Lignocellulosic Biomass III

Lignocellulosic Biomass III

Guest Editors

Alejandro Rodríguez Pascual
Eduardo Espinosa Víctor
Carlos Orestes Martin Medina

MDPI

Basel • Beijing • Wuhan • Barcelona • Belgrade • Novi Sad • Cluj • Manchester

Guest Editors
Alejandro Rodríguez Pascual
Chemical Engineering
Department
Universidad de Córdoba
Córdoba
Spain

Eduardo Espinosa Víctor
Chemical Engineering
Department
Universidad de Córdoba
Córdoba
Spain

Carlos Orestes Martin Medina
Department of Biotechnology
Inland Norway University of
Applied Sciences
Hamar
Norway

Editorial Office
MDPI AG
Grosspeteranlage 5
4052 Basel, Switzerland

This is a reprint of the Special Issue, published open access by the journal *Molecules* (ISSN 1420-3049), freely accessible at: https://www.mdpi.com/journal/molecules/special_issues/2T1B5D7WG6.

For citation purposes, cite each article independently as indicated on the article page online and as indicated below:

Lastname, A.A.; Lastname, B.B. Article Title. *Journal Name* **Year**, *Volume Number*, Page Range.

ISBN 978-3-7258-7863-5 (Hbk)
ISBN 978-3-7258-7864-2 (PDF)
https://doi.org/10.3390/books978-3-7258-7864-2

Contents

About the Editors

Alejandro Rodríguez Pascual

Alejandro Rodríguez Pascual is a Full Professor in the Chemical Engineering Department. He has supervised 12 doctoral theses, and is currently supervising a further 6. Scholar ranking: Google Scholar h = 50, i10 = 122, total cites 7537, Scopus h = 43, total cites 5451. Alejandro has published 152 JCR articles in total. 100% of his published works since 2015 have been cited. 75.8% (47 documents) of his papers are in the top 25% most cited documents worldwide. His Documents and Field-Weighted Citation Impact is 1.95, and he ranks in the Stanford University-World's Top 2% Scientists in Biotechnology. Alejando's normalised impact (Scopus FWCI) is 1.95 (2020–2024). BioPrEn Group (RNM-940), Chemical Engineering Department, Instituto Químico para la Energía y el Medioambiente (IQUEMA), Faculty of Science, Universidad de Córdoba, 14014 Cordoba, Spain.

Eduardo Espinosa Víctor

Eduardo Espinosa Víctor is an Associate Professor working in the Chemical Engineering Area at the Universidad de Córdoba (UCO), and a member of the Bioproducts and Process Engineering group (BIOPREN). He obtained his PhD in Biosciences and Agri-Food Sciences in 2019 (Extraordinary PhD Award), after graduating in Environmental Sciences (Extraordinary Bachelor's Award). His research focuses on the valorization of agri-food by-products through sustainable biorefinery processes for the production of high-value-added bioproducts including bioactive compounds, nanocellulose, and cellulose for applications in paper, food packaging, and biomedicine. His scientific output includes 67 articles (45 Q1, 22 D1). According to Google Scholar, he has 2975 citations (h-index 29, i10-index 49), while Scopus reports 2,327 citations and an h-index of 25. All publications in the 2019–2023 period have been cited, with 70.8% ranked among the top 25% most cited worldwide. Eduardo's normalised impact (Scopus FWCI) is 1.61 (2020–2024). BioPrEn Group (RNM-940), Chemical Engineering Department, Instituto Químico para la Energía y el Medioambiente (IQUEMA), Faculty of Science, Universidad de Córdoba, 14014 Cordoba, Spain.

Carlos Orestes Martin Medina

Carlos Orestes Martin Medina is a Full Professor at the Department of Biotechnology, University of Inland Norway, and Visiting professor at the Department of Chemistry, Umeå University, Sweden.

According to Google Scholar, he has a total of 9049 citations (h = 41, i10 = 76), while Scopus reports 5917 total citations (h = 36). Carlos has a normalised impact (Scopus FWCI) of 1.95 (2020–2024).

molecules

MDPI

Opinion

Exploiting the Biorefinery Potential of Spent Mushroom Substrate: The Time to Do It Is Now

Carlos Martín [1,2,*], Shaojun Xiong [3] and Georgios I. Zervakis [4]

1 Department of Biotechnology, Inland Norway University of Applied Sciences, N-2317 Hamar, Norway
2 Department of Chemistry, Umeå University, SE-901 87 Umeå, Sweden
3 Department of Forest Biomaterials and Technology, Swedish University of Agricultural Sciences, SE-901 83 Umeå, Sweden; shaojun.xiong@slu.se
4 Laboratory of General and Agricultural Microbiology, Department of Crop Science, Agricultural University of Athens, 11855 Athens, Greece; zervakis@aua.gr
* Correspondence: carlos.medina@inn.no; Tel.: +47-966-18-155

Abstract: Spent mushroom substrate (SMS), the residual material left after mushroom cultivation, represents an abundant yet underutilized bioresource. With global mushroom production generating millions of tons of SMS annually, its disposal constitutes a missed opportunity within the circular bioeconomy. This Opinion article highlights why SMS should be repositioned as a valuable raw material for sustainable biorefineries and outlines the technological, economic, and regulatory steps needed to unlock its potential.

Keywords: spent mushroom substrate; biorefinery; circular bioeconomy; lignocellulosic biomass; bioactive compounds

1. Background

Spent mushroom substrate (SMS) is the growing medium leftover after mushroom cultivation and harvesting. It is estimated that the production of one ton of mushrooms generates up to five tons of SMS, resulting in a global annual output of roughly 250 million tons from the mushroom industry [1]. Based on FAO data on global mushroom production and on the typical SMS-to-mushroom ratio, it can be estimated that over 200 million metric tons of SMS are produced annually in China, the world's largest mushroom producer [2]. The next largest producers are Japan, India, and the USA with an estimated generation of 2.3, 1.6, and 1.5 million metric tons, respectively. Regarding the EU, roughly 6 million metric tons of SMS are generated each year, with Poland, the Netherlands, and Spain as the leading producers with ca. 1.2, 1.0, and 0.8 million metric tons, respectively. Most of the generated SMS remains underutilized. As global market demands for edible mushroom increase, SMS management is becoming increasingly problematic, especially in high-production regions, contributing to rising logistical and environmental concerns. However, we believe that SMS should not be considered as waste, but rather as a valuable bioresource [3]. It contains a unique combination of polymers, bioactive compounds of both fungal and plant origin, and enzymes. This particular composition distinguishes SMS from other biomass residues and positions it as a promising feedstock for biorefineries. A biorefinery is an overall concept of a processing plant where biomass feedstocks are converted into a spectrum of valuable products, such as biobased fuels, chemicals, and materials [4]. In addition to these products, a biorefinery processing SMS can also extract bioactive compounds suitable as functional ingredients for pharmaceutical, nutraceutical, and biomedical applications, provided the purification processes meet the required standards.

The valorization of agricultural and agro-industrial residues within a circular biobased economy model is a major focus of scientific interest. However, despite its potential, SMS is typically excluded from biomass inventories and not prioritized in strategic research programs. This exclusion highlights that neither academia nor industry has yet fully recognized the true potential of SMS, which is a concerning oversight. Hereby, we argue that it is time to reassess the role of SMS as a promising raw material for the sustainable production of valuable biobased products in biorefineries.

2. What Is So Unique About Spent Mushroom Substrate (SMS)?

Although SMS is to a large extent composed of biomass derived from agricultural or forest residues, it is qualitatively different from typical lignocellulosic materials. Due to its dual biological origin, SMS contains not only residual plant components but also a substantial amount of fungal biomass, resulting from mycelium growth on the cultivation substrate. During mushroom cultivation, fungi partially degrade lignin and polysaccharides, leaving behind the non-consumed lignocellulosic fraction along with mycelium and other fungal-derived material (including secondary metabolites and other bioactive compounds). SMS of white-rot fungi, which constitute a major share of cultivated mushrooms, typically contains less lignin and more extractives than other lignocellulosic materials, as exemplified by comparing sugarcane bagasse (a gramineous bioresource), maritime pine (softwood), and eucalyptus (hardwood) on one side, and SMS of shiitake (*Lentinula edodes*) and oyster mushrooms (*Pleurotus* spp.), on the other side (Table 1).

Table 1. Summary of the composition of different types of lignocellulosic biomass, including spent mushroom substrate (SMS). The content of each component is given in % (w/w).

Biomass Material	Cellulose	Hemicelluloses	Lignin	Extractives	Source
Sugarcane bagasse	36.5	21.3	19.1	4.9	[5]
Maritime pine	45.0	22.2	26.8	2.9	[6]
Eucalyptus	42.0	22.2	22.9	4.7	[7]
Shiitake SMS	28.1	12.8	6.6	25.1	[8]
Oyster mushroom SMS	30.6	18.6	12.7	13.0	[8]

The partial degradation of lignin during mushroom cultivation increases the accessibility to hydrolyzing enzymes of the cellulose remaining in the spent substrate, as compared to that in the initial substrate [9]. This is a significant advantage from a biorefinery perspective, as it facilitates biomass deconstruction without the need for conventional pretreatment, which would otherwise increase costs and generate inhibitors of enzymes and microorganisms. Additionally, fungal β-glucans, chitin, ergosterol, proteins, and secondary metabolites are retained in the SMS. Many of these fungal-derived compounds are known to exhibit antioxidant, antimicrobial, or immunomodulatory properties, among other bioactivities [10], which are not to be found in common lignocellulosic biomass.

Hence, SMS can be considered a multifunctional bioresource, enriched with hydrolyzable cellulose and bioactive compounds. This makes it suitable as a source of fermentable sugars for producing biofuels and biomaterials, and of extracts for producing high-value-added products to be used in the nutraceutical, functional food, or cosmetic sectors.

3. Unrealized Potential and Missed Opportunities in SMS Exploitation

Despite its high potential value, SMS is generally downgraded to low-value uses (usually as soil amendment after composting) [11], while it is also disposed of as waste by incineration, burial, and piling [12]. Although existing applications contribute to organic matter recycling, the full potential of SMS bioactive compounds and structural polymers is far from being exploited. Mushroom production plants generally lack the facilities required for applying systematic valorization strategies to SMS. As a rule, the infrastructure and technical insight needed to integrate SMS into biorefinery pipelines are far beyond the scope of today's mushroom industry.

This represents a missed opportunity for several economic sectors. For example, the emerging lignocellulosic biorefinery sector, which is interested in low-cost feedstocks, would benefit from the recognition of SMS as a prioritized raw material for producing biofuels and platform chemicals. Furthermore, the pharmaceutical, cosmetic, and food industries could benefit from the SMS potential as a natural source of antioxidants and other bioactive compounds, provided that its use is promoted by national policies.

Although some lab-scale research has demonstrated the suitability of SMS for bioconversion through saccharification and fermentation [13], extraction of bioactive compounds [14], and thermochemical conversion [15], the results rarely progress beyond the laboratory bench. This fact indicates that SMS remains underutilized not because of limited scientific interest, but because of inadequate institutional support to establish it as a mainstream feedstock. Therefore, the existing gap hinders the translation of scientific knowledge into scalable, innovative processes of interest for investors and entrepreneurs.

4. The Biorefinery Potential of Spent Mushroom Substrate (SMS)

4.1. What Does SMS Offer to Biorefineries?

In lignocellulosic biorefineries, cellulose, hemicelluloses, and lignin are transformed into smaller molecules, which are then valorized into biofuels, chemicals, or other materials. During mushroom production, the substrate polymers are partially degraded, but a substantial portion of them remains in the SMS after cultivation. This remaining fraction of carbohydrates and lignin can be further utilized in biorefinery processes.

Cellulose saccharification is a key step in biorefineries. Enzymatic saccharification requires pretreatment to reduce the interference of lignin and hemicelluloses, thereby enhancing enzyme accessibility. Compared to conventional lignocellulosic materials, SMS exhibits higher cellulose digestibility, as much of the lignin and hemicellulose interference has already been reduced during fungal growth. Therefore, when SMS is used as feedstock, the need for energy- and material-intensive pretreatment is minimized. For example, in studies with shiitake SMS, we have achieved cellulose conversions over 80% (w/w), demonstrating its high enzymatic digestibility [8,9]. SMS enzymatic saccharification results in sugar-rich hydrolysates, which are suitable for fermentation, not only because they provide sugars as a carbon source, but also because they are free from inhibitors typically formed during thermochemical pretreatments, and are rich in nutrients that support microbial growth [13].

SMS is also a source of valuable polymers, such as lignin, chitin, and crystalline cellulose, which can be recovered by chemical processes following biorefinery routes. These polymers can be used to formulate composites, nanoparticles, and other advanced materials [16] with the potential to replace fossil-based products.

Unlike other lignocellulosic residues, SMS offers the added value of bioactive compounds, which are either structural components of the mycelium, products of the fungus' metabolism, phytochemicals from the original substrate, or oligomers formed by partial lignocellulose degradation during cultivation. SMS bioactive compounds can be recovered using green extraction methods and directed toward value-added applications.

4.2. Integrated Processing

A simplified flow diagram of our envisioned biorefinery is shown in Figure 1. The process includes, as the first step, the extraction of bioactive compounds, which can serve as precursors of high-value products, such as nutraceuticals, cosmetics, or functional foods.

The extraction also results in an extract-free residue, which represents approximately 70% or 80% of the initial dry weight of the SMS of shiitake or oyster mushrooms, respectively [14].

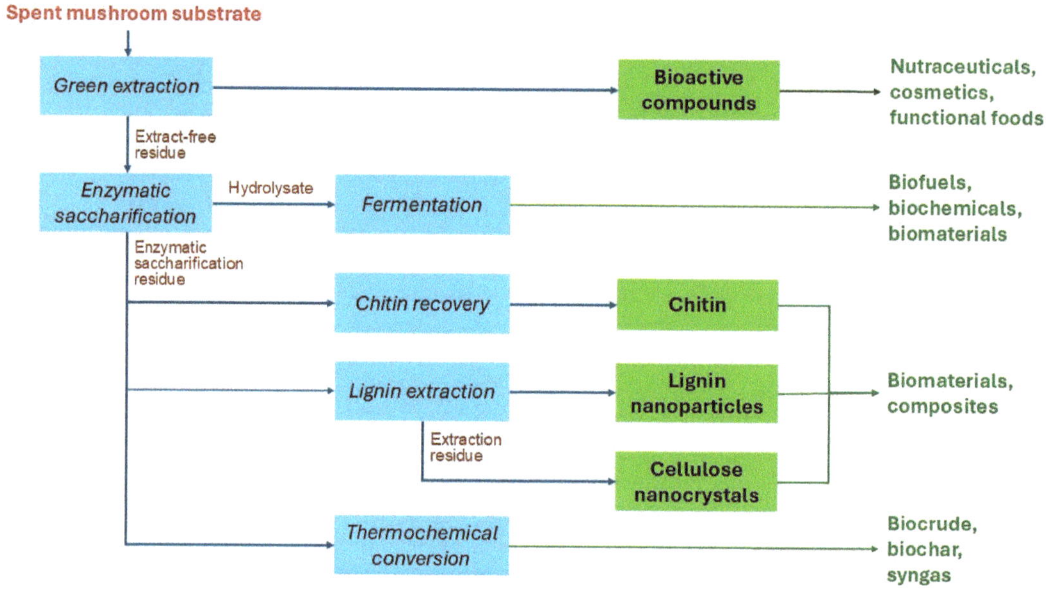

Figure 1. Simplified flow diagram of a biorefinery using spent mushroom substrate as feedstock. Blue boxes represent processes, and green boxes represent value-enabling intermediates. Brown regular text denotes process streams, and green bold text denotes products.

The extract-free residue (EFR) undergoes saccharification to produce a sugar-rich hydrolysate. Enzymatic conversion of over 80% (*w*/*w*) of the cellulose and hemicelluloses can be anticipated in the case of shiitake's SMS [8]. The hydrolysate, typically containing approximately 30–40 g/L glucose and 10–15 g/L xylose, can then be fermented into biofuels, biopolymers, or other products (Figure 1). A residual solid stream, hereafter referred to as enzymatic saccharification residue (ESR), is also formed during saccharification. The ESR yield is expected to be approximately 35% of the weight of the EFR subjected to saccharification, which corresponds to around 25% of the initial SMS dry weight. ESR contains up to 35% (*w*/*w*) lignin, 20–25% cellulose, and some amount of chitin [8]. ESR cellulose is expected to exhibit high crystallinity, given that most of the amorphous cellulose has been removed during enzymatic saccharification. Although data on chitin content in SMS are scarce, our preliminary findings indicate values ranging from 5 to 30 mg/g SMS (Xiong et al., unpublished material). The saccharification residue can be used to recover either chitin or lignin nanoparticles (LNP), while the residue from lignin extraction can subsequently be processed into cellulose nanocrystals (CNC). Chitin, LNP, and CNC can be applied in the formulation of innovative biobased materials. Alternatively, the saccharification residue, as well as any other remaining residue, can be subjected to thermochemical conversion to biocrude, biochar, or syngas, which can serve as renewable fuels or chemical precursors, contributing to closing the carbon loop [15]. This integrated strategy maximizes biomass utilization and supports zero-waste objectives in line with the principles of a circular bioeconomy.

We have partially demonstrated sections of this route by extracting bioactive compounds from shiitake and oyster mushroom SMS [14], running enzymatic saccharification after the extraction of bioactives [17], fermenting the hydrolysates to ethanol [13] and exopolysaccharides [18], and recovering lignin from the saccharification residues [8]. The results of our experiments, together with the insights and outcomes from conceptual studies on specific conversion routes [19], demonstrate that SMS is a versatile and modular feedstock whose various fractions can be selectively recovered and processed through different pathways, depending on the desired product(s) portfolio. Repositioning SMS within the framework of biorefineries enables multi-product valorization according to the cascading concept, whereby extraction, bioconversion, and recycling steps are integrated to maximize both sustainability and economic viability.

5. The Bottlenecks

Despite its promising potential, the widespread adoption of SMS as a biorefinery feedstock remains hindered by several barriers, as discussed below.

5.1. Compositional Variability

One of the key challenges in valorizing SMS is its variable composition, which can change widely depending on the cultivated mushroom species, the initial growth substrate, and the cultivation conditions. The compositional variability of SMS can be illustrated by the wide ranges in the content of cellulose, hemicelluloses, lignin, and extractive compounds reported for shiitake and oyster mushroom SMS (Table 2). Even within the same species, use of different strains and/or small modifications to the substrate recipe can induce metabolic changes, altering carbohydrate and lignin content and bioactive compound profiles [20]. Conversely, deliberate adjustment of substrate composition could be used to enhance the bioactive content of residual fungal biomass, adding value for downstream applications [21,22]. However, this intrinsic compositional variability still hinders the use of SMS in biorefineries, leading to unpredictable yields and product quality. The absence of standardized characterization protocols further complicates SMS integration into industrial value chains. Addressing this bottleneck requires the development of reliable methods for SMS analysis to ensure reproducibility and compatibility with industrial operations.

Table 2. Compositional range of the main components of spent mushroom substrate of shiitake (*Lentinula edodes*) and oyster (*Pleurotus* spp.) mushrooms as reported by different sources. The content of each component is given in % (w/w).

Fungal Species	Cellulose	Hemicelluloses	Lignin	Extractives	Source
Lentinula edodes	24.8–44.3	7.6–17.2	6.6–26.1	22.3–30.0	[8,9,14,23–26]
Pleurotus spp. [1]	27.3–48.7	8.8–22.1	11.0–23.0	13.0–22.6	[8,14,17,25]

[1] Data on *P. ostreatus*, *P. eryngii*, *P. pulmonarius*, and hybrid *P. ostreatus* × *P. eryngii* are compiled.

5.2. Logistical Limitations

The high moisture content of freshly produced SMS presents a major challenge for its transport to industrial facilities. The water content of the initial substrate is typically around 60%. However, controlled irrigation during cultivation, water release from fungal metabolism, condensation of water vapor, and increased water retention capacity -as polymers degrade and the substrate becomes more porous- can raise SMS water content to a level (around 80%) that complicates handling and promotes microbial spoilage.

Furthermore, SMS has a low bulk density, increasing transportation costs. Although dewatering SMS can lower costs, it is costly itself and requires careful evaluation of the adverse effects it might have on efficiency, transportation savings, and SMS quality.

Another logistical challenge is that SMS is often produced in decentralized locations, far from biorefinery facilities. This limits its transportability and shelf life, particularly for bioactive compound extraction, whereby rapid degradation occurring after harvest can compromise extract suitability for further processing into high-value outputs.

Some potential measures to address the logistical challenges include implementing low-cost dewatering and densifying strategies to improve transportability, using mobile extraction units to preserve bioactive compounds, and establishing local biorefinery hubs to improve shelf life and facilitate processing. Additionally, fostering farmer cooperatives could streamline collection and further reduce costs.

5.3. Lack of Processing Infrastructure

Lignocellulosic biorefining is still an emerging industry, and most current plants are pioneering facilities that are equipped to process specific feedstocks, such as certain agricultural or forestry residues, while SMS is so far outside their scope. Adapting existing infrastructure to handle SMS, with its unique composition, moisture content, density, and preprocessing demands, requires substantial investment to implement the necessary technical modifications. Looking forward, we believe that as the market evolves and technologies mature, biorefineries will become more versatile, allowing SMS to be integrated into their feedstock portfolio and play a key role in the global bioeconomy.

5.4. Limited Research

Although SMS has been the focus of some biorefinery-related research, the number of reported studies is considerably lower than that on other biomass materials, such as corn stover, sugarcane bagasse, wheat straw, or forest residues. The lack of comprehensive studies on SMS hinders the development of optimized valorization processes. While early studies have highlighted SMS's potential as a valuable resource, many critical aspects, such as the identification of pre-processing methods, elucidation of microbial degradation pathways, and implementation of cost-effective recovery strategies, remain poorly understood. This research gap prevents developing technologies that could streamline SMS utilization in biorefineries. Consequently, limited research slows down industrial adoption, making it harder for companies to justify the investment needed on a commercial scale.

5.5. Public Perception

Despite the potential of SMS as a valuable bioresource, its categorization as a waste or a compostable by-product remains a major barrier to its broader acceptance. This outdated perception and the general public's limited awareness undermine the perceived economic value of SMS and prevent its recognition as a strategic feedstock. As a consequence, research investment is discouraged, hindering the development of innovative solutions. We believe that a coordinated effort is required to reframe SMS as a value-enabling resource and change the public narrative surrounding it. Without such an effort, the integration of SMS into biorefinery processes will continue to be delayed.

5.6. Policy and Regulatory Gaps

The perception of SMS as an inconsistent bioresource due to the technical limitations discussed in Sections 5.1–5.3 and the underestimation of its true raw material value have discouraged its inclusion in national biomass strategies and policies. For instance, SMS is not explicitly mentioned among the biomass residues eligible for valorization in Poland's Renewable Energy Sources Act [27], the Dutch Climate Agreement [28], or the Spanish Circular Economy Strategy [29], which are the national biomass valorization frameworks of the three largest mushroom-producing countries in Europe. As a consequence, SMS is not included in bioeconomy roadmaps or funding calls. This regulatory ambiguity limits the financial resources available to support SMS research, restricts industrial innovation, and creates uncertainty for investors and entrepreneurs.

5.7. Economic Barriers

The economic feasibility of SMS valorization depends on the cost structure of mushroom production. In Europe, high production costs result in elevated mushroom prices, limiting market growth and, consequently, the availability of SMS for large-scale use. These costs stem mainly from labor- and energy-intensive cultivation methods, and outdated technologies requiring upgrading. Moreover, limited consumer demand in some areas, where acceptance of cultivated mushrooms remains low, reduces incentives for producers and investors. Lowering cultivation costs and increasing consumption would help expand both mushroom production and SMS valorization potential.

6. Why Now Is the Time

Overcoming the above-mentioned bottlenecks requires coordinated efforts between researchers, mushroom producers, biorefinery engineers, policymakers, and funding agencies. These efforts cannot be postponed, as converging trends in policy, sustainability, and innovation only underscore their urgency.

6.1. Circular Bioeconomy Goals and Agricultural Waste Valorization

The EU Green Deal [30], the UN Sustainable Development Goals [31], and national sustainability agendas worldwide emphasize the need to accelerate the transition to a circular bioeconomy. Diversifying biomass feedstocks is a critical step toward that goal. SMS, being renewable, underutilized, and available at scale, aligns well with this vision. Furthermore, valorizing SMS contributes to the sustainability of waste management systems and supports the strategies of governments and environmental agencies in developing waste-to-resource solutions to achieve circular bioeconomy goals.

6.2. Technological Readiness

Innovative techniques in green extraction, pretreatment, saccharification, and fermentation provide solutions to recover valuable intermediates from SMS, amplifying its potential as a biorefinery feedstock. Methods such as supercritical CO_2 extraction and microwave-assisted extraction allow for the efficient and sustainable recovery of bioactive compounds. Similarly, recent advances in pretreatment, saccharification, and fermentation processes—demonstrated at both pilot and commercial scales in 2G ethanol biorefineries—improve the conversion of complex biomass into biobased products. These developments improve the prospects for scalable implementation of SMS integration into sustainable bioeconomy models.

6.3. Emerging Markets for Bio-Based Products

The rising demand for natural antioxidants, prebiotics, and sustainable materials is transforming the food, cosmetic, and pharmaceutical industries, while also driving innovation in functional packaging. Consumers increasingly seek products of non-animal origin that align with ethical, health-conscious, and sustainability-driven values. Natural antioxidants are in high demand for their skin protective, anti-aging, and wellness benefits, while prebiotics are gaining popularity for their digestive health benefits. At the same time, biobased biodegradable packaging offers sustainable alternatives to conventional materials. These trends underscore the SMS potential as a promising feedstock for biorefineries. Rich in bioactive compounds, SMS can be converted into high-value antioxidants, prebiotics, and biopolymers, offering vegan-friendly, sustainable solutions that respond to the growing demand for ethical, eco-conscious, and health-oriented products across these emerging markets.

6.4. Cross-Sector Collaboration Potential

Research on a hypothetical SMS-based biorefinery has great potential to serve as a 'binder' for interdisciplinary collaboration, including projects funded by large research and innovation framework programs such as Horizon Europe. This is because SMS integrates elements from agriculture, biotechnology, and bioenergy, while also offering significant potential for high-value applications in sectors at the interface of food, wellness, pharmaceuticals, and functional materials. Early investment in public–private partnerships could be a visionary step toward launching interdisciplinary innovation across the entire SMS biorefinery value chain.

7. Conclusions

In summary, the scientific rationale is strong, the environmental urgency is clear, and the technological tools are increasingly ready. What is needed now is a strategic commitment to unlock the full potential of spent mushroom substrate (SMS).

This is the moment for coordinated action by researchers, industry, and policymakers to remove existing barriers and bring SMS-based biorefineries to life. Researchers must focus on optimizing scalable processes and fostering interdisciplinary collaboration. Industry players should invest in flexible infrastructure, develop innovative logistics, and strengthen value chain partnerships to ensure consistent feedstock supply and market development. Meanwhile, policymakers need to integrate SMS into circular bioeconomy frameworks, offer targeted incentives and funding mechanisms, and raise public awareness to improve perception and regulatory clarity. With the right support, SMS can become a key component of future sustainable biorefineries.

Author Contributions: The study was conceptualized by C.M., with contributions from S.X. and G.I.Z., C.M. drafted the manuscript, and S.X. and G.I.Z. were actively involved in its revision and edition. All authors have read and agreed to the published version of the manuscript.

Funding: This research was funded by The Research Council of Norway (grant number 342747) and NordForsk (grant number 132066) through the MUSA project (https://www.inn.no/english/research/research-projects/musa/, accessed on 1 November 2025) and by the Bio4Energy strategic research environment (https://www.bio4energy.se, accessed on 19 November 2025).

Institutional Review Board Statement: Not applicable.

Informed Consent Statement: Not applicable.

Data Availability Statement: No new data were created or analyzed in this study. Data sharing is not applicable.

Acknowledgments: The authors acknowledge the administrative support received from their respective universities. During the preparation of the manuscript, ChatGPT (OpenAI, GPT-5) was used to assist in language editing and phrasing improvement. The authors have reviewed and edited the output and take full responsibility for the content of this publication. Diego A. Miranda is thanked for his help with the preparation of the graphical abstract.

Conflicts of Interest: The authors declare no conflicts of interest.

References

1. Ravlikovsky, A.; Pinheiro, M.N.C.; Dinca, L.; Crisan, V.; Symochko, L. Valorization of Spent Mushroom Substrate: Establishing the Foundation for Waste-Free Production. *Recycling* **2024**, *9*, 44. [CrossRef]
2. FAOSTAT. Food and Agriculture Organization of the United Nations Statistics Database. 2025. Available online: https://www.fao.org/faostat/en/#rankings/countries_by_commodity (accessed on 15 November 2025).
3. Martín, C.; Zervakis, G.I.; Xiong, S.; Koutrotsios, G.; Strætkvern, K.O. Spent Substrate from Mushroom Cultivation: Exploitation Potential toward Various Applications and Value-Added Products. *Bioengineered* **2023**, *14*, 2252138. [CrossRef] [PubMed]

4. European Commission. Biorefinery. Knowledge for Policy. Available online: https://knowledge4policy.ec.europa.eu/glossary-item/biorefinery_en (accessed on 19 October 2025).
5. Teófilo, R.F.; Chaves, D.M.; Gonçalves, I.M.; Rufino, A.B.M.X.; Antunes, T.R.; Quintão, A.J.S.; da Silva, R.C.; Sukarno, O.; da Silva, G.A.; Peternelli, L.A.; et al. A Comprehensive Study of Cellulignin Production from Sugarcane Bagasse, Sugarcane Straw, and Energy Cane for Advanced Biofuels: Liquor Composition and Cellulignin Characterizations. *Waste Biomass Valor.* **2025**, *16*, 2195–2208. [CrossRef]
6. López, M.; Vila, C.; Santos, V.; Parajó, J.C. Manufacture of Platform Chemicals from Pine Wood Polysaccharides in Media Containing Acidic Ionic Liquids. *Polymers* **2020**, *12*, 1215. [CrossRef]
7. Penín, L.; Peleteiro, S.; Santos, V.; Alonso, J.L.; Parajó, J.C. Selective Fractionation and Enzymatic Hydrolysis of *Eucalyptus nitens* Wood. *Cellulose* **2019**, *26*, 1125–1139. [CrossRef]
8. Klausen, S.J.; Romero-Soto, L.A.; Liaqat, A.; Dehghanmanshadi, Z.; Strætkvern, K.O.; Xiong, S.; Martín, C. Biorefining Spent Substrates of Shiitake (*Lentinula edodes*) and Oyster Mushroom (*Pleurotus ostreatus*): Enzymatic Saccharification of Cellulose and Xylan, with Lignin Recovery from Residues. *Bioengineered* **2025**, *16*, 2536443. [CrossRef]
9. Chen, F.; Martín, C.; Lestander, T.A.; Grimm, A.; Xiong, S. Shiitake Cultivation as Biological Preprocessing of Lignocellulosic Feedstocks–Substrate Changes in Crystallinity, Syringyl/Guaiacyl Lignin and Degradation-Derived by-Products. *Bioresour. Technol.* **2022**, *344*, 126256. [CrossRef]
10. Gargano, M.L.; van Griensven, L.J.L.D.; Isikhuemhen, O.S.; Lindequist, U.; Venturella, G.; Wasser, S.P.; Zervakis, G.I. Medicinal Mushrooms: Valuable Biological Resources of High Exploitation Potential. *Plant Biosyst.* **2017**, *151*, 548–565. [CrossRef]
11. Lou, Z.; Sun, Y.; Zhou, X.; Baig, S.A.; Hu, B.; Xu, X. Composition Variability of Spent Mushroom Substrates during Continuous Cultivation, Composting Process and Their Effects on Mineral Nitrogen Transformation in Soil. *Geoderma* **2017**, *307*, 30–37. [CrossRef]
12. Luo, J.; Chen, L. Status and Development of Spent Mushroom Substrate Recycling: A Review. *J. Air Waste Manag. Assoc.* **2024**, *74*, 843–860. [CrossRef]
13. Chen, F.; Xiong, S.; Gandla, M.L.; Stagge, S.; Martín, C. Spent Mushroom Substrates for Ethanol Production–Effect of Chemical and Structural Factors on Enzymatic Saccharification and Ethanolic Fermentation of *Lentinula edodes*-Pretreated Hardwood. *Bioresour. Technol.* **2022**, *347*, 126381. [CrossRef]
14. Akter, A.; Klausen, S.J.; Romero-Soto, L.A.; Díaz, F.; Domínguez, H.; Xiong, S.; Strætkvern, K.O.; Martín, C. Comparative Extraction of Bioactive Compounds from Spent Mushroom Substrates of *Lentinula edodes* and *Pleurotus ostreatus* Using Subcritical Water and Pressurized Ethanol. *Ind. Crops Prod.* **2025**, *235*, 121750. [CrossRef]
15. Shahnouri, S.A.; Taghizadeh-Alisaraei, A.; Abbaszadeh-Mayvan, A.; Tatari, A. Catalytic Microwave Pyrolysis of Mushroom Spent Compost (MSC) Biomass for Bio-Oil Production and Its Life Cycle Assessment (LCA). *Biomass Conv. Bioref.* **2024**, *14*, 9949–9965. [CrossRef]
16. Berglund, L.; Rosenstock Völtz, L.; Gehrmann, T.; Antonopoulou, I.; Cristescu, C.; Xiong, S.; Dixit, P.; Martín, C.; Sundman, O.; Oksman, K. The Use of Spent Mushroom Substrate as Biologically Pretreated Wood and Its Fibrillation. *J. Environ. Manag.* **2024**, *372*, 123338. [CrossRef]
17. Klausen, S.J.; Falck-Ytter, A.B.; Strætkvern, K.O.; Martin, C. Evaluation of the Extraction of Bioactive Compounds and the Saccharification of Cellulose as a Route for the Valorization of Spent Mushroom Substrate. *Molecules* **2023**, *28*, 5140. [CrossRef]
18. Miranda, D.A.; Fernández, L.; Strætkvern, K.O.; Carrasco, C.; Martín, C.; Romero-Soto, L.A. Production of Exopolysaccharides by Halotolerant Bacteria Cultivated in Hydrolysates of Spent Mushroom Substrate and Hydrothermally Pretreated Jerusalem Artichoke Stalks. *Ind. Crops Prod.* **2025**, *236*, 121961. [CrossRef]
19. Pandey, A.; Sharma, Y.C. Advancements in Biomass Valorization in Integrated Biorefinery Systems. *Biofuels Bioprod. Biorefin.* **2024**, *18*, 2078–2090. [CrossRef]
20. Koutrotsios, G.; Kalogeropoulos, N.; Stathopoulos, P.; Kaliora, A.C.; Zervakis, G.I. Bioactive Compounds and Antioxidant Activity Exhibit High Intraspecific Variability in *Pleurotus ostreatus* Mushrooms and Correlate Well with Cultivation Performance Parameters. *World J. Microbiol. Biotechnol.* **2017**, *33*, 98. [CrossRef]
21. Koutrotsios, G.; Kalogeropoulos, N.; Kaliora, A.C.; Zervakis, G.I. Toward an Increased Functionality in Oyster (*Pleurotus*) Mushrooms Produced on Grape Marc or Olive Mill Wastes Serving as Sources of Bioactive Compounds. *J. Agric. Food Chem.* **2018**, *66*, 5971–5983. [CrossRef] [PubMed]
22. Koutrotsios, G.; Tagkouli, D.; Bekiaris, G.; Kaliora, A.; Tsiaka, T.; Tsiantas, K.; Chatzipavlidis, I.; Zoumpoulakis, P.; Kalogeropoulos, N.; Zervakis, G.I. Enhancing the Nutritional and Functional Properties of *Pleurotus citrinopileatus* Mushrooms through the Exploitation of Winery and Olive Mill Wastes. *Food Chem.* **2022**, *370*, 131022. [CrossRef] [PubMed]
23. Hiyama, R.; Gisusi, S.; Harada, A. Evaluation of Waste Mushroom Medium from Cultivation of Shiitake Mushroom (*Lentinula edodes*) as Feedstock of Enzymic Saccharification. *J. Wood Sci.* **2011**, *57*, 429–435. [CrossRef]
24. Lin, Y.; Ge, X.; Liu, Z.; Li, Y. Integration of Shiitake Cultivation and Solid-State Anaerobic Digestion for Utilization of Woody Biomass. *Bioresour. Technol.* **2015**, *182*, 128–135. [CrossRef] [PubMed]

25. van Kuijk, S.J.A.; Sonnenberg, A.S.M.; Baars, J.J.P.; Hendriks, W.H.; Cone, J.W. Fungal Treatment of Lignocellulosic Biomass: Importance of Fungal Species, Colonization and Time on Chemical Composition and in Vitro Rumen Degradability. *Anim. Feed Sci. Technol.* **2015**, *209*, 40–50. [CrossRef]
26. Sousa, M.A.C.; Costa, L.M.A.S.; Pereira, T.S.; Zied, D.C.; Rinker, D.L.; Dias, E.S. Enzyme Activity and Biochemical Changes during Production of *Lentinula edodes* (Berk.) Pegler. *Food Sci. Technol.* **2016**, *39*, 774–780. [CrossRef]
27. Act on Renewable Energy Sources ("RES Act", Dz.U. 2015 Poz. 478)-Climate Change Laws of the World. Available online: https://climate-laws.org/document/act-on-renewable-energy-sources-res-act-dz-u-2015-poz-478_5b4d (accessed on 19 October 2025).
28. Ministerie van Economische Zaken en Klimaat. en National Climate Agreement-The Netherlands-Publicatie-Klimaatakkoord. Available online: https://www.klimaatakkoord.nl/documenten/publicaties/2019/06/28/national-climate-agreement-the-netherlands (accessed on 19 October 2025).
29. Spanish Strategy for Circular Economy (España Circular 2030)-Climate Change Laws of the World. Available online: https://climate-laws.org/document/spanish-strategy-for-circular-economy-espana-circular-2030_87c4 (accessed on 19 October 2025).
30. The European Green Deal-European Commission. Available online: https://commission.europa.eu/strategy-and-policy/priorities-2019-2024/european-green-deal_en (accessed on 19 October 2025).
31. Transforming Our World: The 2030 Agenda for Sustainable Development | Department of Economic and Social Affairs. Available online: https://sdgs.un.org/2030agenda (accessed on 19 October 2025).

molecules

MDPI

Review

The Future of Graphene: Preparation from Biomass Waste and Sports Applications

Yueting Wu [1], Yanlong Li [2,*] and Xiangyang Zhang [1]

[1] Graduate School, Harbin Sport University, Harbin 150008, China; wuyueting@hrbipe.edu.cn (Y.W.)
[2] Academic Theory Research Department, Harbin Sport University, Harbin 150008, China
* Correspondence: liyanlong@hrbipe.edu.cn

Abstract: At present, the main raw material for producing graphene is graphite ore. However, researchers actively seek alternative resources due to their high cost and environmental problems. Biomass waste has attracted much attention due to its carbon-rich structure and renewability, emerging as a potential raw material for graphene production to be used in sports equipment. However, further progress is required on the quality of graphene produced from waste biomass. This paper, therefore, summarizes the properties, structures, and production processes of graphene and its derivatives, as well as the inherent advantages of biomass waste-derived graphene. Finally, this paper reviews graphene's importance and application prospects in sports since this wonder material has made sports equipment available with high-strength and lightweight quality. Moreover, its outstanding thermal and electrical conductivity is exploited to prepare wearable sensors to collect more accurate sports data, thus helping to improve athletes' training levels and competitive performance. Although the large-scale production of biomass waste-derived graphene has yet to be realized, it is expected that its application will expand to various other fields due to the associated low cost and environmental friendliness of the preparation technique.

Keywords: graphene; biomass waste; sports equipment; graphene derivatives; biocomposites

1. Introduction

In recent years, with the continuous advancement of a new round of global scientific and technological revolution and industrial upgradation, the requirements for new materials in various fields have become more stringent and diversified. The traditional crystals are too unstable to exist as two-dimensional (2D) structures at non-zero absolute temperatures, which has long plagued researchers. However, the discovery of graphene has successfully filled this gap and completed the graphitic materials system. Graphene—a nanomaterial—consists of a single layer of sp^2-hybridized carbon atoms arranged in two dimensions with a hexagonal honeycomb crystal structure. So far, graphene is the thinnest nanomaterial known and is considered the "magic material of the 21st century". It possesses many excellent properties, such as a large surface area, high conductivity, good chemical stability, and excellent mechanical strength [1–4], which are vital for applications spanning electronics, aerospace, renewable energy systems, sports equipment, and other technologies. In addition, the emergence of this material has spawned a series of new research fields targeted at future needs.

Presently, the proportion of downstream application enterprises of graphene is gradually increasing, and its market is slowly unfolding. Figure 1 shows a new wave of upsurge in the graphene market, which is expected to reach USD 147.9 billion by 2025 [5]. Currently, the primary precursor materials for graphene are mainly metal-based, coal-based, or petroleum coke-based [6–10]. However, these materials are associated with many problems, such as soil and water pollution, releasing harmful gases and particulate matter, and atmospheric pollution, significantly limiting their application [11].

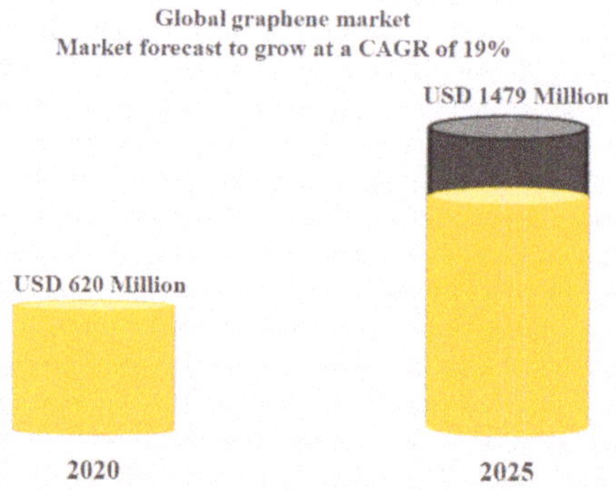

Figure 1. Estimated CAGR of the Global Graphene Market (adapted from Ref. [5]).

The biomass waste generated in the world is increasing year by year, most of which is buried and incinerated [12,13]. This type of treatment not only causes the waste of resources but also leads to environmental pollution. To overcome the challenges associated with the traditional methods of graphene production, such as environmental problems, resource scarcity, and high costs, the research on graphene production from biomass materials has attracted much attention in recent years. Scholars actively seek greener, cleaner, and more efficient materials, including different biomass materials, such as biomass waste resources (composed of plants and animal waste) [14] in agriculture, forestry, and municipal solid waste (MSW), to reduce the economic cost and environmental impact of graphene production. This report reviews the latest progress in the research and development of graphene prepared from biomass waste. The basic structure, characteristics, derivatives, and production methods of graphene are summarized with a detailed discussion of the inherent advantages of selecting biomass waste as a precursor to its preparation. Although many relevant studies on graphene have already been conducted, a detailed focus on its application in sports equipment is still lacking in the literature. Therefore, this report fills the literature gap by focusing on the application prospects of graphene and its derivatives in the sports industry. Finally, the challenges and future outlook of biomass waste-derived graphene are presented from the perspective of the sports industry.

2. Graphene Overview

Graphene is a 2D single-atomic-layer material with a honeycomb lattice structure of closely arranged carbon atoms. Due to its thermodynamic instability, graphene has been used as a theoretical model to describe the structure of other carbon materials before it was discovered [1]. The relevant academic research on graphene can be traced back to 1947 when Wallace et al. [2] proposed a theoretical model of graphene. This breakthrough model established the starting point of academic research on graphite electrical properties. In 2004, physicists Novoselov et al. [15] of the University of Manchester in the United Kingdom practically obtained single-layer graphene for the first time using a tape-stripping method. This breakthrough attracted much attention and inspired scholars to conduct in-depth research on graphene's structure, properties, and derivatives.

2.1. Graphene Structure

Figure 2 shows the structure of graphene and its derivatives, which exhibit a honeycomb-like shape with individual carbon atoms evenly distributed in a 2D lattice, each having 2 s and 2 p electrons in its outer shell [16,17]. The s and p orbitals are sp^2-hybridized to make a typical hexagonal benzene ring structure, which provides the material with significant mechanical stability [18]. In addition, there are p_z orbitals perpendicular to the direction of the plane, forming π chemical bonds. The electrons involved in these π bonds are crucial for the conductivity of graphene, which exhibits different electronic band structures depending upon the structure of the edge carbon chain and the number of staked layers in graphene [19,20].

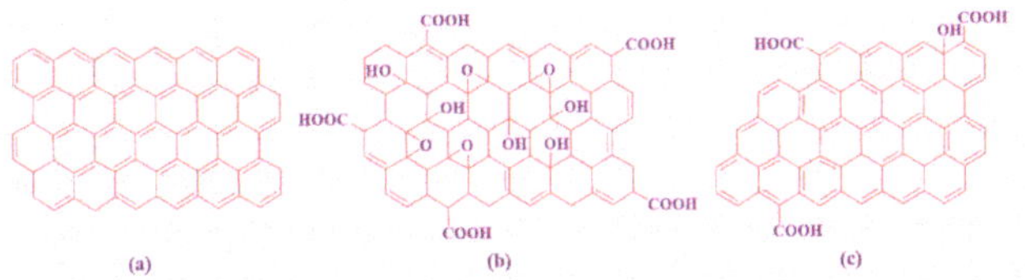

Figure 2. The structure of (**a**) graphene honeycomb, (**b**) graphene oxide (GO), and (**c**) reduced graphene oxide (RGO) (adapted from Ref. [21]).

Graphene has a mechanically stable chemical structure of short C-C bonds (0.142 nm length) connected in a solid fashion [22]. Consequently, when an external force is applied to the sheet, the bond lengths deform to offset the external stress without rearranging the individual carbon atoms, thus maintaining a stable structure [23]. Moreover, scanning tunneling microscopy (STM) shows that the surface of graphene possesses some nanoscale wrinkles [17], which helps overcome the limitations of thermodynamic fluctuations, enabling the stable existence of the 2D lattice [24].

2.2. Graphene Properties

Carbon has achieved a quantum leap from graphite to graphene with a fully exposed bulk and edge atomic arrangement on its surface, which endows it with excellent mechanical, electrical, optical, and other properties. Researchers have conducted in-depth studies on graphene properties and have continuously identified surprising discoveries, which have aroused the interest of many interdisciplinary scholars.

Graphene exhibits high mechanical strength, good toughness, and excellent flexibility. The strong σ and π covalent bonds formed between the graphene carbon atoms are mainly responsible for the superior mechanical properties and structural rigidity [25]. Since graphene is extremely thin in atomic order, its thickness is challenging to measure directly through experimentations. Consequently, determining Young's modulus of graphene is more challenging than that of bulk materials. Lee et al. [26,27] pioneered the study of graphene monolayers' elastic properties and inherent strength using nanoindentation technology based on atomic force microscopy (AFM). The results showed that Young's modulus of graphene monolayers was as high as 1.1 TPa, and its breaking strength could reach up to 42 N m^{-1}. Interestingly, graphite paper composed of graphene exhibited weak toughness initially, but its toughness was significantly enhanced following oxidation and functionalization. In addition, Lee et al. [28] demonstrated graphene balloon technology by preparing graphene through chemical vapor deposition (CVD), transferring it to a circular hole, and placing the sample in a vacuum. Due to the airtightness of graphene, air could not penetrate it, resulting in a higher air pressure inside the balloon. The researchers used Raman spectroscopy to study the balloon and compared the results with numerical simulations. The study concluded that Young's modulus of monolayer graphene was about 2.4 ± 0.4 TPa.

Graphene is very sensitive to its environment due to its unique 2D structure. Its interactions with other materials may affect its electrical and optical properties and trigger new physical phenomena. Most applications based on graphene rely on its high electronic mobility. Electrons in graphene are free to move owing to the extended delocalized bonds. However, the mobility is hindered by internal structural defects and scattering due to lattice vibrations [29], which are almost independent of temperature [30]. Therefore, at room temperature, when the carrier concentration is about 10^{12} cm^{-2}, due to scattering by phonons, the charge mobility can be as high as 200,000 $cm^2V^{-1}s^{-1}$, which is almost 140 times that of commercial crystalline silicon [31]. Moreover, graphene has excellent optical properties, especially high optical transmittance. When white light is irradiated vertically on single-layer graphene, only about 2.3% of the light is absorbed, resulting in a transmittance of 97.7% [32]. As the number of layers increases, the visible light transmittance decreases by 2.3% for each successive layer. Non-interactive Dirac–Fermi simulations showed that the transmittance in graphene was affected by its thickness, and the number of graphene layers could be estimated from the value of transmittance [33,34]. In addition, the optical transitions in graphene could be modulated by changing its Fermi level with an applied gate voltage. For instance, when graphene absorbs photons, the generated electron-hole pairs can recombine in a very short time of picosecond. However, when an external electric field is applied, the excited electron-hole pairs are separated, resulting in a photocurrent flow [35–37]. Due to the unique zero-bandgap structure of graphene, the energy required for electrons to transition from the valence band to the conduction band is almost zero (theoretically calculated). Consequently, the majority spectrum of incident light can generate a certain photocurrent intensity in graphene. Furthermore, graphene can also be excited by appropriately modifying its zero bandgap [38].

2.3. Graphene Derivatives

Graphene derivatives are chemical modifications based on graphene, imparting novel properties to the material, thus significantly expanding its application fields [39,40]. Table 1 compares the mechanical properties of the graphene family.

Table 1. Mechanical properties of the graphene family.

Materials	Experimental Technique	Young's Modulus	Tensile Strength	Refs.
Monolayer graphene	Nano-indentation in AFM	1 TPa	130 GPa	[41]
Free standing GO	Nano-indentation on a dynamic contact tool	697 ± 15 GPa	3–33 GPa	[41]
RGO paper	-	41.8 GPa	293.3 MPa	[42]

Graphene oxide (GO) is the prime representative of graphene-derived materials, also known as graphite oxide. It is prepared by the chemical oxidation of graphite powder. Due to variations in the oxidation conditions, the structure of GO has not been understood in detail until recently. As per the most widely accepted model of GO, the most active functional groups, such as epoxy (-O-) and hydroxyl (-OH), are mainly distributed on its surface. In contrast, functional groups, such as carboxyl (-COOH), carbonyl (-C=O), and phenol (Ar-OH), are connected to the edges of the GO sheet [43]. Moreover, the number of functional groups on GO depends on the preparation method employed.

GO exhibits properties utterly different from those of graphene due to its many oxygen functional groups and its layer structure, which markedly differs from the latter. Compared with graphene, GO exhibits stronger hydrophilicity because of its ability to make hydrogen and polar bonds with water molecules. This property makes GO easy to exfoliate and disperse stably in water and suitable organic solvents [44,45]. However, compared with graphene, oxygen-containing functional groups can lead to structural defects in the 2D atomic structure of GO, resulting in a corresponding loss of mechanical and electrical properties [46,47].

GO can easily disperse in water to make a stable solution that can facilitate its deposition on suitable substrates through the standard deposition techniques of drop casting, spraying, and spin coating [17]. At the same time, these oxygen functional groups can facilitate the covalent or non-covalent binding of the material to other substances, further expanding GO applications [48]. However, the large number of oxygen-containing functional groups on the surface of GO disrupts the continuous conjugated structure of graphene, resulting in the loss of the characteristic high thermal and electrical conductivity of the material. To re-establish the conjugated structure and unique properties of graphene, researchers chemically reduce GO to obtain reduced graphene oxide (rGO) with partial recovery of the conjugated structure associated with unavoidable defects.

rGO is, therefore, also a graphene derivative, which has properties similar to graphene, particularly excellent conductivity and mechanical properties [49]. The GO reduction process removes the oxygen-containing groups and restores the conjugated structure of graphene. Different reduction methods (including thermal, chemical, microwave, photochemical, and hydrothermal reduction) lead to significantly different quality of the final rGO. Chemical and thermal reductions are the two most widely used methods. Chemical reduction usually employs chemical reducing agents, including hydrazine, ascorbic acid, hydroiodic acid, and sodium borohydride, to treat GO in rGO. Among the various reducing agents, ascorbic acid has gained particular interest for the scalable production of rGO without generating toxic gases [50]. Thermal reduction, on the other hand, employs high-temperature conditions to remove oxygen-containing functional groups on the GO surface. Theoretically, GO can be completely deoxidized under extreme temperatures of more than 2000 °C.

Following the elimination of the oxygen-containing groups, the rGO shows completely different properties to GO in some respects [51]. For instance, rGO exhibits better chemical stability and electronic conductivity than GO. Moreover, rGO is prone to aggregation in aqueous and related media, thus controlling its shape and specific surface area [52]. Conversely, oxygen-containing groups significantly enhance the hydrophilicity of GO, making it more stable to disperse uniformly in water [53]. These characteristics open up new opportunities for rGO and rGO-based devices in various fields.

Three-dimensional graphene (3D graphene) is a derivative of graphene consisting of longitudinally grown stacks of 2D graphene, preserving many of its inherent properties, such as high electrical conductivity. In addition, it possesses unique structural/electrical characteristics, resulting in the increased absorption of near-infrared (NIR) to infrared (IR) radiations by up to 16% [54]. The other most common chemical derivative of graphene is nitrogen-doped graphene. Nitrogen is usually present in various carbon-based materials; however, its content is low, with no well-defined stoichiometry. The presence of heteroatoms in the aromatic carbon rings of graphene interferes with its aromaticity, resulting in various effects, including bandgap opening [55,56], charge separation, ferromagnetic ordering, different interlayer forces, and catalytic potential [57]. Subsequently, the properties of pristine graphene are altered, providing broad prospects for its application in specific fields, such as nanoelectronics, electrochemical biosensors, energy storage, and other fields.

3. Preparation Method of Graphene

The preparation of graphene plays a crucial role in advancing its scientific research and wide applications. Different preparation methods vary in the yield and quality of the resulting graphene, which are suitable for different applications [58]. Achieving the large-scale, automated, low-cost production of high-quality graphene is essential for manufacturing high-performance graphene devices. The preparation methods reported in the literature can be classified into the following two categories: top-down and bottom-up approaches [59–61]. The top-down approach aims to break down bulk graphite into lamellar graphene sheets through exfoliation. Conversely, the bottom-up approach assembles graphene from individual carbon atoms using precursors such as methane or polymers with benzene rings and is advantageous for high-yield production. Table 2 compares the different preparation methods of graphene, including the typical micromechanical lift-off, CVD, redox synthesis, and epitaxial growth methods.

Table 2. Comparison of different preparation methods for graphene.

Method	Substrate	Temperature (°C)	Yield	Refs.
Micromechanical exfoliation	SiO_2/Si	Room temperature	Difficult to generate production output	[62]
Chemical vapor deposition	Cu, Pt, Ni, Ru, Ir	>1000	Can be produced on a large scale	[63]
Oxidation reduction	-	<500	Can be produced on a large scale	[64]
Epitaxial growth	SiC	1200–1600	Suitable for small-scale production	[65]

3.1. Micromechanical Lift-Off

The mechanical exfoliation method was the earliest method used to prepare multilayer graphene. This process of graphite delamination was achieved by applying a mechanical force. Geim et al. [66,67] successfully separated graphene nanosheets by repeatedly sticking adhesive tape to highly oriented pyrolytic graphite (HOPG) to obtain layered graphene, combined with applying oxygen plasma etching. The principle of this preparation method is based on the fact that van der Waals forces between graphite sheets are weak and, thus, can easily be separated by applying an external force. Paton et al. [68] reported a simple and efficient method for producing graphene using liquid-phase exfoliation. Graphite powder was placed in an organic solvent containing a specific surfactant and subjected to high-energy shear using a stator-rotor homogenizer (made by Silverson Machine, East Longmeadow, MA, USA). This method effectively improved the speed of graphene production via mechanical exfoliation. Furthermore, the automated exfoliation method has significant advantages, including the simplicity, low cost, and high quality of the resulting graphene. However, the graphene prepared through this method is highly laborious and time-consuming. Additionally, there are uncontrollable factors causing difficulty controlling wrinkles and the number of layers. As a result, this method is limited to laboratory research and is challenging to meet the demands of scaled-up production [69].

3.2. Chemical Vapor Deposition

The CVD technology originated in the 1960s and was initially used for thin film preparation in the semiconductor industry. Later, it gradually expanded to the field of nanomaterials. In recent years, this technology has played an essential role in promoting the development of high-quality graphene. Before introducing the CVD technology to prepare graphene, researchers found that layered graphitic materials can be formed on nickel (Ni) substrates through high-vacuum annealing [70,71]. The CVD preparation is mainly divided into two types based on the growth mechanism: (1) the dissolution precipitation mechanism, employing Ni, cobalt (Co), and other metals with high carbon dissolution ability, and (2) the surface catalytic mechanism, utilizing metals with low dissolved carbon content, such as copper (Cu), molybdenum (Mo), and platinum (Pt). In the former mechanism, the dehydrogenation of the carbon source produces carbon atoms that penetrate the metal substrate at high temperatures. When rapidly cooled, the carbon atoms precipitate out of the interior of the metal to nucleate and grow into graphene on the surface of the substrate [72–76]. On the other hand, in the latter mechanism, the carbon source is dehydrogenated at high temperatures to generate activated carbon atoms, which, after reaching a certain supersaturation on the metal surface, leads to nucleation and growth to form the graphene crystal domains. Finally, continuous graphene is obtained through 2D growth and merging of the crystal domains precipitated on the surface of the substrate [77,78].

CVD technology can prepare single-crystal graphene films with sizes ranging from 10 microns to 1.5 inches [79,80]. Park et al. [81] used CVD technology to prepare few-layer graphene (FLG) on nickel-coated substrates and found that chemical bonds connected different crystal domains of the FLG. Although the FLG samples were not uniform in macrostructure, they demonstrated excellent mechanical properties. The advantages of CVD technology include the preparation of graphene with a large 2D area, superior quality, and complete lattice structure. However, this technology is challenged by the difficulty of separating graphene from the metal substrate and the possible product contamination of the organic molecules used for its separation.

3.3. Oxidation–Reduction Method

rGO is prepared by the oxidation–reduction method. Generally, graphite is first oxidized using a strong acid and oxidant and then exfoliated into GO by ultrasonic treatment. Finally, the oxygen-containing functional groups on the surface of the GO are reduced to obtain rGO. Various oxidation strategies, including Brodie, Hummer, and Staudenmaier methods, are commonly found in the literature [82–84]. Since the synthesized GO is rich in polar functional groups, it can be dispersed in water or alkaline solutions by ultrasonication. Then, GO flakes are reduced using various methods, such as microwave reduction, solvothermal reduction, vapor phase reduction, and chemical reduction [51,85,86]. It is noted that rGO prepared through the redox method destroys the π-electrons conjugated structure, resulting in more functional groups and other types of defects in the synthesized rGO, thus compromising its performance [87,88]. However, a large amount of only a few layers of rGO could be obtained by this method, offering a low-cost production [89–92] demanded by various fields of application.

3.4. Epitaxial Growth Method

In the early 1970s, graphene growth on a nickel single crystal (111) surface was studied [93]. A graphene-decorated SiC substrate was obtained by heating a silicon carbide (SiC) single crystal at high temperatures for the silicon atoms to evaporate, which is associated with the reconstruction of the carbon atoms through self-assembly. Initially, the growth conditions for graphene on a nickel single crystal (111) surface required an ultra-high vacuum (UHV). When a clean Ni (111) surface is exposed to hydrocarbons, surface carbide or graphene is produced. Additionally, it has been confirmed that the thermal decomposition of hydrocarbons at low temperatures leads to the formation of surface carbide, while graphene typically forms at temperatures between 500 °C and 700 °C [94]. However, recent studies showed that an atmosphere of argon or small amounts of disilanes can reduce the rate of silicon sublimation, allowing better-quality graphene to be prepared at higher temperatures [95]. Hass et al. achieved controlled single and multilayered graphene growth using the SiC decomposition method [96]; however, multilayer graphene up to 100 layers is relatively easy to obtain. The quality of graphene prepared through this method is closely related to the heating temperature, reaction pressure, and protective gas type. Therefore, high-quality graphene can be obtained by adjusting the reaction parameters using the epitaxial growth method. However, the graphene yield is low while the reaction temperature is high, limiting its adaptation for the large-scale synthesis of single-layered graphene [97,98].

4. Biomass Waste-Derived Graphene

With the increasing emphasis on resource sustainability and environmental protection, people are gradually becoming aware of the broad prospects of biomass resources [99,100]. Biomass waste, including biomass waste from agriculture, forestry, and MSW, constitutes an integral part of biomass resources, which is cheap and widely available [101]. Figure 3 shows the raw materials and advantages of green synthetic graphene.

The treatment and recycling of biomass waste directly impact the comprehensive utilization and efficiency of natural resources. Following proper treatment and processing, biomass waste has the potential to produce high-value-added chemicals and carbon materials. Therefore, biomass waste recycling is expected to promote the synergistic development of environmental protection and resource utilization. Figure 4 shows various strategies for utilizing biomass to produce graphene.

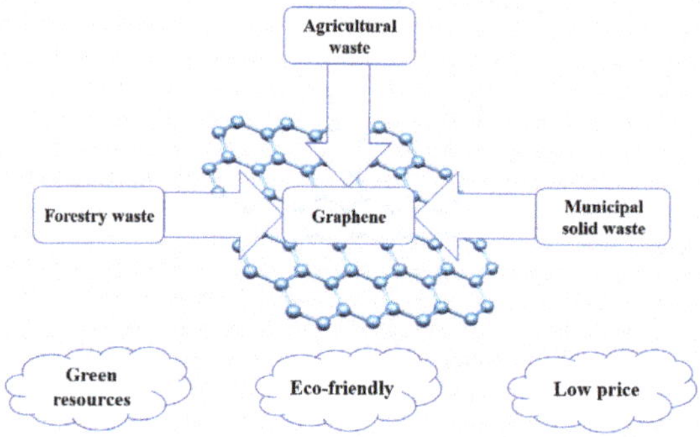

Figure 3. Raw materials and advantages of green synthesis of graphene (adapted from Ref. [102]).

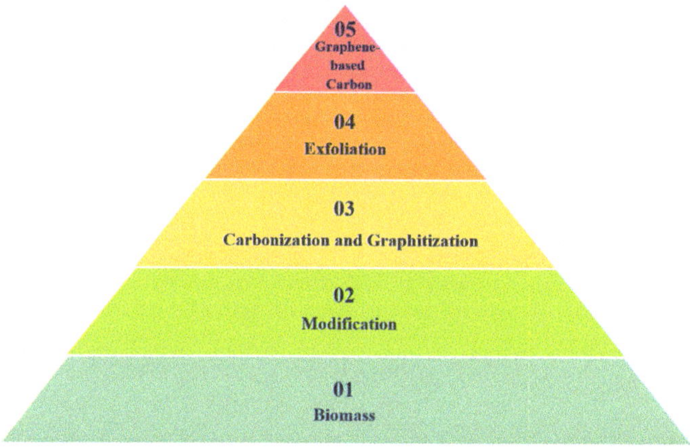

Figure 4. Strategies for using biomass to produce graphene. (adapted from Ref. [103]).

4.1. *Agricultural Waste*

Globally, large quantities of agricultural products are produced daily to meet the needs of a growing population, leading to the large-scale incineration of agricultural waste [104–106]. The indiscriminate incineration of these wastes releases harmful substances, such as carbon monoxide, sulfur dioxide, and dust, which are significant sources of air pollution and pose a severe threat to respiratory health [107]. At the same time, the disposal of agricultural waste without proper treatment may also cause various hazards, such as the spread of infectious diseases, water pollution, and ecosystem damage.

Agricultural waste mainly includes three organic substances: cellulose, hemicellulose, and lignin. Hemicellulose is a heteropolysaccharide whose specific composition varies according to its source. For instance, hemicelluloses from angiosperms and gymnosperms, composed of xylans and glucomannan, respectively [108], are excellent carbon sources for synthesizing carbon materials. In addition, agricultural waste provides abundant raw materials for producing high-performance graphene, with significant advantages in adsorption performance and the quality of the obtained graphene [109].

Rice is one of the major sources of food. The rice husk is the outer covering of the rice grains. During growth, rigid silica in the husk protects rice from insects and bacteria and ensures it can obtain the necessary water and nutrients. To achieve dual functions, silica in rice husks has evolved naturally into a unique 3D porous nanostructure [110–112]. Wang et al. [113] reported biocompatible graphene quantum dots derived from rice husks with a yield of 15% and obtained mesoporous silica nanoparticles as a by-product of the preparation, realizing the comprehensive and effective utilization of agri waste.

Wheat straw is the stalk left after wheat harvesting, and its main components include cellulose, hemicellulose, lignin, trace metals, and crude proteins [114]. Chen et al. [115] used potassium hydroxide (KOH) to dissolve hemicellulose and lignin in wheat straw. After multiple heat treatments to overcome its complex structure, wheat straw was finally converted into cellulose fibers, which, like other lignocellulosic materials, could be converted into graphene using high-temperature carbonization (HTC). Zhou et al. [116] successfully synthesized interconnected high-graphite carbon nanosheets (HGCNSs) using wheat straw as a precursor and combining the hydrothermal and graphitization processes. The experimental results showed that HGCNS possessed an interconnected two-dimensional nanostructure capable of providing multiple storage sites for lithium ions and facilitating the rapid transport of electrons and ions. Additionally, due to its high degree of graphitization, HGCNS significantly reduced voltage hysteresis, demonstrating excellent cycling and rate performance. Graphene prepared from agricultural waste has significant sustainability, cost-effectiveness, and environmental advantages, reducing resource waste and environmental pollution while simultaneously achieving multifunctional applications.

4.2. Forestry Waste

Forestry waste is a vital biomass resource. Energy conversion utilizing forestry waste is one of the research hotspots in renewable energy. Forestry waste, mainly including branches, sawdust, rolled bark, shavings, and scraps, is continuously generated in forestry processing. Whether woody plants or herbs, their cell walls are primarily composed of cellulose, hemicellulose, and lignin [117], indicating that their structure has a high carbon content. Eucalyptus bark extract is rich in 29 polyphenol compounds, which are considered bioactive due to their excellent antioxidant and anticancer properties. Manchala et al. [118] synthesized soluble graphene using a eucalyptus polyphenol solution obtained from the Eucalyptus bark extract. The results showed that the polyphenol compounds in Eucalyptus bark extract had a reducing ability; therefore, exfoliated GO could be reduced to soluble graphene under reflux in an aqueous medium.

Wood processing produces a tremendous quantity of waste wood. Although the wood-based panel processing industry consumes many wood processing residues, wood chips are more cumbersome. Therefore, these biomass resources have not been recycled and reused effectively as fuel. Although sawdust has not been widely used for fuels due to its chemical composition, it still has the potential to produce high-quality, high-value-added carbonaceous materials such as graphene. Severo et al. [119] successfully prepared 3D graphene sponges from sawdust by pre-carbonization and KOH chemical activation. In addition, this method achieves the high value-added utilization of sawdust, helping to reduce biomass waste generation and protect the environment.

4.3. Municipal Solid Waste

MSW is a heterogeneous waste stream, which is an inevitable part of daily life [120], characterized by complex components and a high organic content. On average, 0.74 kg of waste is generated per person per day globally. The per capita waste generation is closely related to the local economic level and the degree of urbanization. Generally, the higher the income, the higher the per capita waste generation. Moreover, global waste generation is growing more than twice the population growth rate and is projected to reach 3.4 billion tons by 2050 [121].

Disposable paper cups (DPCs) are widely used in many families, enterprises, and public places due to their relative hygiene, low cost, convenient use, and mass production [122]. However, most DPCs have a short service life and are difficult to segregate from waste after use. They are eventually incinerated or buried in landfills, posing a severe threat to human ecosystems. DPCs usually comprise high-grade cardboard and an inner polyethene coating. Wang et al. [113] proposed a new route to prepare Fe/graphene sheets using Fe^{2+}

catalysts with DPCs as a carbon source. It was demonstrated that the proposed synthesis strategy could produce graphene sheets with high yield and high quality along with two additional products, Fe/graphene and Pt/graphene, where the latter showed high catalytic activity for an oxygen reduction reaction in fuel cells. Ruan et al. [123] used inexpensive carbonaceous raw materials without pre-purification, including biscuits, chocolate, grass, plastic, cockroaches, and dog feces, to grow graphene directly on copper foil using CVD technology. The results showed high-quality graphene monolayers prepared from these carbon sources. Graphene production using MSW has potential in terms of quality; however, it currently faces challenges regarding yield. With process optimization, this method could become a competitive route to produce high-quality graphene.

5. Graphene Applications in Sports Equipment

In today's rapidly advancing society of science and technology, as we address the contradiction between pushing the limits of human athletic abilities and sports training, the performance of sports equipment consistently plays a crucial role in influencing athletes' performance. To a certain extent, it enhances the ability of the human body to engage in physical activities. According to the specific requirements of different sports events, high-performance sports equipment has unique advantages in material selection, process design, and ergonomics. They also play a significant auxiliary role in improving athletes' experience, performance, and safety. Figure 5 summarizes some applications of graphene in the sports industry.

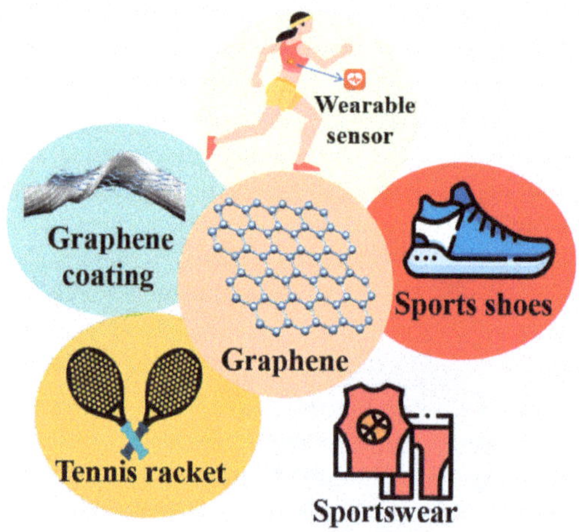

Figure 5. Graphene applications in the sports industry.

5.1. Graphene Wearable Sensors

Scientific training in modern competitive sports relies on intelligent motion monitoring devices that provide athletes with physiological signals during training or competition [124]. Researchers have proposed various types of flexible electronic sensors. Graphene-based flexible wearable electrodes are considered efficient and promising because of their higher elasticity, flexibility, and hydrophobicity, which can better retain contact with human skin and remain stable after long-term use [125–127]. Graphene-coated highly conductive textile electrodes are, therefore, expected to replace the traditional metal-based electrodes in health monitoring devices such as electrocardiograms (ECG), electroencephalograms (EEG), and electromyograms (EMG) [128]. Shanthi et al. [129] successfully developed a highly flexible and washable sports undergarment using graphene-coated textile electrodes by mat curing technology. The study showed that due to the hydrophobic nature of graphene when the textile electrode was exposed to sweat in maximum contact with the skin, its impedance, and conductivity were not disturbed by human sweat. Therefore, this sportswear could monitor high-quality ECG and heart rate under different conditions, such as rest, walking, and running. Graphene-coated textile electrodes exhibit excellent washability and are more suitable for long-term use than metal (Ag/AgCl) ECG electrodes, which are less flexible and stretchable. However, due to the chemical inertness of the graphene surface and the absence of dangling bonds to effectively control the channel carriers, the dielectric layer should be thin by a few nanometers. Li et al. [130] achieved this by optimizing pre-H_2O treatment and employing a two-step temperature growth atomic layer deposition technique. This way, they controlled the gas–solid physical adsorption between water molecules and the graphene surface, directly coating a uniform and dense Al_2O_3 film on graphene with a thickness of 5 nm. Consequently, the quality of the graphene–alumina was comparable to the optimal quality of the silicon–alumina film. Raza et al. [131] designed graphene textile (IGT) sensors by converting different polymer substrates into laser-induced graphene (LIG) and applying them to volleyball sportswear. The results showed that IGT sensors had four functions in volleyball sportswear: catch detection, finger contact foul detection when blocking the ball, spike force measurement, and player position monitoring. It is anticipated that graphene-wearable electronic sensors will improve the consistency between actual sports and virtual activities while enhancing the ability to correct movements in virtual space, thus providing unlimited opportunities for improving competitive levels.

5.2. Graphene Sneakers

Sports shoes can be divided into many types depending on the application requirements. Generally, polymer foams, especially ethylene vinyl acetates (EVA), are widely used as midsole cushioning materials in these articles. Therefore, the key to further improving the performance of sports shoes is to develop polymer foams with the required properties. Graphene is a carbon-based 2D material with potential advantages for improving polymer matrices' mechanical, electrical, thermal, and electromagnetic wave absorption properties [132]. The molecular dynamics (MDs) modeling of the interaction between the graphene and EVA matrix showed that adding graphene increases Young's modulus, the yield strength, and the glass transition temperature of the EVA matrix. When the solid interfacial bonding between the graphene additive and the EVA matrix (the graphene content is 9 wt.%) limits the mobility and flexibility of the EVA chain, the mechanical strength and thermal stability of EVA are improved [133]. Lunchev et al. [134] mixed graphene contents in EVA foam with a twin-screw extruder. After mechanical property analysis, graphene/EVA foam sports shoes were prepared with a sample containing 0.2 phr graphene. The results indicated that the graphene/EVA foam sneakers had a 40% increase in wear resistance and a 30% increase in axial stiffness compared to the pure EVA foam reference sneakers. This increase in the mechanical properties could effectively improve propulsion during running. At the same time, the compression testing revealed that the graphene/EVA foam sneakers were more flexible, which could help enhance their appearance and service life.

5.3. Graphene Tennis Racket

The design of tennis rackets constantly seeks to provide athletes with competitiveness, durability, and comfort; therefore, manufacturers are continually working to improve the materials and designs of their products. Graphene/polymer composites are graphene-dispersed in epoxy resin [135] and have higher strength and elastic modulus than ordinary materials. These composite materials can significantly improve the tennis rackets' stiffness and strength. Moreover, graphene/polymer composites are generally less dense than aluminum, have strength comparable to steel, and have a high elastic modulus and fatigue strength [136], making them more suitable for tennis rackets. In 2013, HEAD launched a new series of tennis rackets (YouTek Graphene Speed series) made of graphene/polymer composites [137]. Young et al. [138] analyzed the microstructure of these rackets through combined optical microscopy and Raman spectroscopy. The study found that graphene-based nanoparticles were utilized in the heads of the tennis rackets and the areas where they connected to the handles, aiming to enhance the mechanical performance of these regions. Thus, the excellent performance of graphene/polymer composites makes rackets lighter and stiffer and lays a solid foundation for athletes to achieve good results.

5.4. Graphene Sportswear

Sports clothing must have a cooling function to rapidly release heat and sweat from the body through different pathways, such as conduction, convection, evaporation, and radiation [139], especially when there is a significant temperature difference between the body and the surrounding environment or surfaces in direct contact [140]. Graphene has become popular in textile applications, especially in lightweight sportswear, due to its excellent properties, high thermal conductivity, mechanical strength, antistatic effect, UV protection, electrical conductivity, and ease of functionalization. It has been used as a conductive filler to increase heat transfer in sportswear textiles [141]. Brazilian sportswear manufacturer BiaBrazil launched sportswear printed with graphene ink that is claimed to improve moisture management and heat transfer by 18% [142].

In boxing, the wrong training method can easily cause physical injury. Therefore, many sports protective clothing use high-performance materials to reduce the negative impact of boxing training. Liu used the orthogonal test method [143] to study the effect of materials and knitting structure on boxing clothing performance. Based on the characteristics of boxing injury, the researcher(s) prepared seamless boxing clothing using a graphene/polymer composite fiber. These graphene-based composite boxing suits were more comfortable and robust than ordinary ones and had excellent impact resistance, protecting the athletes from potential injuries. Li et al. [144] developed a form of compression clothing for running crafted from biomass-derived graphene-modified nylon fibers. The findings of the study revealed that the graphene-modified nylon fibers demonstrated a tensile strength of 4.52 cN/dtex, boasting superior mechanical properties compared to conventional nylon fibers. Additionally, it exhibited an outstanding low-temperature far-infrared (FIR) performance, with the FIR raising its temperature by 2.3 °C/30 s. This feature can enhance blood circulation and improve bodily functions, thereby amplifying athletic performance.

5.5. Graphene Coating

Anterior cruciate ligament (ACL) rupture is a common joint ligament injury, especially in young athletic people aged 16 to 39 [145]. Due to the poor healing ability of ruptured ACLs and the need for injured people to restore their function in time, artificial ligaments are becoming increasingly popular. Graphene can potentially promote cell attachment, proliferation, and differentiation and has shown excellent results in various biological applications in vivo and in vitro. Graphene has been the focus of tissue repair applications due to its excellent mechanical properties. Studies have shown that graphene can promote the differentiation of human bone marrow mesenchymal stem cells into osteoblasts while promoting the differentiation of human neural stem cells into neurons [146,147]. Wang et al. [148] investigated the effect of graphene coating on the bioactivity of PET-based artificial ligaments (PET-ALS). The results showed that bone tunnel healing rates were significantly accelerated after graphene-coated grafts were implanted in bone tunnels. Therefore, graphene-coated grafts can promote early recovery of ACL reconstruction and show great potential in enhancing the biological activity of the material surface.

5.6. Challenges of Biomass Waste Graphene in Sports Applications

Three-dimensional nanostructured graphene not only inherits some unique properties of two-dimensional graphene but also offers numerous novel features, such as a boosted surface area, rapid gas diffusion, and abundant reaction sites [149–151]. Its applications in the sports field are steadily expanding, encompassing enhancements in sports equipment performance and monitoring athletes' physiological indicators. However, certain biomass wastes contain metal ions such as potassium and sodium, which can complicate the exfoliation process when deriving biomass graphene [152]. At the same time, varying temperatures and lignocellulose compositions can induce defects in the biowaste product composition, degradation rate, and other defects. Consequently, ensuring stability in the properties of the biowaste precursor materials becomes a key challenge [153]. In addition, heterogeneous waste precursor materials often contain small amounts of organic or inorganic impurities, necessitating suitable pretreatment [154]. Therefore, at an industrial scale, the costs of pretreatment and the requirements for supply chain stability may impede the large-scale production of biomass waste-derived graphene in the sports sector.

However, although graphene derived from biomass waste has not yet been produced on an industrial scale in the sports sector, its conversion to graphene offers potential benefits over traditional methods. These include reduced emissions of exhaust gases, wastewater, and solid waste to some extent, as well as decreased energy consumption [155], which further promote the development of the green sports industry.

6. Conclusions

This paper discusses the properties, production methods, and different biomass waste sources of graphene. Graphene has excellent electrical conductivity, biocompatibility, and mechanical properties compared to traditional carbon materials. Its application advantages in sports are mainly reflected in manufacturing lightweight and robust sports equipment and improving athletes' comfort. At the same time, it has the potential to study the physiological mechanism of athletes deeply and promote training and rehabilitation. In addition, graphene is expected to shine in ice and snow sports based on its excellent thermal conductivity. Although graphene prepared from biomass waste is similar to graphene from traditional sources, the quality of the graphene produced is compromised due to most of the biological waste being heterogeneous in origin, thus limiting its large-scale application. Future research should focus on the sustainability and scalability of graphene production from natural resources. Graphene and its derivatives with superior performance can be produced on a large scale by improving the processing techniques and standardizing the sources and characteristics of biomass waste to meet the demands of various applications, including sports equipment manufacturing. Although graphene based on biomass waste is of poor quality, it still shows great potential for various applications.

Author Contributions: Writing—original draft, Y.W.; conceptualization, writing—review and editing, project administration, supervision, funding acquisition, Y.L.; investigation, formal analysis, software, visualization, X.Z. All authors have read and agreed to the published version of the manuscript.

Funding: This research was funded by the Harbin Sports University Research Start-up Fund Project, grant number RC20-202112.

Conflicts of Interest: The authors declare no conflicts of interest.

References

1. Geim, A.K.; Novoselov, K.S. The rise of graphene. *Nat. Mater.* **2007**, *6*, 183–191. [CrossRef]
2. Avouris, P.; Dimitrakopoulos, C. Graphene: Synthesis and applications. *Mater. Today* **2012**, *15*, 86–97. [CrossRef]
3. Luo, B.; Liu, S.; Zhi, L. Chemical approaches toward graphene–based nanomaterials and their applications in energy–related areas. *Small* **2012**, *8*, 630–646. [CrossRef]
4. Di Matteo, P.; Petrucci, R.; Curulli, A. Not Only Graphene Two-Dimensional Nanomaterials: Recent Trends in Electrochemical (Bio) sensing Area for Biomedical and Healthcare Applications. *Molecules* **2023**, *29*, 172. [CrossRef]

5. GlobeNewswire. The Worldwide Graphene Industry Is Expected to Reach $1.4 Billion by 2025 at a CAGR of 19% from 2020. 2023. Available online: https://www.globenewswire.com/news-release/2021/04/02/2203853/28124/en/The-Worldwide-Graphene-Industry-is-Expected-to-Reach-1-4-Billion-by-2025-at-a-CAGR-of-19-from-2020.html (accessed on 12 January 2024).
6. Acharya, P.B.; George, A.; Shrivastav, P.S.A. Status Update on the Development of Polymer and Metal-Based Graphene Electrochemical Sensors for Detection and Quantitation of Bisphenol A. *Crit. Rev. Anal. Chem.* **2022**, 1–22. [CrossRef]
7. Sun, L.; Zhao, Z.; Sun, Y.; Wang, X.; Liu, X.; Yang, Y.; Qiu, J. Activated coal-based graphene with hierarchical porous structures for ultra-high energy density supercapacitors. *Diam. Relat. Mater.* **2020**, *106*, 107827. [CrossRef]
8. Li, H.; He, X.; Wu, T.; Jin, B.; Yang, L.; Qiu, J. Synthesis, modification strategies and applications of coal-based carbon materials. *Fuel Process. Technol.* **2022**, *230*, 107203. [CrossRef]
9. Srinivas, A.; Jivanlal Shaha, A.; Jakka, S.C.B. Studies on application of Petcoke-based graphene membranes for CO_2 separation. *Sep. Sci. Technol.* **2023**, *58*, 2738–2747. [CrossRef]
10. Sierra, U.; Álvarez, P.; Blanco, C.; Granda, M.; Santamaría, R.; Menéndez, R. Cokes of different origin as precursors of graphene oxide. *Fuel* **2016**, *166*, 400–403. [CrossRef]
11. Wang, X.; Zhang, L. Green and facile production of high-quality graphene from graphite by the combination of hydroxyl radicals and electrical exfoliation in different electrolyte systems. *RSC Adv.* **2019**, *9*, 3693–3703. [CrossRef]
12. Chew, K.W.; Chia, S.R.; Chia, W.Y.; Cheah, W.Y.; Munawaroh, H.S.H.; Ong, W.J. Abatement of hazardous materials and biomass waste via pyrolysis and co-pyrolysis for environmental sustainability and circular economy. *Environ. Pollut.* **2021**, *278*, 116836. [CrossRef] [PubMed]
13. Zhou, C.; Wang, Y. Recent progress in the conversion of biomass wastes into functional materials for value-added applications. *Sci. Technol. Adv. Mater.* **2020**, *21*, 787–804. [CrossRef] [PubMed]
14. Kalak, T. Potential use of industrial biomass waste as a sustainable energy source in the future. *Energies* **2023**, *16*, 1783. [CrossRef]
15. Pan, H.; Hu, Y.S.; Chen, L. Room-temperature stationary sodium-ion batteries for large-scale electric energy storage. *Energy Environ. Sci.* **2013**, *6*, 2338–2360. [CrossRef]
16. Han, Y.; Ge, Y.; Chao, Y.; Wang, C.; Wallace, G.G. Recent progress in 2D materials for flexible supercapacitors. *J. Energy Chem.* **2018**, *27*, 57–72. [CrossRef]
17. Li, G.; Luican, A.; Andrei, E.Y. Scanning tunneling spectroscopy of graphene on graphite. *Phys. Rev. Lett.* **2009**, *102*, 176804. [CrossRef] [PubMed]
18. Lin, S.; Dong, L.; Zhang, J.; Lu, H. Room-temperature intercalation and ~1000-fold chemical expansion for scalable preparation of high-quality graphene. *Chem. Mater.* **2016**, *28*, 2138–2146. [CrossRef]
19. Kang, F.; Leng, Y.; Zhang, T.Y. Influences of H_2O_2 on synthesis of H_2SO_4-GICs. *J. Phys. Chem. Solids* **1996**, *57*, 889–892. [CrossRef]
20. Tryba, B.; Morawski, A.W.; Inagaki, M. Preparation of exfoliated graphite by microwave irradiation. *Carbon* **2005**, *43*, 2417–2419. [CrossRef]
21. Safian, M.T.U.; Haron, U.S.; Ibrahim, M.M. A review on bio-based graphene derived from biomass wastes. *Bioresources* **2020**, *15*, 9756. [CrossRef]
22. Zhang, Y.; Tan, Y.W.; Stormer, H.L.; Kim, P. Experimental observation of the quantum Hall effect and Berry's phase in graphene. *Nature* **2005**, *438*, 201–204. [CrossRef] [PubMed]
23. Novoselov, K.S.; Jiang, Z.F.; Zhang, Y.S.; Morozov, S.V.; Stormer, H.L.; Zeitler, U.; Maan, J.C.; Boebinger, G.S.; Kim, P.; Geim, A.K. Room-temperature quantum Hall effect in graphene. *Science* **2007**, *315*, 1379. [CrossRef] [PubMed]
24. Fasolino, A.; Los, J.H.; Katsnelson, M.I. Intrinsic ripples in graphene. *Nat. Mater.* **2007**, *6*, 858–861. [CrossRef] [PubMed]
25. Novoselov, K.S.; Jiang, D.; Schedin, F.; Booth, T.J.; Khotkevich, V.V.; Morozov, S.V.; Geim, A.K. Two-dimensional atomic crystals. *Proc. Natl. Acad. Sci. USA* **2005**, *102*, 10451–10453. [CrossRef] [PubMed]
26. Lee, C.; Wei, X.; Kysar, J.W.; Hone, J. Measurement of the elastic properties and intrinsic strength of monolayer graphene. *Science* **2008**, *321*, 385–388. [CrossRef]
27. Shang, H.; Wang, W. Hypervelocity impact properties of graphene armor via molecular dynamics simulations. *EPJ Web Conf.* **2012**, *26*, 04027. [CrossRef]
28. Lee, J.U.; Yoon, D.; Cheong, H. Estimation of Young's modulus of graphene by Raman spectroscopy. *Nano Lett.* **2012**, *12*, 4444–4448. [CrossRef]
29. Wallace, P.R. The band theory of graphite. *Phys. Rev.* **1947**, *71*, 622. [CrossRef]
30. Novoselov, K.S.; Geim, A.K.; Morozov, S.V.; Jiang, D.; Katsnelson, M.I.; Grigorieva, I.V.; Dubonos, S.V.; Firsov, A.A. Two-dimensional gas of massless Dirac fermions in graphene. *Nature* **2005**, *438*, 197–200. [CrossRef]
31. Chen, J.H.; Jang, C.; Xiao, S.; Ishigami, M.; Fuhrer, M.S. Intrinsic and extrinsic performance limits of graphene devices on SiO_2. *Nat. Nanotechnol.* **2008**, *3*, 206–209. [CrossRef]
32. Nair, R.R.; Blake, P.; Grigorenko, A.N.; Novoselov, K.S.; Booth, T.J.; Stauber, T.; Peres, N.M.R.; Geim, A.K. Fine structure constant defines visual transparency of graphene. *Science* **2008**, *320*, 1308. [CrossRef] [PubMed]
33. Bae, S.; Kim, H.; Lee, Y.; Xu, X.; Park, J.S.; Zheng, Y.; Balakrishnan, J.; Lei, T.; Kim, H.R.; Song, Y.L.; et al. Roll-to-roll production of 30-inch graphene films for transparent electrodes. *Nat. Nanotechnol.* **2010**, *5*, 574–578. [CrossRef] [PubMed]
34. Eda, G.; Fanchini, G.; Chhowalla, M. Large-area ultrathin films of reduced graphene oxide as a transparent and flexible electronic material. *Nat. Nanotechnol.* **2008**, *3*, 270–274. [CrossRef]

35. Wang, F.; Zhang, Y.; Tian, C.; Girit, C.; Zettl, A.; Crommie, M.; Shen, Y.R. Gate-variable optical transitions in graphene. *Science* **2008**, *320*, 206–209. [CrossRef] [PubMed]
36. Li, Z.; Henriksen, E.A.; Jiang, Z.; Hao, Z.; Martin, M.C.; Kim, P.; Stormer, H.L.; Basov, D.N. Dirac charge dynamics in graphene by infrared spectroscopy. *Nat. Phys.* **2008**, *4*, 532–535. [CrossRef]
37. Rana, F.; George, P.A.; Strait, J.H.; Dawlaty, J.; Shivaraman, S.; Chandrashekhar, M.; Spencer, M.G. Carrier recombination and generation rates for intravalley and intervalley phonon scattering in graphene. *Phys. Rev. B* **2009**, *79*, 115447. [CrossRef]
38. Gokus, T.; Nair, R.R.; Bonetti, A.; Bohmler, M.; Lombardo, A.; Novoselov, K.S.; Geimt, A.K.; Ferrari, A.C.; Hartschuh, A. Making graphene luminescent by oxygen plasma treatment. *ACS Nano* **2009**, *3*, 3963–3968. [CrossRef] [PubMed]
39. Pumera, M.; Sofer, Z. Towards stoichiometric analogues of graphene: Graphane, fluorographene, graphol, graphene acid and others. *Chem. Soc. Rev.* **2017**, *46*, 4450–4463. [CrossRef]
40. Ho, K.I.; Boutchich, M.; Su, C.Y.; Moreddu, R.; Marianathan, E.S.R.; Montes, L.; Lai, C.S. A self–aligned high–mobility graphene transistor: Decoupling the channel with fluorographene to reduce scattering. *Adv. Mater.* **2015**, *27*, 6519–6525. [CrossRef]
41. Kumar, A.; Sharma, K.; Dixit, A.R. A review on the mechanical properties of polymer composites reinforced by carbon nanotubes and graphene. *Carbon Lett.* **2021**, *31*, 149–165. [CrossRef]
42. Liu, X.; Yang, Q. Molecular dynamic simulation of mechanical behaviour of RGO produced by thermal reduction method. *Micro Nano Lett.* **2017**, *12*, 638–642. [CrossRef]
43. Eda, G.; Chhowalla, M. Chemically derived graphene oxide: Towards large–area thin–film electronics and optoelectronics. *Adv. Mater.* **2010**, *22*, 2392–2415. [CrossRef]
44. Moggio, B.C.; Bergamasco, R.; Andrade, C.M.G.; Aylon, L.B.R. On the Analysis of Cryogels and Xerogels Using Cellulose Nanofibers and Graphene Oxide. *Polymers* **2023**, *15*, 3833. [CrossRef]
45. Andonegi, M.; Correia, D.M.; Pereira, N.; Costa, C.M.; Lanceros-Mendez, S.; Caba, K.; Guerrero, P. Sustainable Collagen Composites with Graphene Oxide for Bending Resistive Sensing. *Polymers* **2023**, *15*, 3855. [CrossRef]
46. Yadav, S.; Singh Raman, A.P.; Meena, H.; Goswami, A.G.; Bhawna; Kumar, V.; Jain, P.; Kumar, G.; Sagar, M.; Rana, D.K.; et al. An update on graphene oxide: Applications and toxicity. *ACS Omega* **2022**, *7*, 35387–35445. [CrossRef]
47. Chen, D.; Feng, H.; Li, J. Graphene oxide: Preparation, functionalization, and electrochemical applications. *Chem. Rev.* **2012**, *112*, 6027–6053. [CrossRef] [PubMed]
48. Shen, J.; Yan, B.; Li, T.; Long, Y.; Li, N.; Ye, M. Study on graphene-oxide-based polyacrylamide composite hydrogels. *Compos. Part A Appl. Sci. Manuf.* **2012**, *43*, 1476–1481. [CrossRef]
49. Barbosa, M.C.; Razzino, C.D.A.; Stocco, T.D.; Santana, M.D.V.; Ghosh, A.; Pereira, L.F.; Tierra-Criollo, C.J.; Lobo, A.O. Production of rGO-Based Electrospinning Nanocomposites Incorporated in Recycled PET as an Alternative Dry Electrode. *Polymers* **2022**, *14*, 4288. [CrossRef] [PubMed]
50. Alam, S.N.; Sharma, N.; Kumar, L. Synthesis of graphene oxide (GO) by modified hummers method and its thermal reduction to obtain reduced graphene oxide (rGO). *Graphene* **2017**, *6*, 1–18. [CrossRef]
51. Pei, S.; Cheng, H.M. The reduction of graphene oxide. *Carbon* **2012**, *50*, 3210–3228. [CrossRef]
52. Becerril, H.A.; Mao, J.; Liu, Z.; Stoltenberg, R.M.; Bao, Z.; Chen, Y. Evaluation of solution-processed reduced graphene oxide films as transparent conductors. *ACS Nano* **2008**, *2*, 463–470. [CrossRef] [PubMed]
53. Stankovich, S.; Piner, R.D.; Chen, X.; Wu, N.; Nguyen, S.T.; Ruoff, R.S. Stable aqueous dispersions of graphitic nanoplatelets via the reduction of exfoliated graphite oxide in the presence of poly (sodium 4-styrenesulfonate). *J. Mater. Chem. A* **2006**, *16*, 155–158. [CrossRef]
54. Zhang, S.; Zhang, G.; Zheng, L.; Liu, Z.; Wang, B.; Wu, H.; He, Z.; Jin, Z.; Wang, G. High-Performance Near-Infrared Photodetector by Integration of PbS Quantum Dots with 3D-Graphene. *IEEE Electron Device Lett.* **2023**, *44*, 1240–1243. [CrossRef]
55. Denis, P.A. Band gap opening of monolayer and bilayer graphene doped with aluminium, silicon, phosphorus, and sulfur. *Chem. Phys. Lett.* **2010**, *492*, 251–257. [CrossRef]
56. García-Lastra, J.M. Strong dependence of band-gap opening at the Dirac point of graphene upon hydrogen adsorption periodicity. *Phys. Rev. B* **2010**, *82*, 235418. [CrossRef]
57. Chen, S.; Duan, J.; Jaroniec, M.; Qiao, S.Z. Three–dimensional N–doped graphene hydrogel/NiCo double hydroxide electrocatalysts for highly efficient oxygen evolution. Angew. *Chem. Int. Ed.* **2013**, *52*, 13567–13570. [CrossRef] [PubMed]
58. Edwards, R.S.; Coleman, K.S. Graphene synthesis: Relationship to applications. *Nanoscale* **2013**, *5*, 38–51. [CrossRef]
59. Kumar, N.; Salehiyan, R.; Chauke, V.; Botlhoko, O.J.; Setshedi, K.; Scriba, M.; Masukume, M.; Ray, S.S. Top-down synthesis of graphene: A comprehensive review. *FlatChem* **2021**, *27*, 100224. [CrossRef]
60. Saeed, M.; Alshammari, Y.; Majeed, S.A.; Al-Nasrallah, E. Chemical vapour deposition of graphene—Synthesis, characterisation, and applications: A review. *Molecules* **2020**, *25*, 3856. [CrossRef]
61. Gutiérrez-Cruz, A.; Ruiz-Hernández, A.R.; Vega-Clemente, J.F.; Luna-Gazcón, D.G.; Campos-Delgado, J. A review of top-down and bottom-up synthesis methods for the production of graphene, graphene oxide and reduced graphene oxide. *J. Mater. Sci.* **2022**, *57*, 14543–14578. [CrossRef]
62. Wehling, T.O.; Novoselov, K.S.; Morozov, S.V.; Vdovin, E.E.; Katsnelson, M.I.; Geim, A.K.; Lichtenstein, A.I. Molecular Doping of Graphene. *Nano Lett.* **2008**, *8*, 173–177. [CrossRef]
63. Tian, W.; Liu, X.; Yu, W. Research progress of gas sensor based on graphene and its derivatives: A review. *Appl. Sci.* **2018**, *8*, 1118. [CrossRef]

64. Land, T.A.; Michely, T.; Behm, R.J.; Hemminger, J.C.; Comsa, G. STM investigation of single layer graphite structures produced on Pt (111) by hydrocarbon decomposition. *Surf. Sci.* **1992**, *264*, 261–270. [CrossRef]
65. Suk, J.W.; Kitt, A.; Magnuson, C.W.; Hao, Y.; Ahmed, S.; An, J.; Swan, A.K.; Goldberg, B.B.; Ruoff, R.S. Transfer of CVD-grown monolayer graphene onto arbitrary substrates. *ACS Nano* **2011**, *5*, 6916–6924. [CrossRef]
66. Geim, A.K. Graphene: Status and prospects. *Science* **2009**, *324*, 1530–1534. [CrossRef] [PubMed]
67. Novoselov, K.S.; Geim, A.K.; Morozov, S.V.; Jiang, D.E.; Zhang, Y.; Dubonos, S.V.; Grigorieva, I.V.; Firsov, A.A. Electric field effect in atomically thin carbon films. *Science* **2004**, *306*, 666–669. [CrossRef] [PubMed]
68. Paton, K.R.; Varrla, E.; Backes, C.; Smith, R.J.; Khan, U.; O'Neill, A.; Boland, C.; Lotya, M.; Istrate, O.M.; King, P.; et al. Scalable production of large quantities of defect-free few-layer graphene by shear exfoliation in liquids. *Nat. Mater.* **2014**, *13*, 624–630. [CrossRef]
69. Yi, M.; Shen, Z. A review on mechanical exfoliation for the scalable production of graphene. *J. Mater. Chem. A* **2015**, *3*, 11700–11715. [CrossRef]
70. Banerjee, B.C.; Hirt, T.J.; Walker, P.L. Pyrolytic carbon formation from carbon suboxide. *Nature* **1961**, *192*, 450–451. [CrossRef]
71. Robertson, S.D. Graphite formation from low temperature pyrolysis of methane over some transition metal surfaces. *Nature* **1969**, *221*, 1044–1046. [CrossRef]
72. Wang, R.; Hao, Y.; Wang, Z.; Gong, H.; Thong, J.T. Large-diameter graphene nanotubes synthesized using Ni nanowire templates. *Nano Lett.* **2010**, *10*, 4844–4850. [CrossRef]
73. Juang, Z.Y.; Wu, C.Y.; Lu, A.Y.; Su, C.Y.; Leou, K.C.; Chen, F.R.; Tsai, C.H. Graphene synthesis by chemical vapor deposition and transfer by a roll-to-roll process. *Carbon* **2010**, *48*, 3169–3174. [CrossRef]
74. Chen, S.; Bai, Q.; Wang, H.; Dou, Y.; Guo, W. Controlled growth of large-area monolayer graphene on Ni (110) facet: Insight from molecular dynamics simulation. *Phys. E* **2022**, *144*, 115465. [CrossRef]
75. Orofeo, C.M.; Ago, H.; Hu, B.; Tsuji, M. Synthesis of large area, homogeneous, single layer graphene films by annealing amorphous carbon on Co and Ni. *Nano Res.* **2011**, *4*, 531–540. [CrossRef]
76. Wang, H.; Yu, G. Direct CVD graphene growth on semiconductors and dielectrics for transfer–free device fabrication. *Adv. Mater.* **2016**, *28*, 4956–4975. [CrossRef]
77. Losurdo, M.; Giangregorio, M.M.; Capezzuto, P.; Bruno, G. Graphene CVD growth on copper and nickel: Role of hydrogen in kinetics and structure. *Phys. Chem. Chem. Phys.* **2011**, *13*, 20836–20843. [CrossRef]
78. Munoz, R.; Gómez–Aleixandre, C. Review of CVD synthesis of graphene. *Chem. Vap. Depos.* **2013**, *19*, 297–322. [CrossRef]
79. Li, X.; Magnuson, C.W.; Venugopal, A.; Tromp, R.M.; Hannon, J.B.; Vogel, E.M.; Colombo, L.; Ruoff, R.S. Large-area graphene single crystals grown by low-pressure chemical vapor deposition of methane on copper. *J. Am. Chem. Soc.* **2011**, *133*, 2816–2819. [CrossRef]
80. Wu, T.; Zhang, X.; Yuan, Q.; Xue, J.; Lu, G.; Liu, Z.; Wang, H.; Wang, H.; Ding, F.; Yu, Q.; et al. Fast growth of inch-sized single-crystalline graphene from a controlled single nucleus on Cu–Ni alloys. *Nat. Mater.* **2016**, *15*, 43–47. [CrossRef]
81. Park, H.J.; Meyer, J.; Roth, S.; Skákalová, V. Growth and properties of few-layer graphene prepared by chemical vapor deposition. *Carbon* **2010**, *48*, 1088–1094. [CrossRef]
82. Brodie, B.C. XIII. On the atomic weight of graphite. *Philos. Trans. R. Soc. Lond.* **1859**, *149*, 249–259.
83. Hummers, W.S., Jr.; Offeman, R.E. Preparation of graphitic oxide. *J. Am. Chem. Soc.* **1958**, *80*, 1339. [CrossRef]
84. Endemann, H. Berichte der Deutschen Chemischen Gesellschaft. *J. Am. Chem. Soc.* **1880**, *2*, 366–371. [CrossRef]
85. Zhu, Y.; Murali, S.; Stoller, M.D.; Velamakanni, A.; Piner, R.D.; Ruoff, R.S. Microwave assisted exfoliation and reduction of graphite oxide for ultracapacitors. *Carbon* **2010**, *48*, 2118–2122. [CrossRef]
86. Wang, H.; Robinson, J.T.; Li, X.; Dai, H. Solvothermal reduction of chemically exfoliated graphene sheets. *J. Am. Chem. Soc.* **2009**, *131*, 9910–9911. [CrossRef] [PubMed]
87. Hu, H.; Zhao, Z.; Zhou, Q.; Gogotsi, Y.; Qiu, J. The role of microwave absorption on formation of graphene from graphite oxide. *Carbon* **2012**, *50*, 3267–3273. [CrossRef]
88. Tateishi, H.; Koinuma, M.; Miyamoto, S.; Kamei, Y.; Hatakeyama, K.; Ogata, C.; Taniguchi, T.; Funatsu, A.; Matsumoto, Y. Effect of the electrochemical oxidation/reduction cycle on the electrochemical capacitance of graphite oxide. *Carbon* **2014**, *76*, 40–45. [CrossRef]
89. Lung-Hao Hu, B.; Wu, F.Y.; Lin, C.T.; Khlobystov, A.N.; Li, L.J. Graphene-modified $LiFePO_4$ cathode for lithium ion battery beyond theoretical capacity. *Nat. Commun.* **2013**, *4*, 1687. [CrossRef] [PubMed]
90. Ren, S.; Rong, P.; Yu, Q. Preparations, properties and applications of graphene in functional devices: A concise review. *Ceram. Int.* **2018**, *44*, 11940–11955. [CrossRef]
91. Manoratne, C.H.; Rosa, S.R.D.; Kottegoda, I.R.M. XRD-HTA, UV visible, FTIR and SEM interpretation of reduced graphene oxide synthesized from high purity vein graphite. *Mater. Sci. Res. India* **2017**, *14*, 19–30. [CrossRef]
92. Wu, W.; Yu, B. Corn flour nano-graphene prepared by the hummers redox method. *ACS Omega* **2020**, *5*, 30252–30256. [CrossRef] [PubMed]
93. Shelton, J.C.; Patil, H.R.; Blakely, J.M. Equilibrium segregation of carbon to a nickel (111) surface: A surface phase transition. *Surf. Sci.* **1974**, *43*, 493–520. [CrossRef]
94. Lahiri, J.; Miller, T.S.; Ross, A.J.; Adamska, L.; Oleynik, I.I.; Batzill, M. Graphene growth and stability at nickel surfaces. *New J. Phys.* **2011**, *13*, 025001. [CrossRef]

95. Virojanadara, C.; Syväjarvi, M.; Yakimova, R.; Johansson, L.I.; Zakharov, A.A.; Balasubramanian, T. Homogeneous large-area graphene layer growth on 6 H-SiC (0001). *Phys. Rev. B* **2008**, *78*, 245403. [CrossRef]
96. Hass, J.; Varchon, F.; Millan-Otoya, J.E.; Sprinkle, M.; Sharma, N.; Heer, W.A.; Berger, C.; First, P.N.; Magaud, L.; Conrad, E.H. Why multilayer graphene on 4 H–SiC (0001) behaves like a single sheet of graphene. *Phys. Rev. Lett.* **2008**, *100*, 125504. [CrossRef] [PubMed]
97. Gao, M.; Pan, Y.; Huang, L.; Hu, H.; Zhang, L.Z.; Guo, H.M.; Du, S.X.; Gao, H.J. Epitaxial growth and structural property of graphene on Pt (111). *Appl. Phys. Lett.* **2011**, *98*, 033101. [CrossRef]
98. Chen, H.; Zhu, W.; Zhang, Z. Contrasting behavior of carbon nucleation in the initial stages of graphene epitaxial growth on stepped metal surfaces. *Phys. Rev. Lett.* **2010**, *104*, 186101. [CrossRef] [PubMed]
99. Li, X.; Zheng, Y. Lignin-enzyme interaction: Mechanism, mitigation approach, modeling, and research prospects. *Biotechnol. Adv.* **2017**, *35*, 466–489. [CrossRef] [PubMed]
100. Tripathi, M.; Sahu, J.N.; Ganesan, P. Effect of process parameters on production of biochar from biomass waste through pyrolysis: A review. *Renew. Sustain. Energy Rev.* **2016**, *55*, 467–481. [CrossRef]
101. Broda, M.; Yelle, D.J.; Serwańska, K. Bioethanol production from lignocellulosic biomass—Challenges and solutions. *Molecules* **2022**, *27*, 8717. [CrossRef]
102. Kharangarh, P.R.; Ravindra, N.M.; Singh, G.; Umapathy, S. Synthesis of luminescent graphene quantum dots from biomass waste materials for energy–related applications—An Overview. *Energy Storage* **2023**, *5*, e390. [CrossRef]
103. Zhou, Y.; He, J.; Chen, R.; Li, X. Recent advances in biomass-derived graphene and carbon nanotubes. *Mater. Today Sustain.* **2022**, *18*, 100138. [CrossRef]
104. Koul, B.; Yakoob, M.; Shah, M.P. Agricultural waste management strategies for environmental sustainability. *Environ. Res.* **2022**, *206*, 112285. [CrossRef]
105. Plazzotta, S.; Manzocco, L.; Nicoli, M.C. Fruit and vegetable waste management and the challenge of fresh-cut salad. *Trends Food Sci. Technol.* **2017**, *63*, 51–59. [CrossRef]
106. Dijk, C.; Doorn, W.; Alfen, B. Long term plant biomonitoring in the vicinity of waste incinerators in The Netherlands. *Chemosphere* **2015**, *122*, 45–51. [CrossRef]
107. Prasad, R.; Abhijeet, S.; Garg, R.; Hosmane, G.B. Biomass fuel exposure and respiratory diseases in India. *Biosci. Trends* **2012**, *6*, 219–228. [CrossRef]
108. Collard, F.X.; Blin, J. A review on pyrolysis of biomass constituents: Mechanisms and composition of the products obtained from the conversion of cellulose, hemicelluloses and lignin. *Renew. Sustain. Energy Rev.* **2012**, *38*, 594–608. [CrossRef]
109. Li, L.; Tang, Y.; Bao, Z.; Tu, W.; Peng, L.; Zou, L.; Zhao, C.; Li, Q. When graphene meets circular agriculture: Insights into agricultural sustainable development. *Biosyst. Eng.* **2024**, *237*, 92–117. [CrossRef]
110. Hossain, S.S.; Mathur, L.; Roy, P.K. Rice husk/rice husk ash as an alternative source of silica in ceramics: A review. *J. Asian Ceram.* **2018**, *6*, 299–313. [CrossRef]
111. Bakar, R.A.; Yahya, R.; Gan, S.N. Production of high purity amorphous silica from rice husk. *Procedia Chem.* **2016**, *19*, 189–195. [CrossRef]
112. Allegretti, C.; Bellinetto, E.; D'Arrigo, P.; Ferro, M.; Griffini, G.; Rossato, L.A.M.; Ruffini, E.; Schiavi, L.; Serra, S.; Strini, A.; et al. Fractionation of Raw and Parboiled Rice Husks with Deep Eutectic Solvents and Characterization of the Extracted Lignins towards a Circular Economy Perspective. *Molecules* **2022**, *27*, 8879. [CrossRef]
113. Wang, Z.; Yu, J.; Zhang, X.; Li, N.; Liu, B.; Li, Y.; Wang, Y.; Wang, W.; Li, Y.; Zhang, L.; et al. Large-scale and controllable synthesis of graphene quantum dots from rice husk biomass: A comprehensive utilization strategy. *ACS Appl. Mater. Interfaces* **2016**, *8*, 1434–1439. [CrossRef]
114. Chen, L.; Zhang, Y.; Lin, C.; Yang, W.; Meng, Y.; Guo, Y.; Li, M.; Xiao, D. Hierarchically porous nitrogen-rich carbon derived from wheat straw as an ultra-high-rate anode for lithium ion batteries. *J. Mater. Chem. A* **2012**, 2, 9684–9690. [CrossRef]
115. Chen, F.; Yang, J.; Bai, T.; Long, B.; Zhou, X. Facile synthesis of few-layer graphene from biomass waste and its application in lithium ion batteries. *J. Electroanal. Chem.* **2016**, *768*, 18–26. [CrossRef]
116. Zhou, X.; Chen, F.; Bai, T.; Long, B.; Liao, Q.; Ren, Y.; Yang, J. Interconnected highly graphitic carbon nanosheets derived from wheat stalk as high performance anode materials for lithium ion batteries. *Green Chem.* **2016**, *18*, 2078–2088. [CrossRef]
117. Monties, B. Plant cell walls as fibrous lignocellulosic composites: Relations with lignin structure and function. *Anim. Feed Sci. Technol.* **1991**, *32*, 159–175. [CrossRef]
118. Manchala, S.; Tandava, V.S.R.K.; Jampaiah, D.; Bhargava, S.K.; Shanker, V. Novel and highly efficient strategy for the green synthesis of soluble graphene by aqueous polyphenol extracts of eucalyptus bark and its applications in high-performance supercapacitors. *ACS Sustain. Chem. Eng.* **2019**, *7*, 11612–11620. [CrossRef]
119. Severo, L.S.; Thue, P.S.; Lima, D.R.; Didó, C.A.; Vasconcellos, M.A.; Armas, L.E.; Lima, E.C.; Benvenutti, E.V.; Menezes, E.W. 3D graphene sponge biomass-derived with high surface area applied as adsorbent for nitrophenols. *J. Environ. Chem. Eng.* **2023**, *11*, 109924. [CrossRef]
120. Alfè, M.; Gargiulo, V.; Porto, M.; Migliaccio, R.; Le Pera, A.; Sellaro, M.; Pellegrino, C.; Abe, A.A.; Urciuolo, M.; Caputo, P.; et al. Pyrolysis and Gasification of a Real Refuse-Derived Fuel (RDF): The Potential Use of the Products under a Circular Economy Vision. *Molecules* **2022**, *27*, 8114. [CrossRef]

121. Amaral, C.; Pedro, M.I.; Ferreira, D.C.; Marques, R.C. Performance and its determinants in the Portuguese municipal solid waste utilities. *Waste Manag.* **2022**, *139*, 70–84. [CrossRef]
122. Harst, E.; Potting, J. A critical comparison of ten disposable cup LCAs. Environ. *Impact Assess. Rev.* **2012**, *43*, 86–96. [CrossRef]
123. Ruan, G.; Sun, Z.; Peng, Z.; Tour, J.M. Growth of graphene from food, insects, and waste. *ACS Nano* **2011**, *5*, 7601–7607. [CrossRef]
124. Rum, L.; Sten, O.; Vendrame, E.; Belluscio, V.; Camomilla, V.; Vannozzi, G.; Truppa, L.; Notarantonio, M.; Sciarra, T.; Lazich, A.; et al. Wearable sensors in sports for persons with disability: A systematic review. *Sensors* **2021**, *21*, 1858. [CrossRef]
125. Pani, D.; Achilli, A.; Bonfiglio, A. Survey on textile electrode technologies for electrocardiographic (ECG) monitoring, from metal wires to polymers. *Adv. Mater. Technol.* **2018**, *3*, 1800008. [CrossRef]
126. Ge, G.; Huang, W.; Shao, J.; Dong, X. Recent progress of flexible and wearable strain sensors for human-motion monitoring. *J. Semicond.* **2018**, *39*, 011012. [CrossRef]
127. Zou, Y.; Zhong, M.; Li, S.; Qing, Z.; Xing, X.; Gong, G.; Zhou, C. Flexible Wearable Strain Sensors Based on Laser-Induced Graphene for Monitoring Human Physiological Signals. *Polymers* **2023**, *15*, 3553. [CrossRef]
128. Choi, S.; Han, S.I.; Kim, D.; Hyeon, T.; Kim, D.H. High-performance stretchable conductive nanocomposites: Materials, processes, and device applications. *Chem. Soc. Rev.* **2019**, *48*, 1566–1595. [CrossRef]
129. Shathi, M.A.; Chen, M.; Khoso, N.A.; Rahman, M.T.; Bhattacharjee, B. Graphene coated textile based highly flexible and washable sports bra for human health monitoring. *Mater. Des.* **2020**, *193*, 108792. [CrossRef]
130. Zheng, L.; Cheng, X.; Cao, D.; Wang, G.; Wang, Z.; Xu, D.; Xia, C.; Shen, L.; Yu, Y.; Shen, D. Improvement of Al_2O_3 films on graphene grown by atomic layer deposition with pre-H_2O treatment. *ACS Appl. Mater. Interfaces* **2014**, *6*, 7014–7019. [CrossRef]
131. Raza, T.; Tufail, M.K.; Ali, A.; Boakye, A.; Qi, X.; Ma, Y.; Ali, A.; Qu, L.; Tian, M. Wearable and flexible multifunctional sensor based on laser-induced graphene for the sports monitoring system. *ACS Appl. Mater. Interfaces* **2022**, *14*, 54170–54181. [CrossRef]
132. Kinloch, I.A.; Suhr, J.; Lou, J.; Young, R.J.; Ajayan, P.M. Composites with carbon nanotubes and graphene: An outlook. *Science* **2018**, *362*, 547–553. [CrossRef]
133. Shi, J.; Yang, J.; Zhou, J.; Ji, H.; Tang, X.; Gao, T. Effect of graphene on thermal stability and mechanical properties of ethylene-vinyl acetate: A molecular dynamics simulation. *Mater. Res. Express* **2020**, *7*, 035304. [CrossRef]
134. Lunchev, A.V.; Kashcheev, A.; Lipik, V. Mechanical characteristics of poly (ethylene vinyl acetate) foams with graphene for the applications in sport footwear. *Polym. Test.* **2022**, *113*, 107688. [CrossRef]
135. Tang, L.C.; Wan, Y.J.; Yan, D.; Pei, Y.B.; Zhao, L.; Li, Y.B.; Wu, L.B.; Jiang, J.X.; Lai, G.Q. The effect of graphene dispersion on the mechanical properties of graphene/epoxy composites. *Carbon* **2013**, *60*, 16–27. [CrossRef]
136. Fu, B.; Gao, J. Study on the application of graphene polymer material in sports equipment. *J. Phys. Conf. Ser.* **2020**, *1676*, 012044. [CrossRef]
137. HEAD Technology GMBH Ltd. Sporting Goods with Graphene Material. U.S. Patent 2013/0090193, 11 April 2013.
138. Young, R.J.; Liu, M. The microstructure of a graphene-reinforced tennis racquet. *J. Mater. Sci.* **2016**, *51*, 3861–3867. [CrossRef]
139. Kalaoglu-Altan, O.I.; Kayaoglu, B.K.; Trabzon, L. Improving thermal conductivities of textile materials by nanohybrid approaches. *iScience* **2022**, *25*, 103825. [CrossRef]
140. Havenith, G.; Fiala, D. Thermal indices and thermophysiological modeling for heat stress. *Compr. Physiol.* **2011**, *6*, 255–302.
141. Bhattacharjee, S.; Joshi, R.; Chughtai, A.A.; Macintyre, C.R. Graphene modified multifunctional personal protective clothing. *Adv. Mater. Interfaces* **2019**, *6*, 1900622. [CrossRef]
142. Determining the Performance of Graphene-Enhanced Sportswear | Technical Textiles. Available online: https://www.technicaltextiles.net/node/76391 (accessed on 28 December 2023).
143. Liu, T. Properties of graphene composite fiber seamless knitted fabric and its application in boxing training. *Front. Mater.* **2023**, *10*, 1098652. [CrossRef]
144. Li, X.; Liu, L.; Wang, S. Running compression garments design using biomass graphene modified fiber. *J. Fiber Bioeng. Inform.* **2019**, *12*, 155–166. [CrossRef]
145. Widuchowski, W.; Widuchowski, J.; Koczy, B.; Szyluk, K. Untreated asymptomatic deep cartilage lesions associated with anterior cruciate ligament injury: Results at 10-and 15-year follow-up. *Am. J. Sports Med.* **2009**, *37*, 688–692. [CrossRef]
146. Park, J.K.; Jung, J.; Subramaniam, P.; Shah, B.P.; Kim, C.; Lee, J.K.; Cho, J.H.; Lee, C.; Lee, K.B. Graphite-coated magnetic nanoparticles as multimodal imaging probes and cooperative therapeutic agents for tumor cells. *Small* **2011**, *7*, 1647. [CrossRef]
147. Liu, F.; Liu, C.; Zheng, B.; He, J.; Liu, J.; Chen, C.; Lee, I.; Wang, X.; Liu, Y. Synergistic effects on incorporation of β-tricalcium phosphate and graphene oxide nanoparticles to silk fibroin/soy protein isolate scaffolds for bone tissue engineering. *Polymers* **2020**, *12*, 69. [CrossRef]
148. Wang, C.H.; Guo, Z.S.; Pang, F.; Zhang, L.Y.; Yan, M.; Yan, J.H.; Li, K.W.; Li, X.J.; Li, Y.; Bi, L.; et al. Effects of graphene modification on the bioactivation of polyethylene-terephthalate-based artificial ligaments. *ACS Appl. Mater. Interfaces* **2015**, *7*, 15263–15276. [CrossRef]
149. Chen, Z.; Wang, J.; Cao, N.; Wang, Y.; Li, H.; Rooij, N.F.D.; Umar, A.; Feng, Y.; French, P.J.; Zhou, G. Three-dimensional graphene-based foams with "greater electron transferring areas" deriving high gas sensitivity. *ACS Appl. Nano Mater.* **2021**, *4*, 13234–13245. [CrossRef]
150. Chen, Z.; Wang, J.; Pan, D.; Wang, Y.; Noetzel, R.; Li, H.; Xie, P.; Pei, W.; Umar, A.; Jiang, L.; et al. Mimicking a dog's nose: Scrolling graphene nanosheets. *ACS Nano* **2018**, *12*, 2521–2530. [CrossRef]

151. Chen, Z.; Wang, J.; Umar, A.; Wang, Y.; Li, H.; Zhou, G. Three-dimensional crumpled graphene-based nanosheets with ultrahigh NO_2 gas sensibility. *ACS Appl. Mater.* **2017**, *9*, 11819–11827. [CrossRef]
152. Trubetskaya, A.; Larsen Andersen, M.; Talbro Barsberg, S. The nature of stable char radicals: An ESR and DFT study of structural and hydrogen bonding requirements. *ChemPlusChem* **2018**, *83*, 780–786. [CrossRef]
153. Surup, G.R.; Nielsen, H.K.; Heidelmann, M.; Trubetskaya, A. Characterization and reactivity of charcoal from high temperature pyrolysis (800–1600 C). *Fuel* **2019**, *235*, 1544–1554. [CrossRef]
154. Tamuly, J.; Bhattacharjya, D.; Saikia, B.K. Graphene/graphene derivatives from coal, biomass, and wastes: Synthesis, energy applications, and perspectives. *Energy Fuels* **2022**, *36*, 12847–12874. [CrossRef]
155. Ji, Z.; Zhang, G.; Liu, R.; Qu, J.; Liu, H. Potential applications of solid waste-based geopolymer materials: In wastewater treatment and greenhouse gas emission reduction. *J. Clean. Prod.* **2024**, *443*, 141144. [CrossRef]

molecules

MDPI

Article

Mathematical Modelling and Optimization for Facile Synthesis of Structured Activated Carbon (ACs) from *Adansonia kilima* (*Baobab*) Wood Chips Integrating Microwave-Assisted Pyrolysis for the Elimination of Lead (II) Cations from Wastewater Effluents

Santhana Sellamuthu [1], Zaira Zaman Chowdhury [1,*], Khalisanni Khalid [2], Shahjalal Mohd. Shibly [1], Md Mahfujur Rahman [3,*], Masud Rana [1,4], Irfan Anjum Badruddin [5], H. M. T. Khaleed [6], Sarfaraz Kamangar [5], Mohd. Rafie Bin Johan [1], Mohamed Hussein [7], Ajita Mitra [1] and Abu Nasser Faisal [1]

1 Nanotechnology and Catalysis Research Center, University of Malaya, Kuala Lumpur 50603, Malaysia; ajitamitra@gmail.com (A.M.); faisal_dhaka2003@yahoo.com (A.N.F.)
2 Malaysian Agricultural Research & Development Institute (MARDI), Serdang 43000, Malaysia
3 Institute of Halal Management, Islamic Business School, Universiti Utara Malaysia, Kedah 06010, Malaysia
4 Department of Civil Engineering, World University of Bangladesh, Dhaka 1230, Bangladesh
5 Department of Mechanical Engineering, College of Engineering, King Khalid University, P.O. Box 394, Abha 61421, Saudi Arabia
6 Department of Mechanical Engineering, Faculty of Engineering, Islamic University, Madinah Munawwarra 42351, Saudi Arabia
7 Department of Chiemistry, King Khalid University, P.O. Box 9004, Abha 61413, Saudi Arabia
* Correspondence: dr.zaira.chowdhury@um.edu.my or zaira.chowdhury76@gmail.com (Z.Z.C.); dr_mahfuj@yahoo.co.uk (M.M.R.); Tel.: +60-37967-2929 or +60-10267-5621 (Z.Z.C.)

Abstract: In this research, activated carbon (AC) was synthesized from ligno-cellulosic residues of *Adansonia kilima (Baobab)* wood chips (AKTW) using two-step semi-carbonization and subsequent pyrolysis using microwave-induced heating (MWP) in the presence of a mild activating agent of K_2CO_3. The influence of process input variables of microwave power (x_1), residence time (y_1), and amount of K_2CO_3 (z_1) were analysed to yield superior quality carbon having maximum removal efficiencies (R_1) for lead (II) cations from waste effluents, fixed carbon percentages (R_2), and carbon yield percentages (R_3). Analysis of variance (ANOVA) was used to develop relevant mathematical models, with an appropriate statistical assessment of errors. Level factorial response surface methodology (RSM) relying on the Box–Behnken design (BBD) was implemented for the experimental design. The surface area and porous texture of the samples were determined using Brunauer, Emmett, and Teller (BET) adsorption/desorption curves based on the N_2 isotherm. Surface morphological structure was observed using field emission scanning electron microscopic (FESEM) analysis. Thermogravimetric analysis (TGA) was carried out to observe the thermal stability of the sample. Change in the carbon content of the samples was determined using ultimate analysis. X-ray diffraction (XRD) analysis was performed to observe the crystalline and amorphous texture of the samples. The retention of a higher proportion of fixed carbon (80.01%) ensures that the synthesized adsorbent (AKTWAC) will have a greater adsorption capacity while avoiding unwanted catalytic activity for our synthesized final sample.

Keywords: ligno-cellulosic biomass; activated carbon (AC); response surface methodology (RSM); Box–Behnken design (BBD); lead (II) cations (Pb^{+2}); microwave-assisted pyrolysis (MWP); conventional pyrolysis; optimization and analysis of variance (ANOVA)

1. Introduction

Adsorption is the most widely utilized approach for treating wastewater among all known techniques due to its ease of operation, minimal installation costs, and superior effluent treatment effectiveness [1–4]. It is a reasonably environmentally benign and sustainable strategy that utilizes waste materials, such as lignocellulosic wastes, to develop adsorbent materials. Subsequently, it leads to the reduction of environmental pollution [5–7]. According to the adsorption studies, activated carbon (AC) is a commonly utilized adsorbent material due to its excellent textural features; such as enlarged surface area, average pore volume, and well-defined porous structures [8–10].

The morphological, as well as the chemical, features of the synthesized carbon are determined by the lignocellulosic precursors [8,9], including the experimental variables under which the activation procedure has taken place [10]. Physical and chemical processes of activation, or a combination of both techniques, are used to generate activated carbons (ACs) [11,12]. The porous nature of activated carbon (AC) enables it to have a high sorption capacity. It is frequently used in industries to purify wastewater containing metallic/non-metallic impurities, colours, and organic compounds. Additionally, it has exhibited considerable performance as a catalyst support and heating substrate in thermochemical systems.

There are two processes used to produce ACs: carbonization and activation. Lignocellulosic biomass is a suitable precursor for the production of ACs. ACs can be generated using physical, chemical, and physiochemical thermochemical conversion of lignocellulosic biomass. There is one step involved in the physical and chemical process during pyrolysis. The physiochemical method involves two steps of carbonization and activation in the presence of a suitable chemical reacting agent. These processes are carried out independently, having the raw materials first carbonized and pyrolyzed at a higher temperature (400–1000 °C) in the presence or absence of physical or chemical activating agents to enhance their porosity. Nevertheless, carbonization and activation can be accomplished in a single-step process. There are mainly three types of pyrolysis process: traditional, microwave heating, and catalytic pyrolysis. Conventional fixed-bed/tubular pyrolysis with electric heating is more prevalently utilized to manufacture ACs [4]. Due to the extended period required for the pyrolysis in an inert environment, the traditional electric heating technique has economical limitations [13,14]. Additionally, due to the diameter and length of the tubular furnace, a temperature gradient exists between the upper surface of the precursors and the interior side of the sample, which may influence the composition of the activated carbons (ACs) produced [13,14].

The conventional heating method for transferring the heat may be ineffective and slow. The heat must be transmitted from the heat source (e.g., heating surface or filament) to the reactor system, and then to the thermal conductor (e.g., the gas contained within the furnace compartment) before reaching the sample surface which needs to be heated. Additionally, this leads to a considerable loss of energy, requiring additional time to heat the substances to the appropriate temperature for the pyrolysis to commence. Subsequently, microwave-assisted pyrolysis (MWP) has been described for the production of high quality ACs [15–19].

Microwave-induced heating provides an advantage over conventional heating in terms of faster treatment time (10–15 min approximately), which usually result in energy conservation and sustainability of the environment. This leads to a decrease in the processing expenditure, using traditional pyrolysis process [13,14]. Microwave-induced heating generates heat differently than traditional heating in a regular furnace. In a traditional heating process, heat is transferred via conduction and convection, while in MWP heating, heat is effectively produced via ion conduction and molecular dipole–dipole oscillation or rotation of the sample. Consequently, sample temperature rapidly rises and tends to produce a consistent, homogeneous temperature dispersion across the sample [13,14]. MWP heating does not involve direct interaction between the radiation supply and the feedstock. This stimulates friction/interfacial turbulence between the subatomic particles in the substance, which results in a rapid thermal energy transfer across the entire volume of the substance. This can offer additional ultra-fast, 'volumetric' heating to reheat the content in large quantities. Numerous researchers have used MWP to produce ACs from a variety of precursors, including durian shell [16], palm oil shell [13,20,21], banana frond [17], palm shell [18], sugarcane bagasse [15], cocoa shell [19], seed pods of *Albizia* [22,23], rambutan peel [24], peel of pomegranate [25], date stones [26], wastes from microalgae [27], stalks of lotus [28], Crofton weeds [29], endocarp of macadamia [30], sludge of pulp mill [31], industrial waste lignin [32], corn stover [33], peanut shell [34], etc.

Lead is considered as one of the hazardous heavy metals that can be ingested through exposure to water and air [35]. Even at low concentrations, it is injurious to human and aquatic organisms. It is typically found to form divalent cations, which is a highly toxic element that is widely present in surface, ground, and industrial wastewater [36]. It has the potential to be absorbed through the skin, as well as the digestive and respiratory systems, which can have an effect on the body's physiological mechanisms. Furthermore, its deposition can lead to severe environmental and ecological threats for water bodies [37,38]. Wastewater from metal plating, paint, dye, battery production, and glass sectors can release lead cations into water [38,39]. Based on the World Health Organization (WHO), the limit of lead in water should be 0.01 mg/L. Earlier divalent cations of lead (Pb^{2+}) concentration was found to be approximately 0.321 mg/L in open drainage systems in certain industrial sectors [39,40].

In this research, semi-carbonized char obtained from a fixed bed carbonization system, derived from baobab wood chips from the trunk, was activated using microwave-assisted pyrolysis (MWP), where potassium carbonate (K_2CO_3) was used as a chemical activating agent. The process parameters of MWP were optimized in terms of microwave power (x_1), irradiation period (y_1), and impregnation ratio (z_1) using K_2CO_3. By incorporating the Box–Behnken design (BBD), removal percentages of lead, Pb (II) cations (R_1), fixed carbon percentages (R_2), and carbon yield percentages (R_3) of each sample were measured depending on the response surface methodology (RSM). These assessments of input/independent variables are essential in determining the effectiveness of the developed method for recycling and converting the waste chips using a more energy-efficient process to yield high-quality carbon that can be used for the removal of Pb (II) cations from polluted streams. In previous literature, alkaline hydroxide, $ZnCl_2$, or H_3PO_4 acid, which are strong Lewis acid-based activating agents, have been used for pyrolysis using MWP as well as the conventional heating process. Furthermore, $ZnCl_2$ itself can introduce secondary pollutants in aqueous effluents after leaching out from the carbon substrate. *Adansonia kilima* is mainly found in southern and eastern Africa. Later on, it was also found in Kenya, Tanzania, Zambia, South Africa, and Namibia. Previously, *Adansonia digitata*, another kind of Baobab stem powder, and shell were activated using highly concentrated $ZnCl_2$ to produce ACs using conventional heating [41]. *Adansonia digitata* seed hulls, fruit shells, and fibres were also activated using conventional heating using KOH, $ZnCl_2$, H_3PO_4 activation; steam pyrolysis; and two-step conventional pyrolysis, whereby both stages were conducted at a very high temperature of 800–900 °C, respectively [42–44]. The use of harsh, corrosive chemicals (KOH, $ZnCl_2$, H_3PO_4) in the production of ACs is not economically feasible. Nevertheless, the conventional pyrolysis technique is time consuming. In this study, the potential of a relatively mild activating agent (K_2CO_3) is explored for microwave-assisted pyrolysis (MWP) to produce ACs with outstanding physiochemical properties. To the best of our knowledge, no research has been reported regarding the application of the suggested process of microwave-assisted pyrolysis (MWP) using a Box–Behnken design (BBD) for producing ACs from *Adansonia kilima* (*Baobab*) wood chips (AKTW) using the mild activating agent of K_2CO_3.

2. Results

2.1. Regression Model Development and Statistical Analysis

Tables 1 and 2 illustrate the magnitude of operational parameters during the microwave-assisted pyrolysis (MWP) process. The design matrix based on a Box–Behnken design (BBD) design suggested 17 experimental runs under different conditions containing three input variables (power—x_1, residence time—y_1, and ratio—z_1) and three output variables/responses (removal percentages—R_1, fixed carbon content—R_2, and yield percentages—R_3). Five experimental runs under identical conditions were performed for the interpretation of the standard error at the centre/zero position.

Table 1. Input/independent variables for microwave-assisted pyrolysis (MWP) of AKTWC char, including coded and actual level.

Input/Independent Variables	Code	Unit	Variable Levels (Coded and Actual)			Desired Output/Dependent Variables/Responses		
			−1	0	+1	Removal Percentages (%)	Active Carbon Content (%)	Yield Percentages (%)
Microwave Power	x_1	Watt	550	650	750	R_1	R_2	R_3
Radiation/Residence Time	y_1	Min.	10	12.5	15			
Ratio	z_1	-	1	1.5	2			

Table 2. Box–Behnken design (BBD) and experimental responses for microwave-assisted pyrolysis (MWP) of AKTWC char to obtain AKTWAC, including coded and actual Level.

Sample ID	Run	Type of Point	Level (Coded Factors)			Input/Independent Variables (Actual Factors)			Removal Percentage (%)	Fixed Carbon Content (%)	Yield Percentages (%)
						Power x_1, (Watt)	Radiation Time, y_1, (Min.)	Ratio z_3	R_1 (%)	R_2 (%)	R_3 (%)
S-1	1	Center	0	0	0	650	12.50	1.50	85.99	78.77	83.44
S-2	2	Center	0	0	0	650	12.50	1.50	85.98	79.99	84.54
S-3	3	Center	0	0	0	650	12.50	1.50	85.59	78.87	83.33
S-4	4	Center	0	0	0	650	12.50	1.50	85.98	78.32	83.54
S-5	5	Center	0	0	0	650	12.50	1.50	85.77	79.33	82.99
S-6	6	IBFact	0	−1	0	550	10.00	1.50	87.04	80.22	85.88
S-7	7	IBFact	+1	−1	0	750	10.00	1.50	80.99	62.77	80.08
S-8	8	IBFact	+1	0	−1	750	12.50	1.00	84.33	62.76	78.99
S-9	9	IBFact	0	−1	−1	650	10.00	1.00	86.39	76.33	84.06
S-10	10	IBFact	0	+1	+1	650	15.00	2.00	72.33	70.44	76.87
S-11	11	IBFact	−1	0	−1	550	12.50	1.00	86.03	75.66	86.76
S-12	12	IBFact	−1	0	+1	550	12.50	2.00	81.22	70.43	81.09
S-13	13	IBFact	−1	0	0	550	15.00	1.50	82.09	68.44	83.99
S-14	14	IBFact	0	+1	−1	650	15.00	1.00	73.56	70.75	82.45
S-15	15	IBFact	+1	+1	−1	750	15.00	1.50	69.99	61.79	78.89
S-16	16	IBFact	0	0	+1	650	10.00	2.00	74.22	69.87	79.77
S-17	17	IBFact	+1	0	+1	750	12.50	2.00	68.55	61.99	73.32

Three quadratic regression models based on removal percentages—R_1, fixed carbon content—R_2, and percentage yield—R_3 are suggested by the software and represented by Equations (1)–(3).

$$Removal\ Percentages\ of\ Lead,\ Pb(II)Cations\ (R_1) = +85.86 - 4.06x_1 - 3.83y_1 - 4.25z_1 - 1.51x_1y_1 \\ -2.74x_1z_1 + 2.73y_1z_1 - 1.21x_1^2 - 1.62y_1^2 - 1.62z_1^2 \quad (1)$$

$$\text{Fixed Carbon Percentages } (R_2) = +79.06 - 5.68x_1 - 2.22y_1 - 1.60\,z_1 + 2.70x_1y_1 + 1.11\,x_1z_1 + 1.54y_1z_1 - 7.44x_1^2 - 3.31y_1^2 - 3.90z_1^2 \quad (2)$$

$$\text{Carbon Yield Percentages } (R_3) = +83.57 - 3.30\,x_1 - 0.95y_1 - 2.65z_1 + 0.17\,x_1y_1 - 0.77\,x_1z_1 - 0.32\,y_1z_1 - 1.05x_1^2 - 0.31\,y_1^2 - 2.48\,z_1^2 \quad (3)$$

The R^2 values for Equations (1), (2), and (3) are 0.987, 0.981, and 0.972, respectively, which were closer to unity, indicating that the developed models are suitable (Figure 1). The coefficient of variation (CV) and standard deviations found were minimal, indicating that the proposed models are reliable (Table 3). The signal-to-noise ratio was used to estimate the magnitude of adequate precision [45,46]. Adequate precision values need to be higher than 4 for successful simulation of the proposed models. For removal percentages, R_1, fixed carbon content, R_2, and yield, R_3, the adequate precision was 21.76, 18.96, and 20.26, respectively. Therefore, the experimental data set reported here is statistically significant for navigating the design [47].

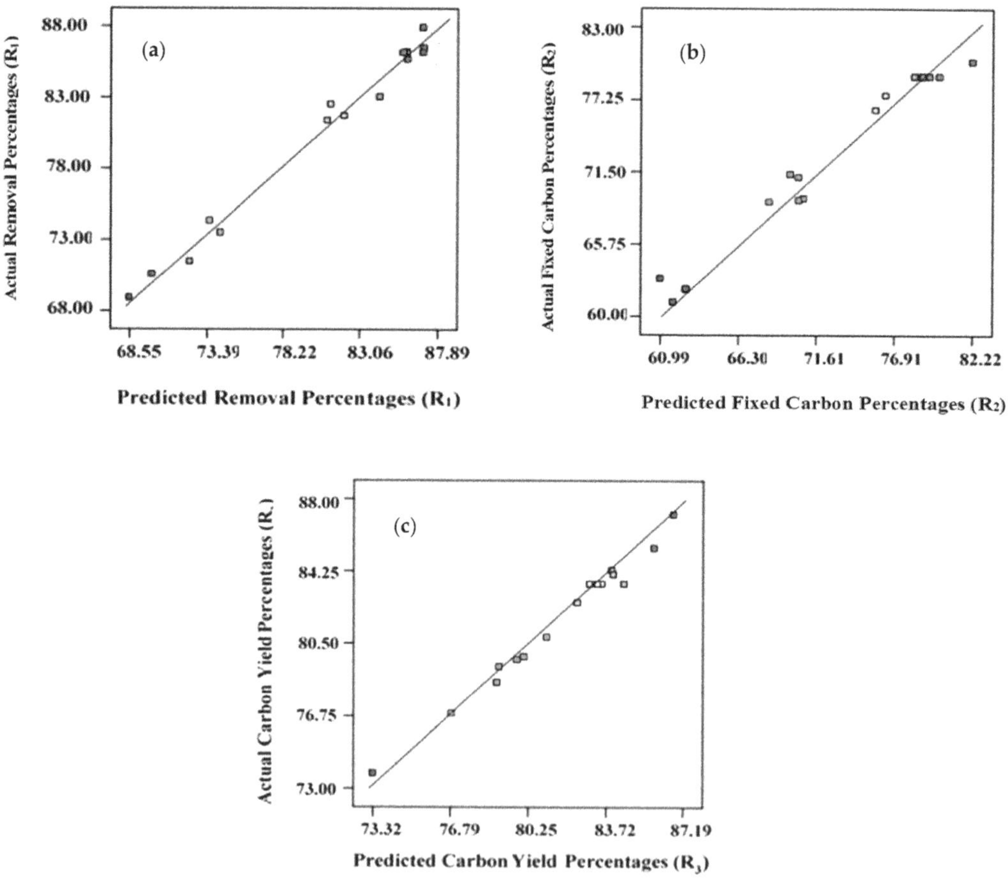

Figure 1. Predicted versus actual/experimental data points: (**a**) removal percentages, R_1; (**b**) fixed carbon percentages, R_2; (**c**) yield percentages, R_3.

Table 3. Statistical tools for model validation.

Statistical Tools	Output Variables/Responses		
	Removal Percentages R_1	Fixed Carbon Content R_2	Percentage Yield R_3
Standard Deviation, SD%	1.11	1.20	0.81
Mean	80.94	72.16	81.76
Correlation Coefficient, R^2	0.98	0.98	0.97
Adjusted R^2	0.97	0.96	0.94
Coefficient of Variation, CV	1.37	1.67	0.99
Adequate Precision	21.76	18.96	20.26

Tables 4–6 present the findings from ANOVA analysis for removal percentage (R_1), carbon content percentages (R_2), and carbon yield percentages (R_3), correspondingly, indicating that these regression models are statistically relevant, having a confidence level at 95%. To assess the level of competence of the developed models, the magnitude of F-test results, reflecting the divergence of the data from the average/mean value, were computed. Figure 1 represents linear plots for predicted versus actual data points for R_1, R_2, and R_3, respectively.

Table 4. Influence of input/independent variables on removal percentages (R_1): analysis of variance (ANOVA) test.

Source	Sum of Squares	Degree of Freedom	Mean Square	F Value	Prob > F	Comments
Model	665.58	9	73.95	60.45	<0.0001	*Significant*
x_1	132.19	1	132.19	108.05	<0.0001	
y_1	117.58	1	117.58	96.11	<0.0001	
z_1	144.42	1	144.42	118.01	0.0003	
x_1^2	6.20	1	6.20	5.07	0.0591	
y_1^2	89.91	1	89.91	73.49	<0.0001	
z_1^2	89.72	1	89.72	73.33	<0.0001	
x_1y_1	9.15	1	9.15	7.48	0.0291	
x_1z_1	30.09	1	30.09	24.59	0.0016	
y_1z_1	29.92	1	29.92	24.46	0.0017	
Residuals	8.56	7	1.22			
Lack of Fit	8.44	3	2.81	88.80	0.0004	
Pure Error	0.137	0.0049				

Table 5. Influence of input/independent variables on fixed carbon content percentages (R_2): analysis of variance (ANOVA) test.

Source	Sum of Squares	Degree of Freedom	Mean Square	F Value	Prob > F	Comments
Model	738.36	9	82.04	56.64	<0.0001	*Significant*
x_1	258.10	1	258.10	178.18	<0.0001	
y_1	39.47	1	39.47	27.25	0.0012	
z_1	20.38	1	20.38	14.07	0.0072	
x_1^2	33.33	1	33.33	168.08	<0.0001	
y_1^2	46.06	1	46.06	31.78	0.0008	
z_1^2	64.10	1	64.10	44.25	0.0003	
x_1y_1	29.16	1	29.16	1.71	0.1063	
x_1z_1	4.97	1	4.97	13.08	0.0378	
y_1z_1	9.46	1	9.46	3.43	0.0028	
Residuals	10.14	7	1.45			
Lack of Fit	8.53	3	2.84	14.67	0.0126	
Pure Error	1.61	4	0.40			

Table 6. Influence of input/independent variables on yield percentages (R_3): analysis of variance (ANOVA) test.

Source	Sum of Squares	Degree of Freedom	Mean Square	F Value	Prob > F	Comments
Model	184.13	9	20.46	31.34	<0.0001	*Significant*
x_1	87.38	1	87.38	133.86	<0.0001	
y_1	7.20	1	7.20	11.03	0.0127	
z_1	56.23	1	56.23	86.14	<0.0001	
x_1^2	4.67	1	4.67	7.15	0.0318	
y_1^2	0.39	1	0.39	0.60	0.4636	
z_1^2	25.80	1	25.80	39.52	0.0004	
x_1y_1	0.10	1	0.10	0.19	0.6774	
x_1z_1	0.30	1	0.30	0.77	0.2987	
y_1z_1	0.42	1	0.42	0.64	0.4509	
Residuals	4.57	7	0.65			
Lack of Fit	3.22	3	1.07	1.10	0.1471	
Pure Error	1.35	4	0.34			

The model F values for Pb (II) cations removal percentages (R_1), fixed carbon content percentages (R_2), and yield percentages (R_3) were 60.45, 56.64, and 31.34, respectively, indicating that these models were credible and significant. Nevertheless, for these output responses, the magnitude of Prob > F values were lower than 0.05, indicating that the model variables included here for the analysis of responses were significant [48].

The linear variables of power (x_1), time (y_1) and ratio (z_1), interaction terms x_1y_1 and y_1z_1, and also their quadratic terms y_1^2 and z_1^2, were significant for the response of removal percentages, R_1. In comparison to power (x_1), other linear factors such as time (y_1) and ratio (z_1) had a moderate effect on the elimination percentages (R_1). The interaction between power and ratio (x_1z_1) had a stronger influence on removal percentages, R_1 (according to the greatest value of F = 30.09) than the remaining interaction variables of y_1z_1 and $z_{1\times 1}$.

For fixed carbon content percentages, R_2, power (x_1), time (y_1), and ratio (z_1), along with the interaction terms of x_1y_1 and y_1z_1, are significant model terms. The quadratic model terms x_1^2, y_1^2, and z_1^2 were also significant (Table 5). The power (x_1), radiation duration (y_1), and ratio (z_1) had a substantial impact on percentage yield of carbon (R_3). The quadratic terms of x_1^2 and z_1^2 were also significant for R_3. The interaction between time and ratio, y_1z_1, had a more prominent influence compared with the other interaction terms (x_1z_1 and x_1y_1) on percentage yield, R_3 (Table 6).

2.2. Process Optimization

The optimal values for the output variables, designated as responses (removal percentages, R_1, fixed carbon content percentages, R_2, and yield percentages, R_3), were determined. Due to the distinct interest areas of input variables (power—x_1, radiation time—y_1, and ratio—z_1), it was difficult to maximize the selected responses (R_1, R_2, and R_3) using the identical operating conditions. Maximum removal percentages, R_1, and fixed carbon percentages, R_2, were observed under identical conditions shown at Run 6 (power, 550 watts, time, 10 min, and ratio 1.50). This is expected, as the maximum amount of fixed carbon content with less ash residues can ensure maximum removal percentages. However, yield percentage, R_3, was maximum under the condition of Run 11 (power, 550 watts, time, 12.5 min, and ratio, 1.00). Two different sets of parameters were observed for maximum values of R_1, R_2, and R_3. Thus, the desirability indicator was implemented using the State Ease Design Expert Software–9. To determine the optimal set of processing variables for the production of ACs, the target criterion for the output variables/responses were fixed to "maximum", while the level of input variables were kept "in the range". The desirability ramp with the graph is shown in Figure 2.

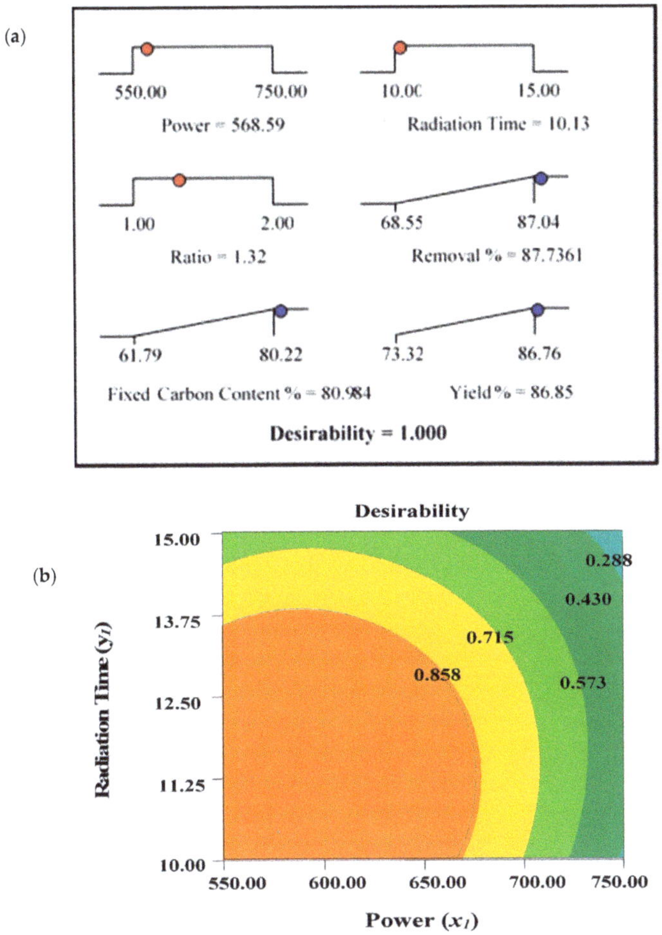

Figure 2. (**a**) Optimization/desirability ramp and (**b**) 2D contour plots for desirability function.

The predicted and actual experimental values for the responses determined under optimum processing conditions are summarised in Table 7. The measured values demonstrate the average of the three experimental results carried out for the assessment of the accuracy of the process.

Table 7. Optimization for the microwave-assisted pyrolysis (MWP) process.

Power	Radiation Time	Ratio	Percentage Removal (R_1)			Fixed Carbon Content (R_2)			Yield Percentages (R_3)		
(Watt)	(Min.)	(-)	Predicted	Experimental	Error	Predicted	Experimental	Error	Predicted	Experimental	Error
568.59	10.13	1.32	87.73	85.66	2.35	80.98	80.02	1.18	86.85	85.90	1.09

2.3. *Effect of Process Variables on Adsorption Performance (R_1) of AKTWAC*

The optimum ACs need to ensure higher removal efficiencies of contaminants from the aqueous phase, having superior physicochemical properties. In such scenarios, the carbon materials (AKTWAC) ought to have significant BET surface area and suitable porosity to enable the aqueous state sorption process. RSM was used to explore the influence of three factors (power—x_1, residence time—y_1, and ratio—z_1) on the removal percentages of

Pb (II) metal ions (R_1), employing a three-dimensional response surface (RSM) mesh and two-dimensional contour plots. According to the RSM analysis, all three factors contributed successively to increasing the percentage removal (R_1) up to a specified level.

Figure 3a illustrates the effect of microwave power (x_1) and radiation duration (y_1) on the percentage removal (R_1) when the other parameters of impregnation ration (z_1) were kept consistent at centre/zero level (1.5). Figure 3b depicts the impact of duration (y_1) and ratio (z_1) on the percentage removal (R_1) when power (x_1) was fixed at the centre point (650 watts). Figure 3c shows the cumulative impact of power (x_1) and ratio (z_1) on the removal percentages (R_1) when radiation time (y_1) was kept constant at the centre level (12.5 min). The values provided along the x- and y-axis are precise, actual values.

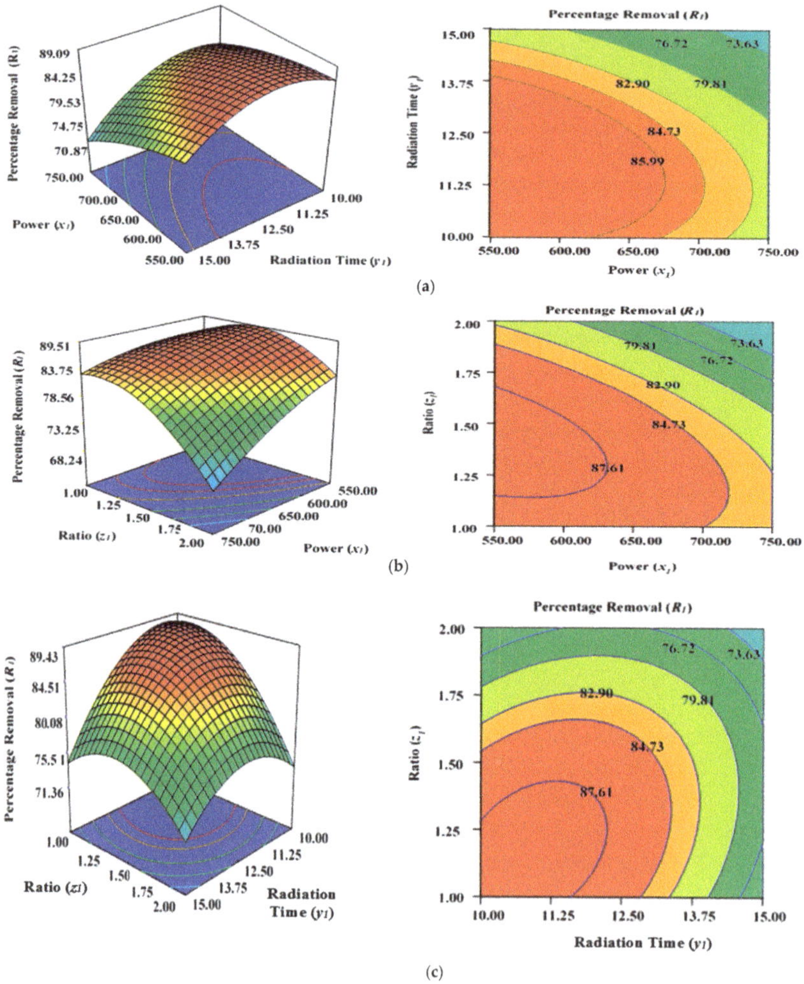

Figure 3. Three-dimensional RSM surface mesh plots with respective 2D contour plots for removal percentages (R_1): (**a**) effect of microwave power (x_1) and radiation duration (y_1) when impregnation ration (z_1) was kept consistent at the centre/zero level (1.5), (**b**) effect of duration (y_1) and ratio (z_1) when power (x_1) was fixed at the centre point (650 watts), (**c**) effect of power (x_1) and ratio (z_1) when radiation time (y_1) was kept constant at the centre level (12.5 min).

Figure 3a–c demonstrates that increasing the radiation power (x_1) as well as extending the retention period (y_1) in the microwave improves the percentage removal (R_1) up to a definite point. This inevitably results in a concave layout of the RSM figures. The greatest percentage removal (R_1) achieved was approximately 87.04% (from experimental Run 6; condition: lower power level (x_1)—550 watts, duration(y_1)—10 min, and ratio (z_1)—1.5 under N_2 flow of 50 mL/min). The application of microwave power (x_1) for a certain duration (y_1) and ratio (z_1) can increase the elimination and dehydration processes during the MWP pyrolysis/activation phase. Enhanced power (x_1) and duration (y_1) of activation can ensure the emission of more volatile chemicals from the interior side of the carbon matrix. To increase the porosity and surface area, an optimal contact period is necessary to allow greater interaction between the semi-carbonized char (AKTWC) and K_2CO_3 in the presence of N_2 gas [49–51]. On the other hand, a too low ratio (z_1) cannot enhance the cracking mechanism necessary for the formation of pores, consequently having no favourable leverage on the sorption performance of the ACs synthesised here.

The lower removal percentages (R_1) at a very high level of MWP power (x_1) could be related to the sintering/thermal decomposition, accompanied by deformation of the semi-carbonized char (AKTWC). This can cause realignment of the synthesised activated carbon structure, leading to reduced surface areas and pore volume. This phenomenon was explained earlier for the sorption of MB using ACs produced from the stalks of cotton using MWP-based KOH activation, whereby 400-watt power was chosen as the optimal level of power for activation [52]. The findings of ANOVA results in Table 3 reveal that ratio (z_1) has the major impact on Pb (II) cation uptake [49,50].

The cumulative effect of power (x_1) and ratio (z_1) is decreasing the removal percentages (R_1), as can be seen by the contours of the graphs (Figure 3b), until a certain range. This indicates that increasing the power (x_1) and ratio (z_1) of activating salt in the environment of N_2 gas would improve the rate of the reaction between K_2CO_3 and semi-carbonized char (AKTWC), leading to more porosity, up to a fixed threshold value. Maximizing the power (x_1) and ratio (z_1) beyond that specific threshold point will have a detrimental impact on the sorption process. This has been anticipated because a very high level of power or ratio of K_2CO_3 can disrupt certain functional groups over the carbon surface. Microwave annealing due to heat may negatively affect the porous structure of the synthesized carbon, incinerating the pore walls and decreasing the overall volume of the pore [44,53,54].

The residence time (y_2) also had a progressive effect on increasing the surface area, with a porous texture of the ACs produced from cassava peel. Meanwhile, the proportion of micro- and mesopores was significantly influenced by the impregnation ratio using the strong base of KOH and temperature [49–51]. Enhanced radiation time (y_1) and power (x_1) decreased the removal percentages (R_1) after a certain level. At a higher power level (x_1) of 650 to 750 watts and residence time (y_1) of approximately 12.5 to 15 min, more radiation takes place, and the semi-carbonized char (AKTWC) is exposed under that condition. This may disrupt the C–C and C–O–C bonds of carbon and can cause burning of the carbon to produce more ash and weight loss. Too much exposure under microwave radiation is not recommended because it may increase ash production, destruction of pores, and functional groups, resulting in a reduction in the removal efficiencies (R_1).

Thereby, it can be inferred that exceeding the optimal limits for radiation power (x_1), duration (y_1), and ratio (z_1) destroys the porous network, which tends to result in a decreased BET surface area and reduced removal rate percentages (R_1). Nevertheless, the removal percentages (R_1) were reduced at extremely minimal levels of ratio (z_1) of approximately 1 at a power (x_1) of 650 or 750 watts, irrespective of duration (y_1) at 10 to 15 min (Table 2). This occurs as a result of inadequate interactions between the semi-carbonized char (AKTWC) and K_2CO_3, potentially degrading the efficiency of the synthesized carbon (AKTWAC) [55].

2.4. *Effect of Process Variables on Fixed Carbon Content (R_2)*

Figure 4a illustrates the incremental effect of two variables, power (x_1) and radiation time (y_1), on the percentage fixed carbon (R_2) inside the AC samples produced using the condition of experimental Runs 1–17 (Table 2), whereby K_2CO_3 ratio (z_1) was set to zero at the centre point (1.25). Figure 4b depicts the simultaneous effect of radiation time (y_1) and K_2CO_3 ratio (z_1) on percentage fixed carbon (R_2), with power (x_1) maintained at its respective centre point (650 watts). Figure 4c shows the collective effect of power (x_1) and K_2CO_3 ratio (z_1) on percentage fixed carbon (R_2) when time (y_1) was maintained at its respective centre point (12.50 min).

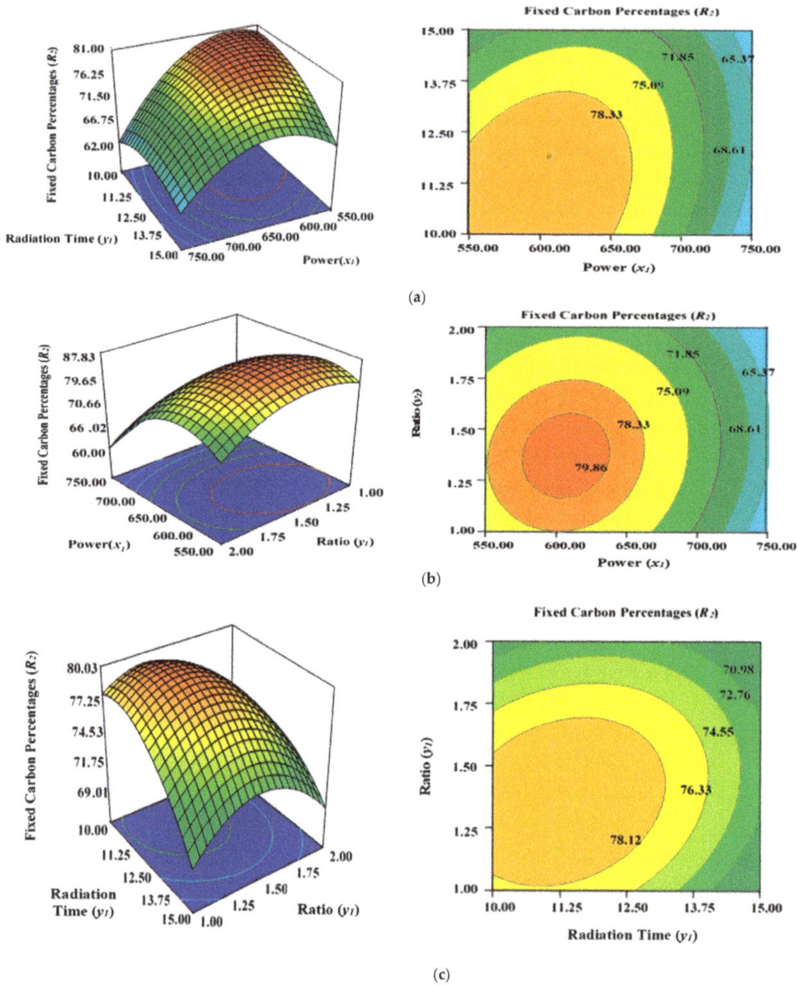

Figure 4. Three-dimensional RSM surface mesh plots with respective 2D contour plots for fixed carbon percentages (R_2): (**a**) effect of microwave power (x_1) and radiation duration (y_1) when impregnation ration (z_1) was kept consistent at the centre/zero level (1.5), (**b**) effect of duration (y_1) and ratio (z_1) when power (x_1) was fixed at the centre point (650 watts), (**c**) effect of power (x_1) and ratio (z_1) when radiation time (y_1) was kept constant at the centre level (12.5 min).

In this study, all three factors examined had a stimulating effect on the fixed carbon content percentages (R_2). It was discovered that increasing the power (x_1), duration (y_1), and ratio (z_1) significantly lowered the carbon content (R_2) of the sample after the optimal limit. Power (x_1) had the greatest effect on fixed carbon content (R_2), as demonstrated before by the maximum F value of 258.10 in Table 4, but ratio (z_1) was relatively less significant (F value is 20.38) in comparison to power (x_1) and time (y_1). The lowest carbon content was observed at 750 watts for 15 min, with a K_2CO_3 ratio of 1.50 (Run 15), as indicated by Table 2. Increasing the power (x_1), radiation duration (y_1), and ratio (z_1) after a certain level will accelerate the diffusion K_2CO_3 within the char matrix, leading to the formation of ash. This would lower the fixed carbon percentages (R_2) due to enhanced de-volatilization and burning of the carbon to ash residues. This phenomenon is consistent with previous findings when producing ACs from orange peel using microwave-induced KOH and NaOH activation [56].

2.5. *Effect of Process Variables on Yield Percentages of AKTWAC Carbon (R_3)*

Figure 5a demonstrates the combined effect of two variables, power (x_1) and radiation time (y_1), on the percentage yield of carbon (R_3) inside the AC samples produced using the condition of experimental Runs 1–17 (Table 2), whereby the K_2CO_3 ratio (z_1) was fixed to the centre at 1.25. Figure 5b represents the concurrent effect of radiation time (y_1) and K_2CO_3 ratio (z_1) on percentage carbon yield (R_2), with power (x_1) kept at its respective centre point (650 watts). Figure 5c shows the collective effect of power (x_1) and K_2CO_3 ratio (z_1) on percentage carbon yield (R_3) when time (y_1) was retained at its respective centre point (12.50 min).

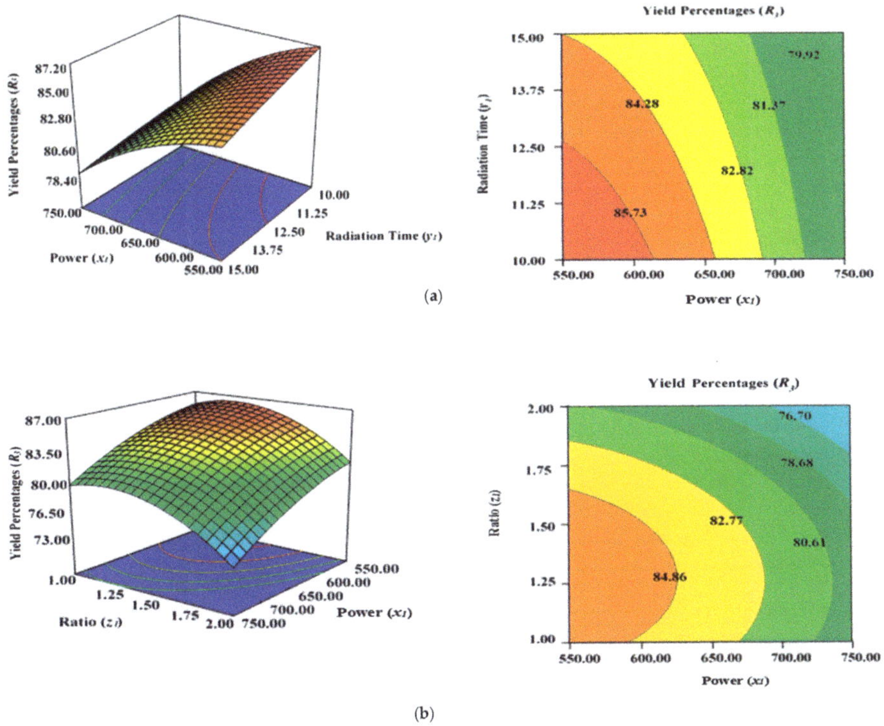

Figure 5. *Cont.*

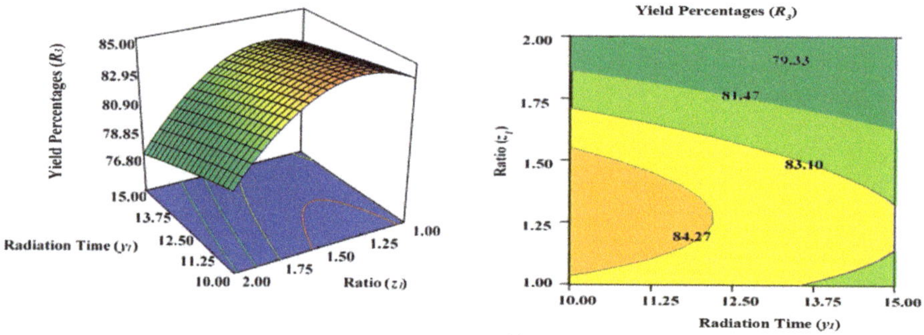

(c)

Figure 5. Three-dimensional RSM surface mesh plots with respective 2D contour plots for yield percentages (R_3): (**a**) effect of microwave power (x_1) and radiation duration (y_1) when impregnation ration (z_1) was kept consistent at the centre/zero level (1.5), (**b**) effect of power (x_1) and ratio (z_1) when radiation time (y_1) was kept constant at the centre level (12.5 min), (**c**) effect of duration (y_1) and ratio (z_1) when power (x_1) was fixed at the centre point (650 watts).

All three variables considered in this study had a significant impact on the carbon yield percentages (R_3). It was observed that raising the power (x_1), time (y_1), and ratio (z_1) of the MWP pyrolysis of the sample resulted in a substantial decrease in the yield percentages (R_3) after the optimum limit. As previously indicated by the maximum F value of 87.38 in Table 5, power (x_1) had the highest impact on carbon yield (R_3), while radiation time (y_1) was considerably less significant (F value of 7.20). Ratio (z_1) had a stronger effect on yield percentages (R_3) compared with time (y_1). As stated in Table 2, the lowest carbon yield was found at 750 watts for 12.5 min with a K_2CO_3 ratio of 2.00 (Run 15).

A higher magnitude of power (x_1) in conjunction with a greater ratio (y_1) results in lower carbon yield percentages (R_3) (Figure 5a). The rise in power (x_1) will induce more heat energy and will eventually intensify the elimination and dehydration activities inside the sample, resulting in the formation of more gaseous and liquid fractions rather than solid ACs with a higher proportion of ash residues. Thus, yield percentage will drop at higher power levels with enhanced radiation time. This observation was previously reported in the literature when oil palm fronds were used to prepare ACs using the conventional heating approach for the elimination of zinc (Zn (II)) cations [57]. Overall, the yield percentages (R_3) of ACs dropped when the level of the process parameters was increased. Several studies have documented a similar tendency [58–60]. The negative relationship between microwave radiation duration and AC production can be explained by the increasing amount of breakdown of organic residues by prolonged radiation time (heating) [60]. As a result, more volatile content was discharged from the partially carbonized char, resulting in a reduced yield of ACs. The greater proportion of K_2CO_3, along with the increased thermal energy generated by the higher microwave power, likely aided the burning of the semi-carbonized char, resulting in a reduced conversion of char to AC substrate.

2.6. *Physio-Chemical Characterizations*

2.6.1. Surface Morphological Analysis

FESEM images of untreated AKTW, AKTWC, and AKTWAC produced under optimal circumstances are shown in Figure 6a–c. As can be seen from Figure 6a, the untreated precursor of AKTW has a relatively smooth surface with few cracks and craters and a negligible amount of pores. It is almost non-porous before the semi-carbonization process using the fixed bed reactor. The semi-carbonization process in the presence of N_2 gas at 650 °C has created some pores over the surface of AKTWC (Figure 6b). Microwave-assisted pyrolysis (MWP) in the presence of K_2CO_3 leads to the formation of many distorted pores. However, the majority of these pores have a circular shape (Figure 6c)

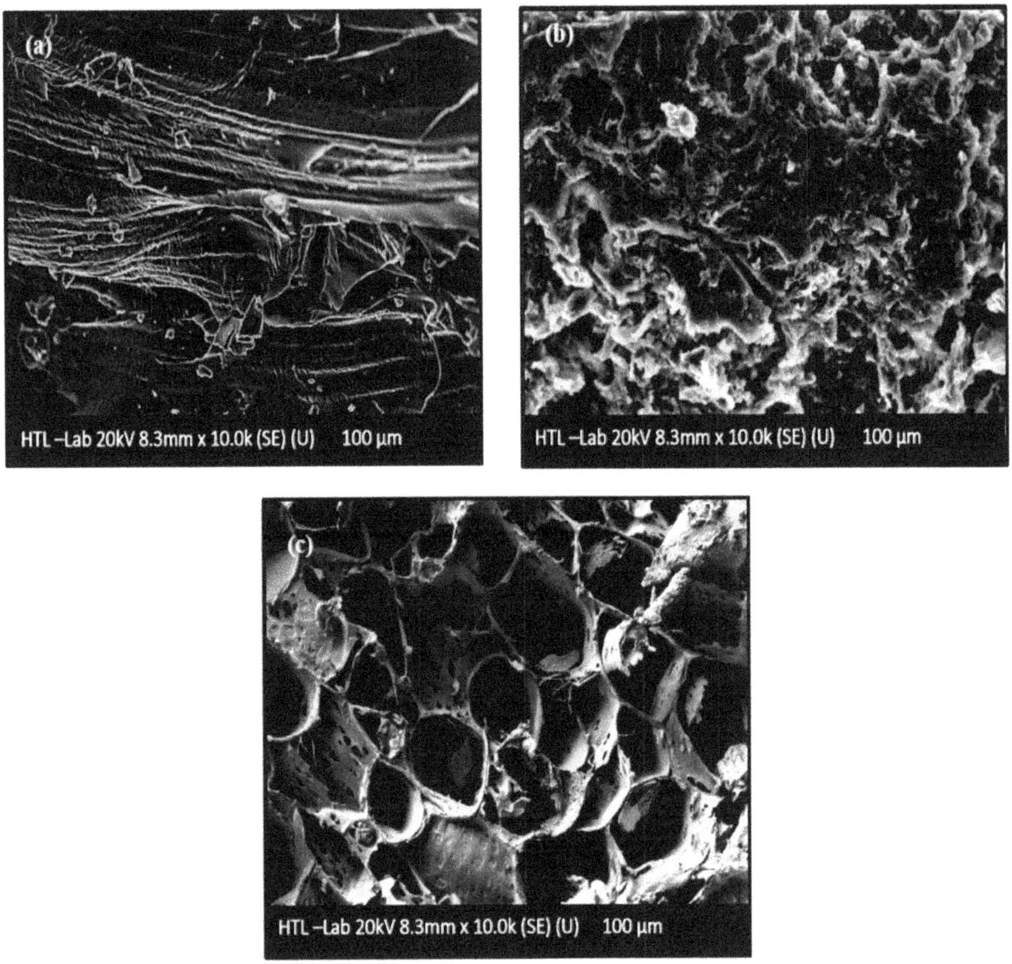

Figure 6. Surface morphological analysis using field emission scanning electron microscopic (FESEM) images: (**a**) AKTW, (**b**) AKTWC, (**c**) AKTWAC.

The gasification process in the presence of K_2CO_3 is responsible for the development of the porous texture of the carbon and can be represented by the following reactions [61]. Due to the reducing effect, the carbon can generate K_2O, K, CO, and CO_2 gas from K_2CO_3 based on the following reactions [61]:

$$K_2CO_3 + 2C \rightarrow 2K + 3CO$$

$$K_2CO_3 \rightarrow K_2O + CO_2$$

$$K_2O + C \rightarrow 2K + CO$$

It is expected that metallic potassium (K) produced during the gasification will penetrate into the internal surface of the char cavity, enlarging the existing holes, which can form new pores. Pore diameter can be regulated by specific surface interactions and the quantity and rate of volatile compounds emitted. Evidently, the volume and geometry of the pores are highly dependent on, not only the magnitude of radiation time, ratio, reactor design, flow rate, and types of the activating gas, but also on the sort (conventional or microwave-assisted) and rate of heating. After microwave-assisted pyrolysis (MWP), the microstructure of pores was clearly apparent (Figure 6c). This indicates that K_2CO_3 impregnation and soaking had annihilated the undesirable particles that had been clogging the pores after the semi-carbonization process (Figure 6b) [56,57]. As a result, the surface area and volume of the pores were significantly enhanced after the microwave-assisted pyrolysis process. This phenomenon is further supported by BET analysis of the AKTWC and AKTWAC sample. The final sample of AKTWAC was washed several times to ensure the removal of unreacted K_2CO_3 from sample. As depicted earlier, metallic potassium intercalated during microwave-assisted pyrolysis to enlarge the surface area according to mechanisms (I), (II), and (III). EDX analysis was carried out, and after final washing in AKTWAC, the potassium content was relatively small, approximately 1.02%, with carbon at 83.23%, oxygen at 5.57%, silicon at 6.34%, aluminium at 3.54%, and a trace amount of iron content at 0.30%. Other elements, such as silicon, aluminium, and iron, were found due to the presence of ash, as impurities inside the carbon matrix [35]. The lower content of potassium is due to the vigorous washing of the final sample (AKTWAC) until the pH became neutral.

2.6.2. Surface Area and Porous Texture Analysis

N_2 adsorption–desorption curves reveal qualitative feedback on the adsorption capacity and porosity of the carbonaceous substrate. The N_2 adsorption isotherm assessment (Figure 7a) revealed that the isotherm obtained here for AKTWAC can be classified as an intermediate between Category I and Category II isotherms, designated by the International Union of Pure and Applied Chemistry (IUPAC) association. This adsorption pattern indicates a hybrid structure containing both micro- and meso-porous texture. At relatively higher pressures, the isotherm curve exhibits a small hysteresis loop, indicating a meso-porous texture.

Table 8 provides detailed information regarding the surface area and porous texture of the AKTWC and AKTWAC samples. Mesopores have encompassed approximately 49.06% of the total volume of the pores in AKTWAC. This reflects an adequate porous texture suitable for aqueous phase adsorption.

Table 8. Surface textural properties analysis.

Sample	S_{BET} (m^2/g)	S_{mic} (m^2/g)	Langmuir Surface Area (m^2/g)	External Surface Area (m^2/g)	S_{mic}/S_{BET} (%)	V_{mic} (cm^3/g)	V_{tot} (cm^3/g)	Diameter (nm)	Bulk Density (g/mL)	Iodine Number (mg/g)	P_{ZPC}
AKTWC	448.79	209.99	699.89	238.81	46.79	0.07	0.16	4.76	-	-	-
AKTWAC	1390.76	798.67	1882.91	592.09	55.26	0.49	0.96	6.73	0.07	1297.87	4.6

When AKTWAC was compared to carbonized AKTWC, it was observed that the BET surface area, micropore surface area, Langmuir surface area, and total pore volume all increased significantly, confirming pore formation and enlargement of the existing ones during the microwave pyrolysis step (Table 8). The AKTWAC produced in this work had a relatively higher BET surface area of 1390.76 m^2/g and total pore volume of 0.9643 cm^3/g, which is efficient enough for the removal of Pb (II) cations from waste water.

The pore size distribution is a representation of the intricate hollow areas inside the interior surface of the solid substrate. The pores are classified as micropores (diameter less than 2 nm), mesopores (diameter equal to 2–50 nm), and macropores (diameter greater than 50 nm) according to the IUPAC classification. The pore size distribution of AKTWAC was determined using density functional theory (DFT) modelling. The highest peak with adsorbed volume was observed between 5 and 7 nm, with an average diameter of the pore of approximately 6.73 nm (Figure 8b), indicating that the largest proportion of pores developed inside the AKTWAC were mesopores.

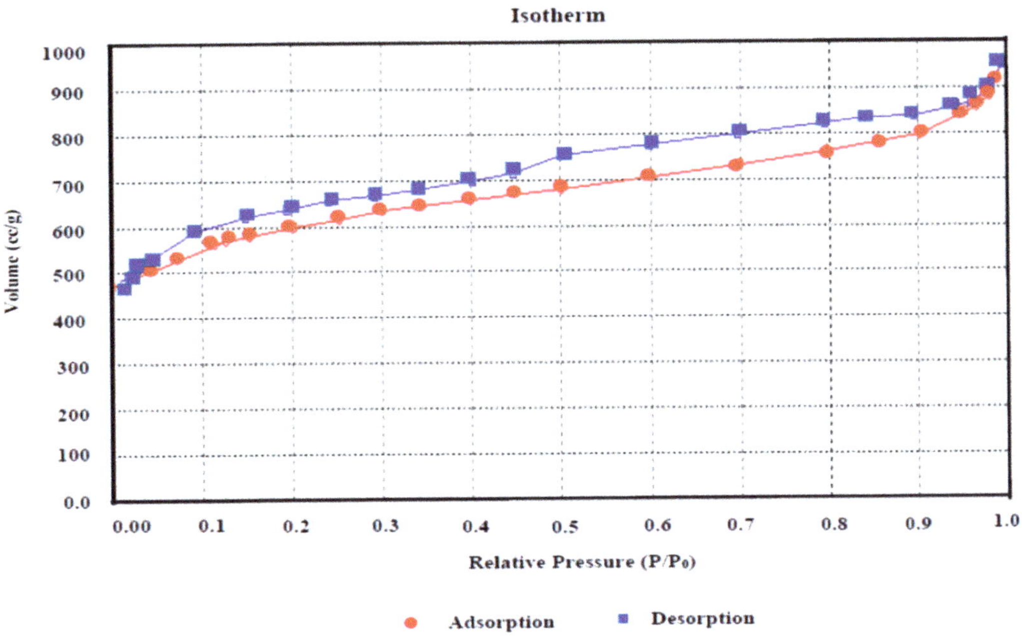

(**a**)

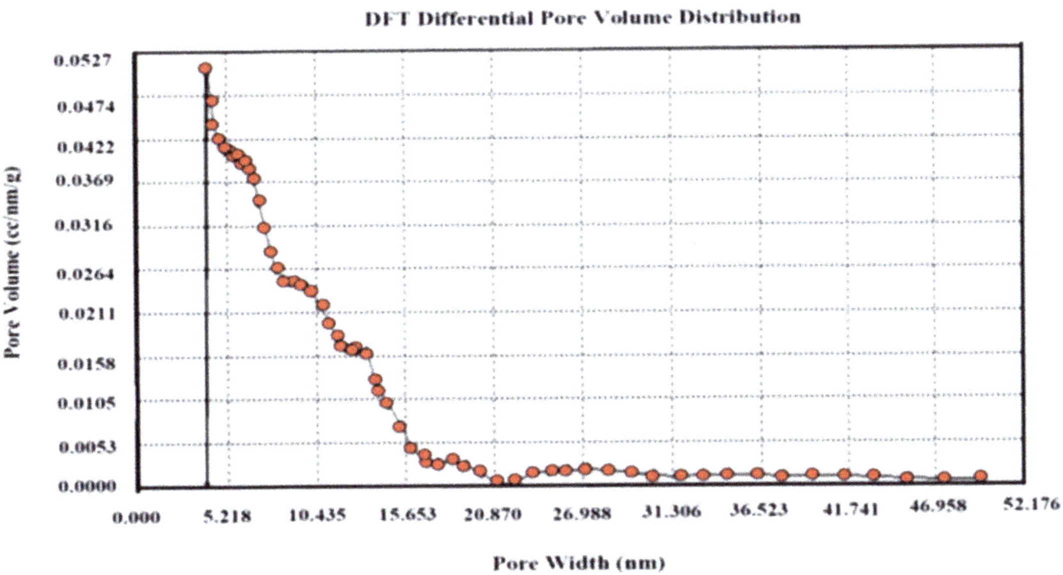

(**b**)

Figure 7. (**a**) BET adsorption isotherm and (**b**) pore size distribution of AKTWAC sample.

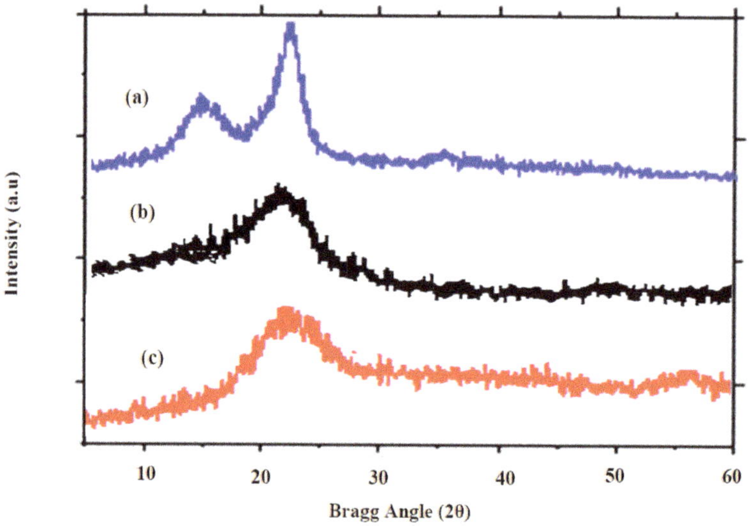

Figure 8. X-ray Diffraction Spectra for (**a**) AKTW (**b**) AKTWC (**c**) AKTWAC.

2.6.3. Thermogravimetric and Elemental Analysis

Thermo-gravimetric (TGA) analysis was used to determine the thermal stability of AKTW, AKTWC, and AKTWAC. AKTWAC showed increased thermal stability compared with AKTWC and AKTW. As shown in Table 9, the dtg_{max} values for AKTW, AKTWC, and AKTWAC were 347.38 °C, 371.22 °C, and 378.87 °C, respectively. Between the three primary components of ligno-cellulosic wastes, hemicelluloses had the minimum thermal stability [62–64]. Semi-carbonization as well as microwave-assisted pyrolysis of AKTW eliminated hemicelluloses and the amorphous portion of the cellulose. This finally increased the thermal stability of the AKTWAC sample. Thus, AKTWAC contained a greater portion of stabilized carbon fragments and was more resistant to heat. The first degrading phase, which tends to occur between 70 and 130 °C, is due to the moisture evaporation. The second degradation phase for AKTW occurs between 200 °C and 300 °C, which reflects the degradation of hemicellulose. At 300 to 400 °C, degradation of cellulose takes place. It was previously established that hemicellulose degradation occurs concurrently with degradation of cellulose in ligno-cellulosic residues. However, degradation of lignin takes place between 200 and 800 °C [62–64].

Table 9. Thermo-gravimetric (proximate analysis—TGA) of AKTW, AKTWC, and AKTWAC.

Proximate Analysis (TGA)	AKTW	AKTWC	AKTWAC
Carbon Content (%)	51.92	70.70	80.02
Moisture (%)	3.88	2.87	1.76
Volatile Matter (%)	36.98	14.55	2.07
Ash Residues (%)	7.22	11.88	16.21
dtg_{max}	347.38	371.22	378.87
Elemental Analysis			
C	53.98	68.78	81.03
H	6.87	4.89	1.12
N	2.12	0.78	0.24
O	35.05	24.52	17.29
S	1.98	1.03	0.32

This effect is also noticeable from ultimate analysis, wherein the percentage carbon in AKTW was originally 50.87% but was increased to 81.99% after the consecutive process of semi-carbonization and microwave-assisted activation (MWP). Elemental analysis was carried out and is tabulated in Table 9. The CHNOS analysis revealed that successive treatment with pyrolysis increased the carbon content of the sample, but the proportion of hydrogen and oxygen decreased. Enhancement of carbon content in activated AKTWAC illustrated that microwave-assisted pyrolysis after K_2CO_3 impregnation was appropriate enough to enhance the quality of the final activated carbon required for sufficient removal of lead cations from water. Pyrolysis under microwave caused vigorous reactions to take place due to metallic potassium intercalation inside the carbon matrix, releasing a significant amount of volatile organic compounds as gaseous and liquid products. This finally reduced the amount of hydrogen and oxygen content inside the sample [51,64].

2.6.4. X-ray Diffraction Analysis

The technique of X-ray diffraction analysis (XRD) is used in materials research to identify the crystalline properties of a material [65–67]. The structural variations of the AKTW, AKTWC, and AKTWAC were investigated using XRD analysis. The XRD patterns for the samples are shown in Figure 8. Owing to the stacking arrangement of the aromatic layers, a blunt diffraction peak appeared at 2θ = 22–25° for AKTWC and AKTWAC, corresponding to the diffraction of (002). This represents the amorphous texture of the AKTWC and AKTWAC samples. However, the AKTW sample contains two diffraction peaks at 16° and 22°, reflecting the presence of a crystalline and amorphous region of cellulose. After semi-carbonization and microwave-assisted pyrolysis, the microcrystalline framework of the cellulose in untreated AKTW has been disrupted. A similar trend was observed for ACs obtained from lotus stalk after the activation process using phosphorous oxy-acids [68].

3. Materials and Methods

3.1. Materials

3.1.1. Preparation of Feedstock

AKTW trunk chips were obtained from a local furniture factory in Africa. The chips were washed under running water to avoid contamination by dust and dried for 24 h at 120 °C. The dried chips were crushed and ground to powder and stored in plastic vials prior to semi-carbonization using a fixed-bed reactor under nitrogen (N_2) flow at a temperature of 650 °C for 2 h. The temperature ramping was kept constant at 10 °C per minute to avoid excessive burn off to have a lower amount of ash residues and prevent the weight loss of activated carbon (ACs). The semi-carbonized char designated as AKTWC was impregnated with K_2CO_3 at different ratios based on the BBD design matrix, as illustrated by Tables 1 and 2. With that impregnated sample, 500 mL of distilled water was added and stirred at room temperature for 8 h. The mixture was kept for 24 h after the stirring to ensure adequate soaking of the sample. It was filtered and dried in a vacuum oven overnight before the microwave-assisted pyrolysis (MWP). An impregnated and soaked sample (AKTWC) was placed inside the quartz reactor of the microwave equipped with a thermocouple by changing the power (x_1), residence time (y_1), and ratio (z_1), as suggested by Table 2. Inside the microwave, N_2 gas flow was maintained at 50 mL/min. Each experimental run carried out under different conditions yielded different types of activated carbon (AC) samples. A synthesized AKTWAC sample after the microwave-assisted heating process was washed vigorously with distilled water to remove the residual salt. Washing was carried out until the sample pH was neutral. The final sample (AKTWAC) was dried overnight in a vacuum oven at 45 °C and sent for necessary characterizations. Lead (II) cations, Pb (II) removal percentages (R_1), fixed carbon content (R_2), and yield percentages of carbon (R_3) were determined and are tabulated in Table 2. For further mathematical modelling, ANOVA analysis and process optimization were performed on the final sample of AKTWAC.

3.1.2. Preparation of Adsorbate Solution

Lead nitrate, $Pb(NO_3)_2$, salt, and K_2CO_3 were purchased from Sigma–Aldrich, Japan. A stock solution of lead (II) cations with a concentration of 1000 ppm was prepared using distilled water (DI). The stock solution was diluted to 100 ppm before observing the performance of ACs prepared using two-step microwave-assisted (MWP) pyrolysis of semi-carbonized char (AKTWC).

3.2. Methodology

3.2.1. Adsorption Studies

The concentration of 100 mg/L was agitated with 0.25 g of the AKTWAC sample to calculate the removal percentages, R_1. (Table 2). At 30 °C and 200 rpm rotation speed, the adsorption experiment was carried out. The pH of the pollutant Pb (II) solution was fixed to 5.5 before agitation, and volume treated with 0.25 g of AKTWAC containing Pb (II) cations with a concentration of 100 mg/L was 50 mL until equilibrium time (6 h—after which carbon became exhausted and no further adsorption took place). The following equation [51,64,69,70] was used to determine the quantity of Pb (II) ions adsorbed onto the solid surface of ATWAC:

$$q_e = \frac{(C_0 - C_e)\,V}{W} \tag{4}$$

where he quantity of cation loaded after the equilibration time is denoted by q_e (mg/gm); C_0 represents the initial concentration Pb (II) cation; C_e (mg/L) is the remaining liquid phase concentrations present after equilibrium is achieved; V (L) is the pollutant volume; and W (gm) denotes the weight of AKTWAC used. Based on Table 2, the subsequent equation was used to calculate removal percentages under various experimental conditions [51,64,69,70]:

$$Removal\ Percentages = \frac{(C_0 - C_e)}{C_0} \times 100 \tag{5}$$

3.2.2. Design of Experiment Using Response Surface Methodology (RSM)

To investigate the influence of several effective variables on Pb (II) cation removal efficiency (R_1) from aqueous solution, the Box–Behnken design (BBD) approach, which relies on the RSM technique, was used. The BBD design is suitable for identifying the most optimal permutations of factors; at the same time, it can ensure maximum fixed carbon content (R_2) with maximum yield (R_3) while maintaining the highest adsorption capacity (R_1). It is employed to generate a second order polynomial model including linear, interaction, and quadratic terms (Equation (6)) [66]. The independent/input variables used here for microwave-assisted pyrolysis (MWP) were power (x_1), time (y_1), and ratio (z_1). The percentage removal of Pb (II) ions (R_1), fixed carbon content (R_2), and yield (R_3) were selected as responses (dependent variable). Fixed carbon content was measured using TGA analysis.

To conclude, the optimization technique consists of three key steps: statistically planning and conducting the experiments, developing the suitable mathematical models, and assessing the appropriateness of the developed models [47,48,62,71,72].

$$R = (f(x_1\,, x_2,\, x_3,\, x_4 \ldots x_n) \tag{6}$$

Here, R represents the responses and x_i represents the input variables under consideration. For each input factor, the low, middle, and high values were expressed by the coded numbers 1, 0, and +1, respectively. It is significant to ensure an accurate estimation of the appropriate functional correlation between the input/independent factors and the predefined responses in order to optimize the system [71,72].

After the experimental runs were completed, the responses were used to develop qualitative models that could be used to correspond the responses of removal percentages

(R_1) with fixed carbon (R_2) and yield (R_3) using a second-degree mathematical model, as expressed by Equation (7) [72]:

$$R = b_0 + \sum_{i=0}^{n} b_1 x_1 + \sum_{i=1}^{n} b_{ii} x_1^2 + \sum_{i=0}^{n} \times \sum_{j>1}^{n} b_{ij} x_i x_j \quad (7)$$

R stands for the anticipated responses, b_0 designates the constant coefficient, b_1 for the linear coefficients, b_{ij} for the interaction coefficients, b_{ii} for the quadratic coefficients, and x_i, x_j for the adsorption in coded terms [66]. The proposed number of trials at the centre point for three factors is 5, and the overall number of experiments (N) necessary is 17 (Table 2). The coded and actual levels of three factors, as well as the responses acquired for the BBD experimental design matrix, are presented in Table 2.

3.2.3. Analytical Methods and Physio-Chemical Characterization

To determine the surface characteristics of the produced samples, different analytical methods were used. A field emission scanning electron microscope (FE-SEM) connected with EDX analysis was used to examine the surface morphological variations of final derivatives (SUPRA Zeiss 35-VP; Zeiss Group, Berlin, Germany).

A BET Surface Area Analyser (TriStar II; Micrometrics, Berlin, Germany) was used to determine the BET surface area, micropore, mesopore area, volume, and porosity of the samples. The samples were outgassed under vacuum at 400 °C for 6 h before N_2 gas adsorption to eliminate moisture content. The Brunauer–Emmett–Teller (BET) method was employed to estimate the surface area and diameter of the pores, while the t-plot approach was used to quantify micropore volume. Iodine number and bulk density, along with P_{ZPC} of the synthesized AKTWAC, were measured [52]. X-ray diffraction (XRD, Bruker AXS-D8 Advance; Bruker Corporation, Kuala Lumpur, Malaysia) operating at 40 kV and 40 mA with a Cu-Kα radiation source was used. XRD analysis was carried out to examine the crystalline phase of the samples. To assess the thermal stability of the materials, thermo-gravimetric analysis (Star Mettler Toledo, Mettler-Toledo (M) Sdn Bhd, Selangor, Malaysia) was carried out. In the TGA study, 5 mg of each sample was heated at 1000 °C, with a heating rate of approximately 5 °C/min under N_2 gas flow. The fixed carbon (R_2) content, moisture, volatile materials, and ash content were determined using TGA analysis. The percentages of elemental carbon (C), oxygen (O), hydrogen (H), nitrogen (N), and sulphur (S) in AKTW, AKTWC, and AKTWAC, were calculated using an ultimate analysis (Perkin-Elmer; Series II: 2600; Malvern Panalytical, Tokio, Japan). Yield percentage (%) is a physical parameter, and after each experimental run based on Table 2 under different conditions, it was calculated using Equation (8).

$$Yield\ \% = \frac{Initial\ Weight\ of\ the\ Char - Final\ Microwave\ Assisted\ Activated\ Carbon\ Weight}{Initial\ weight\ of\ the\ Char} \times 100 \quad (8)$$

All the experiments were triplicated, and average results are tabulated in Table 2. The overall experimental steps are shown in Figure 9.

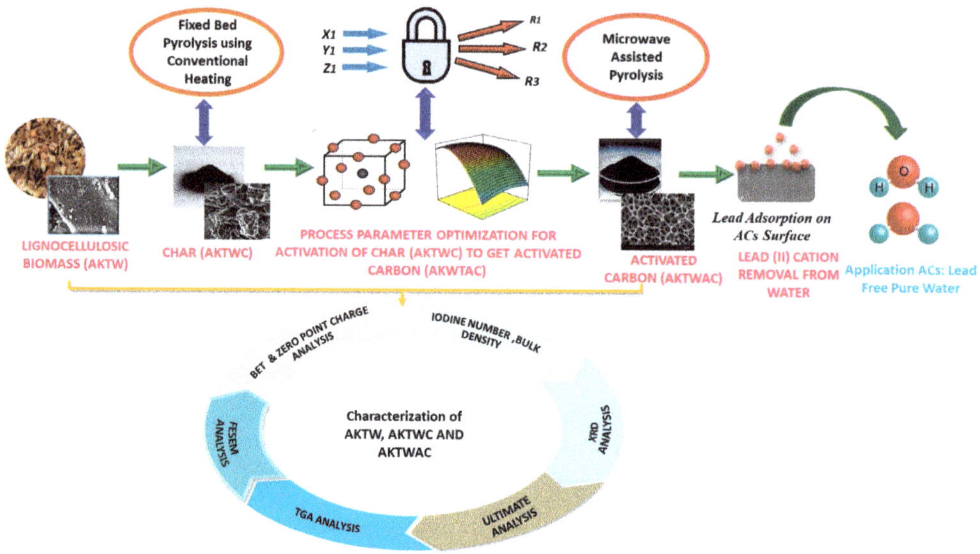

Figure 9. Simplified flow sheet for experimental steps.

4. Conclusions

The use of microwave heating during the pyrolysis process (MWP) shortened the production time while also generating activated carbon by expanding inaccessible pores and forming new ones. This is due to the bulk, volumetric, internal heating approach induced by the microwave-assisted pyrolysis (MWP) process. The findings indicate a unique technique for producing activated carbon (AC) which is cost-effective and energy-efficient.

The removal percentages of Pb (II) cations (R_1), fixed carbon content percentages (R_2), and yield percentages (R_3) are all controlled by input variables such as power (x_1), radiation time (y_1), and activating agent (K_2CO_3) ratio (z_1). Consequently, the level of input variables and the targeted output responses should be assessed initially before designing the AC manufacturing process.

The extent to which the cross-linking and elimination processes occur during the carbonisation and activation process controls the porosity and surface area of the activated carbon (Ass). The results indicate that the optimum process condition leads to the formation of new cavities due to multiple chemical interactions between the AKTWC and K_2CO_3, which makes it an efficient adsorbent for elimination of Pb (II) cations from waste effluents. Th Box–Behnken design (BBD) methodology applied in this research illustrates a reasonable and competent strategy to produce activated carbon from *Adansonia kilima (Baobab)* wood chips (AKTW) for removing Pb (II) cations from wastewater using a reduced number of experiments. AKTW has the potential to be used as a feedstock for the production of superior quality ACs, having an enlarged surface area and porous texture.

Author Contributions: Conceptualization, Investigation, Writing original draft S.S.; Development of methodology, Formal Analysis, Supervision, Writing original draft Z.Z.C.; Resource provided, Data curation, Software analysis K.K.; Formal Analysis, Validation, Data curation S.M.S.; Funding, Resource, Validation M.M.R.; Investigation, original draft preparation, M.R.; investigation, Funding, Project Administration I.A.B.; Formal analysis, Funding acquisition Conceptualization H.M.T.K.; Data curation, Funding S.K.; Supervision and Conceptualization M.R.B.J.; Data curation, Methodology and Funding M.H.; Software analysis and Writing—review & editing A.M. and A.N.F. All authors have read and agreed to the published version of the manuscript.

Funding: The authors are thankful for the funding provided by ST-077-2022 and IIRG003A-2022IISS under the University of Malaya, Kuala Lumpur 50603, Malaysia. The authors extend their appreciation to the Deanship of Scientific Research at King Khalid University for funding this work through the large group project under grant number R.G.P 2/36/44. The authors would like to acknowledge Doctors Feed Ltd., Bangladesh, and Universiti Utara Malaysia, research opportunity with grant number RI 10010030/SO 21153 for their support in this work.

Data Availability Statement: The data presented in this study are available on request from the corresponding author. The data are not publicly available due to privacy.

Acknowledgments: The authors extend their appreciation to the Deanship of Scientific Research at King Khalid University for funding this work through the large group project under grant number R.G.P 2/36/44. The authors are thankful for the funding provided by ST-077-2022 and IIRG003A-2022IISS under the University of Malaya, Kuala Lumpur 50603, Malaysia. The authors would like to acknowledge Doctors Feed Ltd., Bangladesh, and Universiti Utara Malaysia, research opportunity with grant number RI 10010030/SO 21153 for their support in this work.

Conflicts of Interest: The authors declare no conflict of interest.

Sample Availability: Not applicable.

References

1. Foroutan, R.; Peighambardoust, S.J.; Peighambardoust, S.H.; Pateiro, M.; Lorenzo, J.M. Adsorption of Crystal Violet Dye Using Activated Carbon of Lemon Wood and Activated Carbon/Fe_3O_4 Magnetic Nanocomposite from Aqueous Solutions: A Kinetic, Equilibrium and Thermodynamic Study. *Molecules* **2021**, *26*, 2241. [CrossRef] [PubMed]
2. Azam, M.; Wabaidur, S.M.; Khan, M.R.; Al-Resayes, S.I.; Islam, M.S. Removal of Chromium(III) and Cadmium(II) Heavy Metal Ions from Aqueous Solutions Using Treated Date Seeds: An Eco-Friendly Method. *Molecules* **2021**, *26*, 3718. [CrossRef] [PubMed]
3. Kacan, E. Optimum BET surface areas for activated carbon produced from textile sewage sludges and its application as dye removal. *J. Environ. Manag.* **2016**, *166*, 116–123. [CrossRef] [PubMed]
4. Tomic, N.M.; Dohcevic-Mitrovic, Z.D.; Paunovic, N.M.; Mijin, D.Z.; Radic, N.D.; Grbic, B.V.; Askrabic, S.M.; Babic, B.M.; Bajuk-Bogdanovic, D.V. NanocrystallineCeO_2—As effective adsorbent of azo dyes. *Langmuir* **2014**, *30*, 11582–11590. [CrossRef]
5. Das, S.K.; Khan, M.M.R.; Parandhaman, T.; Laffir, F.; Guha, A.K.; Sekarana, G.; Mandal, A.B. Nano-silica fabricated with silver nanoparticles: Antifouling adsorbent for efficient dye removal, effective water disinfection and biofouling control. *Nanoscale* **2013**, *5*, 5549–5560. [CrossRef]
6. Da Silva, L.G.; Ruggiero, R.; Gontijo, P.M.; Pinto, R.B.; Royer, B.; Lima, E.C.; Fernandes, T.; Calvete, T.H.M. Adsorption of Brilliant Red 2BE dye from watersolutions by a chemically modified sugarcane bagasse lignin. *Chem. Eng. J.* **2011**, *168*, 620–628. [CrossRef]
7. Chowdhury, Z.Z.; Abd Hamid, S.B.; Das, R.; Hasan, M.R.; Zain, S.M.; Khalid, K.; Uddin, M.N. Preparation of carbonaceous adsorbents from lignocellulosic biomass and their use in removal of contaminants from aqueous solution. *BioResources* **2013**, *8*, 6523–6555. [CrossRef]
8. Chowdhury, Z.Z.; Krishnan, B.; Sagadevan, S.; Rafique, R.F.; Hamizi, N.A.B.; Abdul Wahab, Y.; Khan, A.A.; Johan, R.B.; Al-douri, Y.; Kazi, S.N.; et al. Effect of Temperature on the Physical, Electro-Chemical and Adsorption Properties of Carbon Micro-Spheres Using Hydrothermal Carbonization Process. *Nanomaterials* **2018**, *8*, 597. [CrossRef]
9. Akinpelu, A.A.; Chowdhury, Z.Z.; Shibly, S.M.; Faisal, A.N.M.; Badruddin, I.A.; Rahman, M.M.; Amin, M.A.; Sagadevan, S.; Akbarzadeh, O.; Khan, T.M.Y.; et al. Adsorption Studies of Volatile Organic Compound (Naphthalene) from Aqueous Effluents: Chemical Activation Process Using Weak Lewis Acid, Equilibrium Kinetics and Isotherm Modelling. *Int. J. Mol. Sci.* **2021**, *22*, 2090. [CrossRef]
10. Chowdhury, Z.Z.; Zain, S.M.; Khan, R.A.; Ahmad, A.A.; Khalid, K. Application of response surface methodology (RSM) for optimizing production condition for removal of Pb (II) and Cu (II) onto kenaf fiber based activated carbon. *Res. J. Appl.* **2011**, *4*, 458–465.
11. Xu, J.; Chen, L.; Qu, H.; Jiao, Y.; Xie, J.; Xing, G. Preparation and characterization ofactivated carbon from reedy grass leaves by chemical activation with H_3PO_4. *Appl. Surf. Sci.* **2014**, *320*, 674–680. [CrossRef]
12. Ribas, M.; Adebayo, M.A.; Prola, L.D.T.; Lima, E.C.; Cataluña, R.; Feris, L.A.; Puchana-Rosero, M.J.; Machado, F.M.; Pavan, F.A.; Calvete, T. Comparison of a home-made cocoa shell activated carbon with commercial activated carbon for the removal of reactive violet 5 dye from aqueous solutions. *Chem. Eng. J.* **2014**, *248*, 315–326. [CrossRef]
13. Hesas, R.H.; Arami-Niya, A.; Daud, W.M.A.W.; Sahu, J.N. Preparation of granular activated carbon from oil palm shell by microwave-induced chemical activation: Optimisation using surface response methodology. *Chem. Eng. Res. Des.* **2013**, *91*, 2447–2456. [CrossRef]
14. Hesas, R.H.; Daud, W.M.A.W.; Sahu, J.N.; Arami-Niya, A. The effects of a microwave heating method on the production of activated carbon from agricultural waste: A review. *J. Anal. Appl. Pyrol.* **2013**, *100*, 1–11. [CrossRef]
15. Chayid, M.A.; Ahmed, M.J. Amoxicillin adsorption on microwave prepared activated carbon from *Arundo donax* Linn: Isotherms kinetics, and thermodynamics studies. *J. Environ. Chem. Eng.* **2015**, *3*, 1592–1601. [CrossRef]

16. Foo, K.Y.; Hameed, B.H. Textural porosity, surface chemistry and adsorptive properties of durian shell derived activated carbon prepared by microwave assisted NaOH activation. *Chem. Eng. J.* **2012**, *187*, 53–62. [CrossRef]
17. Foo, K.Y.; Lee, L.K.; Hameed, B.H. Preparation of banana frond activated carbon by microwave induced activation for the removal of boron and total iron from landfill leachate. *Chem. Eng. J.* **2013**, *223*, 604–610. [CrossRef]
18. Al-mahbashi, N.M.Y.; Kutty, S.R.M.; Jagaba, A.H.; Al-nini, A.; Sholagberu, A.T.; Baker, N.S.; Aldhawi, B.N.S.; Rathnayake, U. Sustainable sewage sludge biosorbent activated carbon for remediation of heavy metals: Optimization by response surface methodology. *Case Stud. Chem. Environ. Eng.* **2023**, *8*, 100437. [CrossRef]
19. Saucier, C.; Adebayo, M.; Lima, E.C.; Cataluña, R.; Thue, P.; Prola, L.D.T.; Puchana-Rosero, M.J.; Machado, F.; Pavan, F.; Dotto, G.L. Microwaved-assisted activated carbon from cocoa shell as adsorbents for removal of sodium diclofenac and nimesulide from aqueous effluents. *J. Hazard. Mater.* **2015**, *289*, 18–27. [CrossRef]
20. Foo, K.Y.; Hameed, B.H. Microwave-assisted preparation of oil palm fiber activated carbon for methylene blue adsorption. *Chem. Eng. J.* **2011**, *166*, 792–795. [CrossRef]
21. Salema, A.A.; Ani, F.N. Microwave-assisted pyrolysis of oil palm shell biomass using an overhead stirrer. *J. Anal. Appl. Pyrol.* **2012**, *96*, 162–172. [CrossRef]
22. Ahmed, M.J.; Theydan, S.K. Fluoro-quinolones antibiotics adsorption onto microporous activated carbon from lignocellulosic biomass by microwave pyrolysis. *J. Taiwan Inst. Chem. Eng.* **2014**, *45*, 219–226. [CrossRef]
23. Ahmed, M.J.; Theydan, S.K. Optimization of microwave preparation conditions for activated carbon from *Albizia lebbeck* seed pods for methylene blue dye adsorption. *J. Anal. Appl. Pyrol.* **2014**, *105*, 199–208. [CrossRef]
24. Njoku, V.O.; Foo, K.Y.; Asif, M.; Hameed, B.H. Preparation of activated carbonsfrom rambutan (*Nephelium lappaceum*) peel by microwave-Induced KOH activation for acid yellow 17 dye adsorption. *Chem. Eng. J.* **2014**, *250*, 198–204. [CrossRef]
25. Ahmad, M.A.; Puad, N.A.A.; Bello, O.S. Kinetic: Equilibrium and thermodynamic studies of synthetic dye removal using pomegranate peel activated carbon prepared by microwave-induced KOH activation. *Water Resour. Ind.* **2014**, *6*, 18–35. [CrossRef]
26. Abbas, A.F.; Ahmed, M.J. Mesoporous activated carbon from date stones (*Phoenix dactylifera* L.) by one-step microwave assisted K_2CO_3 pyrolysis. *J. Water Proc. Eng.* **2016**, *9*, 201–207. [CrossRef]
27. Ferrera-Lorenzo, N.; Fuente, E.; Suárez-Ruiz, I.; Ruiz, B. KOH activated carbon from conventional and microwave heating system of a macro-algae waste from the Agar–Agar industry. *Fuel Process. Technol.* **2014**, *121*, 25–31. [CrossRef]
28. Huang, L.; Sun, Y.; Wang, W.; Yue, Q.; Yang, T. Comparative study on characterization of activated carbons prepared by microwave and conventional heating methods and application in removal of oxy-tetracycline(OTC). *Chem. Eng. J.* **2011**, *171*, 1446–1453. [CrossRef]
29. Qiang, Z.Z.; Ying, X.H.; Srinivasakannan, C.; Hui, P.J.; Bo, Z.L. Utilization of Crofton weed for preparation of activated carbon by microwave induced CO_2 activation. *Chem. Eng. Process.* **2014**, *82*, 1–8.
30. Pezoti, O., Jr.; Cazetta, A.L.; Gomes, R.; Barizão, É.O.; Souza, I.P.A.F.; Martins, A.C.; Asefa, T.; Almeida, V.C. Synthesis of $ZnCl_2$-activated carbon from macadamia nut endocarp (*Macadamia integrifolia*) by microwave-assisted pyrolysis: Optimization using RSM and methylene blue adsorption. *J. Anal. Appl. Pyrol.* **2014**, *105*, 166–176. [CrossRef]
31. Namazi, A.B.; Allen, D.G.; Jia, C.Q. Microwave-assisted pyrolysis and activation of pulp mill sludge. *Biomass Bioenergy* **2015**, *73*, 217–224. [CrossRef]
32. Maldhure, A.V.; Ekhe, J.D. Preparation and characterizations of microwave assisted activated carbons from industrial waste lignin for Cu(II) sorption. *Chem. Eng. J.* **2011**, *168*, 1103–1111. [CrossRef]
33. Zhu, L.; Lei, H.; Wang, L.; Yadavalli, G.; Zhang, X.; Wei, Y.; Liu, Y.; Yan, D.; Chen, S.; Ahring, B. Biochar of corn stover: Microwave-assisted pyrolysis condition induced changes in surface functional groups and characteristics. *J. Anal. Appl. Pyrol.* **2015**, *115*, 149–156. [CrossRef]
34. Georgin, J.; Dotto, G.L.; Mazutti, M.A.; Foletto, E.L. Preparation of activatedcarbon from peanut shell by conventional pyrolysis and microwaveirradiation-pyrolysis to remove organic dyes from aqueous solution. *J. Environ. Chem. Eng.* **2016**, *4*, 266–275. [CrossRef]
35. Jaber, L.; Ihsanullah, I.; Almanassra, I.W.; Backer, S.N.; Abushawish, A.; Khalil, A.K.A.; Alawadhi, H.; Shanableh, A.; Atieh, M.A. Adsorptive Removal of Lead and Chromate Ions from Water by Using Iron-Doped Granular Activated Carbon Obtained from Coconut Shells. *Sustainability* **2022**, *14*, 10877. [CrossRef]
36. Blanco, A.; Pignata, M.L.; Lascano, H.R.; Salazar, M.J.; Rodriguez, J.H. Lead uptake and translocation pathways in soybean seedlings: The role of ion competition and transpiration rates. *Environ. Sci. Pollut. Res.* **2021**, *28*, 20624–20636. [CrossRef]
37. Kumar, A.; Kumar, A.; Cabral-Pinto, M.; Chaturvedi, A.K.; Shabnam, A.A.; Subrahmanyam, G.; Mondal, R.; Gupta, D.K.; Malyan, S.K.; Kumar, S.S.; et al. Lead Toxicity: Health Hazards, Influence on Food Chain, and Sustainable Remediation Approaches. *Int. J. Environ. Res. Public Health* **2020**, *17*, 2179. [CrossRef]
38. Almanassra, I.W.; Khan, M.I.; Atieh, M.A.; Shanableh, A. Adsorption of Lead Ions from an Aqueous Solution onto NaOH-Modified Rice Husk. *Desalin. Water Treat.* **2022**, *254*, 104–115. [CrossRef]
39. Otieno, J.; Kowal, P.; Mąkinia, J. Monitoring Lead Concentration in the Surrounding Environmental Components of a Lead Battery Company: Plants, Air and Effluents—Case Study. Kenya. *Int. J. Environ. Res. Public Health* **2022**, *19*, 5195. [CrossRef]
40. Ahmad, K.; Shah, H.B.; Khan, M.S.; Iqbal, A.; Potrich, E.; Amaral, L.S.; Rasheed, S.; Nawaz, H.; Ayub, A.; Naseem, K.; et al. Lead In drinking water: Adsorption method and role of zeolitic imidazolate frameworks for its remediation: A Review. *J. Clean. Prod.* **2022**, *368*, 133010. [CrossRef]

41. Agwogie, A.B. The Potentials of *Adansonia digitata* Root and Stem Powders and Stem Activated Carbon as Low-Cost Adsorbents for the Removal of Heavy Metals from Aqueous Solutions. Masters's Thesis, University of Nigeria, Nsukka, Nigeria, 2014.
42. Ndjientcheu, L.M.; Yossa, S.K.; Ouiminga, S.S.; Sidibe, I.; Ouedraogo, W.K. Synthesis of a cleaner potassium hydroxide-activated carbon from baobab seeds hulls and investigation of adsorption mechanisms for diuron: Chemical activation as alternative route for preparation of activated carbon from baobab seeds hulls and adsorption of diuron. *Sci. Afr.* **2020**, *9*, e00476. [CrossRef]
43. Nedjai, R.; Alkhatib, M.; Kabbashi, N.; Alam, M.Z.; Ahmed Kabbashi, N. Removal of Phenol from Aqueous Solution by Adsorption onto Baobab Fruit Shell Activated Carbon: Equilibrium and Kinetics Studies. *J. Environ. Treat. Tech.* **2021**, *9*, 686–697. [CrossRef]
44. Tchikuala, E.; Mourão, P.; Nabais, J. Valorisation of Natural Fibres from African Baobab Wastes by the Production of Activated Carbons for Adsorption of Diuron. *Procedia Eng.* **2017**, *200*, 399–407. [CrossRef]
45. Marcos, J.C.; Fonseca, L.P.; Ramalho, M.T.; Cabral, J.M.S. Application of surface response analysis to the optimization of penicilin acylase purification in aqueous two-phase system. *Enzym. Microb. Technol.* **2002**, *31*, 1006–1014. [CrossRef]
46. Karacan, F.; Ozden, U.; Karacan, S. Optimization of manufacturing conditions for activated carbon from Turkish by chemical activation using response surface methodology. *Appl. Therm. Eng.* **2007**, *27*, 1212–1218. [CrossRef]
47. Karim, M.Z.; Chowdhury, Z.Z.; Abd Hamid, S.B.; Ali, M.E. Optimizing the pre-treatment condition for a-cellulose using Lewis acid catalyst. *Sci. Adv. Mater.* **2016**, *8*, 534–544. [CrossRef]
48. Chowdhury, Z.Z.; Zain, S.M.; Rashid, A.K.; Khalid, K. Process Variables Optimization for Preparation and Characterization of Novel Adsorbent from Lignocellulosic Waste. *BioResources* **2012**, *7*, 3732–3754. [CrossRef]
49. Sudaryanto, Y.; Hartono, S.B.; Irawaty, W.; Hindarso, H.; Ismadji, S. High surface area activated carbon prepared from cassava peel by chemical activation. *Bioresour. Technol.* **2006**, *97*, 734–739. [CrossRef]
50. Cao, Q.; Xie, K.C.; Lv, Y.K.; Bao, W.R. Process effects on activated carbon with large specific surface area from corn cob. *Bioresour. Technol.* **2006**, *97*, 110–115. [CrossRef]
51. Chowdhury, Z.Z.; Hasan, M.D.; Abd Hamid, S.B.; Samsudin, E.M.; Zain, S.M.; Khalid, K. Catalytic pre-treatment of biochar residues derived from lignocellulosic feedstock for equilibrium studies of manganese, Mn(II) cations from aqueous solution. *RSC Adv.* **2015**, *5*, 6345–6356. [CrossRef]
52. Deng, H.; Li, G.X.; Yang, H.B.; Tang, J.P.; Tang, J.Y. Preparation of activated carbons from cotton stalk by microwave assisted KOH and K_2CO_3 activation. *Chem. Eng. J.* **2010**, *163*, 373–381. [CrossRef]
53. Arenas, E.; Chejne, F. The effect of the activating agent and temperature on the porosity development of physically activated coal chars. *Carbon* **2004**, *42*, 2451–2455. [CrossRef]
54. Zolin, A.; Jensen, A.; Dam-Johansen, K.; Jensen, L.S. Influence of experimental protocol on activation energy in char gasification: The effect of thermal annealing. *Fuel* **2001**, *80*, 1029–1032. [CrossRef]
55. Lua, A.C.; Guo, J. Activated carbon prepared from oil palm stone by one-step CO_2 activation for gaseous pollutant removal. *Carbon* **2000**, *38*, 1089–1097. [CrossRef]
56. Lam, S.S.; Liew, R.K.; Wong, Y.M.; Yek, P.Y.; Ma, N.; Lee, C.L.; Chase, H. Microwave-assisted pyrolysis with chemical activation, an innovative method to convert orange peel into activated carbon with improved properties as dye adsorbent. *J. Clean. Prod.* **2017**, *162*, 1376–1387. [CrossRef]
57. Zahangir, M.A.; Suleyman, A.M.; Noraini, K. Production of activated carbon from oil palm empty fruit bunches for removal of Zinc. In Proceedings of the Twelfth International Water Technology Conference, IWTC12, Alexandria, Egypt, 1 January 2008; pp. 373–383.
58. Kundu, A.; Gupta, B.S.; Hashim, M.; Redzwan, G. Taguchi optimization approach for production of activated carbon from phosphoric acid impregnated palm kernel shell by microwave heating. *J. Clean. Prod.* **2015**, *105*, 420–427. [CrossRef]
59. Kundu, A.; Gupta, B.S.; Hashim, M.; Sahu, J.; Mujawar, M.; Redzwan, G. Optimisation of the process variables in production of activated carbon by microwave heating. *RSC Adv.* **2015**, *5*, 35899–35908. [CrossRef]
60. Duan, X.; Chandrasekar, S.; Peng, J.-h.; Zhang, L.-b.; Zhang, Z.-y. Preparation of activated carbon from Jatropha hull with microwave heating: Optimization using response surface methodology. *Fuel Process. Technol.* **2011**, *92*, 394–400.
61. Basta, A.H.; Fierro, V.; El-Saied, H.; Elzard, C.A. 2-steps KOH activation of rice straw: An efficient method for preparing high performance activated carbons. *Bioresour. Technol.* **2009**, *100*, 3941–3947. [CrossRef]
62. Chowdhury, Z.Z.; Julkapli, N.M.; Ali, A.M.; Abdul, R.A.M.A.H.; Wageeh, A.Y. Microwave Assisted Synthesis, Characterization and Application of Graphitic Bio-carbon using Two Level Factorial Designs. *BioResources* **2016**, *11*, 3637–3659. [CrossRef]
63. Chowdhury, Z.Z.; Karim, M.Z.; Ashraf, M.A. Influence of Carbonization Temperature on Physicochemical Properties of Biochar derived from Slow Pyrolysis of Durian wood sawdust (*Durio zibethinus*). *BioResources* **2016**, *11*, 3356–3372. [CrossRef]
64. Chowdhury, Z.Z.; Abd Hamid, S.B.; Zain, S.M. Base Catalytic Approach: A Promising Technique for Activation of Bio Char for Equilibrium Sorption Studies of Copper, Cu(II) ions in Single Solute System. *Materials* **2014**, *7*, 2815–2832.
65. Sagadevan, S.; Pal, K.; Chowdhury, Z.Z. Fabrication of CuO nanoparticles for structural, optical and dielectric analysis using chemical precipitation method. *J. Mater. Sci. Mater. Electron.* **2017**, *28*, 12591–12597. [CrossRef]
66. Sagadevan, S.; Pal, K.; Chowdhury, Z.Z.; Foley, M. Controllable synthesis of Graphene/ZnO-nanocomposite for novel switching. *J. Alloys Compd.* **2017**, *728*, 645–654. [CrossRef]
67. Hamid, S.A.; Chowdhury, Z.Z.; Karim, M.; Ali, M. Catalytic Isolation and Physicochemical Properties of Nanocrystalline Cellulose (NCC) using HCl-FeCl3 System Combined with Ultrasonication. *BioResources* **2016**, *11*, 3840–3855. [CrossRef]

68. Liu, H.; Jian Zhang, J.; Bao, N.; Cheng, C.; Ren, L.; Zhang, C. Textural properties and surface chemistry of lotus stalk-derived activated carbons prepared using different phosphorus oxyacids: Adsorption of trimethoprim. *J. Hazard Mater.* **2012**, *235–236*, 367–375. [CrossRef]
69. Adebisi, G.A.; Chowdhury, Z.Z.; Alaba, P.A. Equilibrium, kinetic, and thermodynamic studies of lead ion and zinc ion adsorption from aqueous solution onto activated carbon prepared from palm oil mill effluent. *J. Clean. Prod.* **2017**, *148*, 958–968. [CrossRef]
70. Chowdhury, Z.Z.; Zain, S.M.; Khan, R.A.; Ahmed, A.A. Equilibrium kinetics and isotherm studies of Cu (II) adsorption from waste water onto alkali activated oil palm ash. *Am. J. Appl. Sci.* **2011**, *8*, 230–237. [CrossRef]
71. Kaveh, M.; Abbaspour-Gilandeh, Y.; Taghinezhad, E.; Witrowa-Rajchert, D.; Nowacka, M. The Quality of Infrared Rotary Dried Terebinth (*Pistacia atlantica* L.)-Optimization and Prediction Approach Using Response Surface Methodology. *Molecules* **2021**, *26*, 1999. [CrossRef]
72. Chowdhury, Z.Z.; Hamid, S.B.A.; Rahman, M.M.; Rafique, R.F. Catalytic activation and application of micro-spherical carbon derived from hydrothermal carbonization of lignocellulosic biomass: Statistical analysis using Box–Behnken design. *RSC Adv.* **2016**, *6*, 102680–102694. [CrossRef]

molecules

Article

Açai Seeds (*Euterpe oleracea* Mart) Are Agroindustrial Waste with High Potential to Produce Low-Cost Substrates after Acid Hydrolysis

Willen Silva Igreja [1], Luiza Helena da Silva Martins [1,2], Rafaela Rodrigues de Almeida [3], Johnatt Allan Rocha de Oliveira [1,4], Alessandra Santos Lopes [1,3] and Renan Campos Chisté [1,3,*]

1 Postgraduate Program of Food Science and Technology (PPGCTA), Institute of Technology (ITEC), Federal University of Pará (UFPA), Belém 66075-110, PA, Brazil
2 Instituto de Saúde e Produção Animal (ISPA), Universidade Federal Rural da Amazônia (UFRA), Belém 66075-900, PA, Brazil
3 School of Food Engineering (FEA), Institute of Technology (ITEC), Federal University of Pará (UFPA), Belém 66075-110, PA, Brazil
4 Nutrition College, Federal University of Pará (UFPA), Belém 66075-110, PA, Brazil
* Correspondence: rcchiste@ufpa.br

Abstract: Açai seeds have been discarded improperly around the Amazonia region, but they can be seen as promising low-cost substrates for fermentation processes. The structural carbohydrates and physicochemical characterization of açai seeds from the Amazonia were assessed followed by the determination of the optimal hydrolysis conditions using H_3PO_4 (phosphoric acid) and H_2SO_4 (sulfuric acid) to obtain a liquor with high contents of simple carbohydrates and low levels of potential microbial inhibitors usually generated during acid hydrolysis of carbohydrates. A central composite rotational design was carried out varying the concentrations of diluted acid (0–5%, *w*/*v*), solids (0.1–25%, *w*/*v*), and hydrolysis time (9.5–110 min). Acid hydrolysis with H_2SO_4 was more effective in producing reducing sugars (15.9–103.1 g/L) than H_3PO_4 (2.9–33.9 g/L) during optimization. The optimal hydrolysis conditions with H_2SO_4 were 3.5% of acid (*w*/*v*), 25% of solids during 70 min at 121 °C, which provided a liquor with 55 g/L of reducing sugars and low levels of microbial inhibitors: acetic acid (1.8 g/L), hydroxymethyl furfural (338 mg/L), and furfural (10 mg/L). Thus, açai seeds were characterized as promising agroindustrial waste with high potential to be used as a low-cost substrate in biotechnological processes, comprising relevant environmental and bioeconomic aspects for the development of the Amazonia.

Keywords: Amazonian fruits; byproduct; reducing sugars; fiber seeds; bioeconomy

1. Introduction

The world economy has improved in production systems with increasingly higher productivity yields and lower operating costs that provide increasingly profitable agroindustrial production. In line with environmental and bioeconomic aspects, it is of paramount importance that raw materials, and derived wastes, be used in any technological processes in its entirety. In this sense, the Amazon highlights its contribution in the agroindustrial market with açai fruits being one of the main exploited products, yielding more than 540 million USD in 2019 [1].

Despite the expressive numbers, during açai processing, to obtain the juice, only 15% of the fruit is used in the process (epicarp and mesocarp), with the remaining 85% composed by fibers (endocarp), which are improperly discarded around Amazonia [2]. Waste with fibrous characteristics formed by cellulose, lignin, and hemicellulose are susceptible to pretreatments to release fermentable sugars [3], which can further be used as a culture media in biotechnological processes, with the aim of promoting and produce high value-added products from low-cost substrates, providing cost savings in the process.

The study of the use of new industrial fermentation means from agribusiness waste has been intensified in recent years [3,4], and it can be produced from various lignocellulose complexes representing a cheap source, renewable and plentiful, enabling sustainable and profitable production. In view of this, several pre-treatments have been reported in the literature with the aim of fractionating, solubilizing, hydrolyzing, and separating the biomass components for adequate purposes, with acid pre-treatment being one of the most promising in deconstructing the lignocellulose chain and changing the conformation structure of lignins [5].

At the moment, there is a scarcity of data in the literature that assess the potential application of agroindustrial wastes, namely açai seeds, as a substrate for the production of high-added value products. Therefore, the objective of this work was to investigate the acid hydrolysis treatment of açai seeds to produce a liquor with high contents of simple carbohydrates and low levels of potential microbial inhibitors usually generated during acid hydrolysis of carbohydrates, which were monitored for the first time, for the application as low-cost substrate in further biotechnological processes. The innovative nature of this work is that it provides a potential socioeconomic and environmental alternative through the waste generated by the açai production chain, making it possible to cheapen several fermentable processes to produce economically viable products.

2. Results and Discussion

2.1. Characterization of the Dried Açai Seeds

Açai seeds are agroindustrial wastes that are still little explored, with few scientific data available in the literature, and their physicochemical composition is essential to enable their application in a variety of investigations, including as a substrate in biotechnological approaches. As can be seen in Figure 1, the major constituent of the dried seeds was accounted for as total carbohydrates (≈87%), and according to the literature [6–9], due to the high content of lignocellulosic material, açai seeds can be considered a promising low-cost substrate for the production of derived compounds with high added value by biotechnological means.

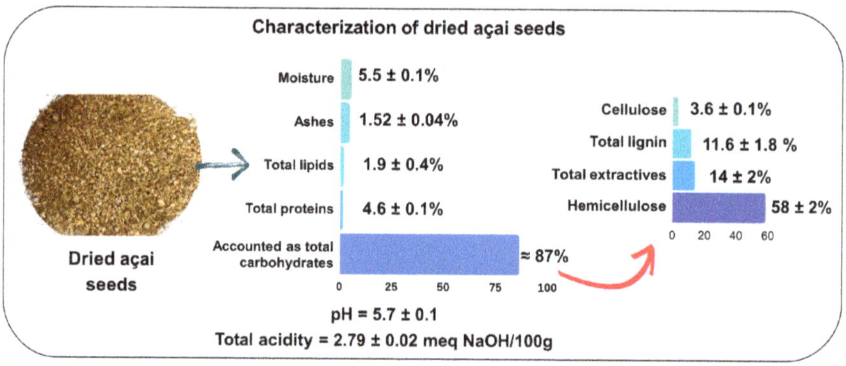

Figure 1. Physicochemical and structural carbohydrates characterization of the dried açai seeds used in this work.

However, the determination of total carbohydrates, by difference, overestimated the total carbohydrate values in the dried açai seeds since it also included the contents of total extractives and total lignin. Thus, after determining the contents of total extractives, structural carbohydrates, and total lignin, the dried açai seeds exhibited low contents of cellulose (3.6%) and total lignin (11.68%), 14% of total extractives and high hemicellulose contents (58%) (Figure 1). According to Oliveira et al. [6], the fibers of *E. oleracea* seeds presented 38% cellulose, 18% lignin, and 25% hemicellulose; and in another study, Martins et al. [10] reported that the fibers of the açai seeds were composed of 37% hemicellulose, 33%

lignin, and 33% cellulose. A recent study showed that the fibers of açai seeds are mainly composed of hemicellulose (50%, dry basis) [11], which supported our findings. Based on these values, açai seeds presented a differentiated composition of lignocellulosic material to be used directly as a substrate for fermentation processes, requiring pre-treatment strategies such as acid hydrolysis to release simple fermentable sugars.

In relation to the other constituents (Figure 1), the moisture content of the dried açai seeds (5.5%) was lower than those found in other studies with dehydrated açai seeds (8 and 13%) [6,7], but it clearly depends on the chosen drying parameters. Importantly, when a dried material is used for the purpose of releasing fermentable sugars through acid hydrolysis, moisture becomes a preponderant factor, as high values may dilute the acid concentration, decreasing the hydrolysis efficiency, and values < 10% are the most applicable [12,13]. The total protein values found in this work (4.6%) were similar to those found in other studies (1.7–4.3%) [6–9], and the total lipid values (1.9%) were similar to those found by Ferreira et al. (1.70%) [8]. Additionally, based on the pH and total acidity values (Figure 1), the dried açai seeds used in this study can be considered a low acidic product. In general, the variations in the contents of moisture, total proteins, total lipids, and ashes in the seeds are usually attributed to differences in planting conditions, ripening stage, climate, geographic location, harvest period, time of sun exposure, and its own genetic variability.

2.2. Effect of the Acid Hydrolysis Treatments on the Dried Açai Seeds

During acid hydrolysis with diluted acid, the lignocellulosic material is hydrolyzed causing the release of sugar monomers [14]. Although H_3PO_4 has a higher cost than other acids, such as H_2SO_4 and hydrochloric (HCl), it is considered a weaker acid (and consequently, lower combined severity factor) as compared to the previously mentioned acids, and its use implies lower environmental impact, in addition to being less corrosive and toxic. Moreover, H_3PO_4 has phosphorus in its composition, which can be incorporated as a micronutrient by microorganisms after hydrolysis [15,16].

In our study, although H_3PO_4 presented lower combined severity factors than H_2SO_4 for the selected acid hydrolysis conditions in the central composite rotational design (CCRD) runs, the range of total reducing sugar contents in the produced liquor (2.93–33.97 g/L) (Supplementary Table S1) was much lower than the contents obtained by H_2SO_4 (6.93–103.12 g/L) (Table 1). Thus, we decided to focus our investigation on only the H_2SO_4 treatment, but all the results for the acid hydrolysis with H_3PO_4 can also be found as Supplementary Material.

According to Table 1, in general, the contents of total reducing sugars increased as the combined severity factor increased during the acid hydrolysis of dried açai seeds, and the same tendency was observed for H_3PO_4 (Supplementary Table S1). For H_2SO_4, the increase in the acid concentration from 0 (run 9) to 2.5% (runs 15, 16, and 17) increased by about seven times the contents of total reducing sugars. If the acid treatments are carried out under mild conditions of acid concentration and temperature, the hemicellulose fraction may be extracted without significantly affecting cellulose and lignin contents [14]. Furthermore, hydrolysis time also has relevant effects to be considered since the increase in time from 60 (run 12) to 110 min (runs 15, 16, and 17), at the same acid and solid concentrations, did not increase the contents of reducing sugars in the liquor. This behavior can be explained due to the higher severity of the treatment at 110 min than 60 min that probably degraded sugars resulting in the formation of derived products [5,17]. Thus, monitoring the combined severity factor and time during acid hydrolysis may ensure the release of glycoside monomers while avoiding or minimizing the accumulation of sugar degradation products, such as hydroxymethylfurfural (HMF) [5].

The highest content of total reducing sugar in the liquor after the H_2SO_4 treatments (103.12 g/L) was obtained at 4% acid and 20% solids after 90 min of hydrolysis (Table 1). Interestingly, these same conditions also yielded the highest contents of total reducing sugars to the liquor after the hydrolysis using H_3PO_4 (33.97 g/L) (Supplementary Table S1).

Table 1. Combined severity factor (CSF), contents of total reducing sugars (TRS), hydroxymethylfurfural (HMF) and solubilized mass obtained during the central composite rotational design (CCRD) runs for the acid hydrolysis of açai seeds using sulfuric acid (H_2SO_4).

	Level of Independent Variables				Dependent Variables		
Runs	X_1	X_2	X_3	CSF	TRS (g/L)	HMF (mg/L)	Solubilized Mass (%)
1	1 (−1)	30 (−1)	5 (−1)	1.06	12.25	5.60	27.93
2	4 (+1)	30 (−1)	5 (−1)	1.21	17.65	33.68	38.58
3	1 (−1)	90 (+1)	5 (−1)	1.53	14.76	12.82	31.94
4	4 (+1)	90 (+1)	5 (−1)	1.68	25.40	90.30	48.83
5	1 (−1)	30 (−1)	20 (+1)	1.06	28.98	19.92	21.53
6	4 (+1)	30 (−1)	20 (+1)	1.21	59.60	82.06	35.23
7	1 (−1)	90 (+1)	20 (+1)	1.53	57.54	14.79	29.25
8	4 (+1)	90 (+1)	20 (+1)	1.68	103.12	448.83	48.49
9	0 (−1.68)	60 (0)	12.5 (0)	NC1	6.93	1.49	11.56
10	5 (+1.68)	60 (0)	12.5 (0)	1.62	44.70	148.90	49.58
11	2.5 (0)	9.54 (−1.68)	12.5 (0)	0.69	15.93	<LOD	23.52
12	2.5 (0)	110 (+1.68)	12.5 (0)	1.75	51.34	191.34	41.22
13	2.5 (0)	60 (0)	0.1 (−1.68)	1.49	<LOD	0.59	NC2
14	2.5 (0)	60 (0)	25 (+1.68)	1.49	63.95	87.01	33.28
15	2.5 (0)	60 (0)	12.5 (0)	1.49	49.06	70.05	37.13
16	2.5 (0)	60 (0)	12.5 (0)	1.49	48.16	70.58	38.50
17	2.5 (0)	60 (0)	12.5 (0)	1.49	48.92	72.53	37.18

X_1: Acid concentration (%, w/v). X_2: Hydrolysis time (min). X_3: Solids concentration (%, w/v). <LOD: below the limit of detection (equal 0 for the prediction purpose). NC1: Combined severity factor not calculated given the 0% diluted acid concentration. NC2: Not calculated (equals to 0) due to the low solid concentration (0.1%, w/v).

According to the Pareto charts (Figure 2), the contents of solids and acid concentration were the variables that most affected ($p \leq 0.1$) the contents of total reducing sugars, with positive linear effects, followed by the hydrolysis time, suggesting that the higher the percentage of dried açai seeds and the concentration of H_2SO_4, the higher the total reducing sugar contents in the liquor. Furthermore, the treatments with H_2SO_4 resulted in a very low variation in the contents of total reducing sugars at the CCRD central points (relative standard deviation (RSD) = 1%), which indicated high repeatability of the hydrolysis procedure. Statistical details regarding the values of the (t and (p tests carried out for the experimental design for all the parameters observed in this study can be found as Supplementary Table S2.

As the acid treatment acts mainly in the solubilization of hemicelluloses, very severe treatments degrade sugars during hydrolysis, generating microbiological inhibitors, such as HMF, which may affect fermentation processes [5]. As can be seen in Table 1, the increase in the combined severity factor contributed to the increase in the HMF contents in the hydrolyzed liquor, with the highest concentration found in run 8 (448 mg/L), which corresponds to the highest factor (1.68). Oliveira et al. [6] carried out pre-treatment of açai seeds using diluted H_2SO_4 under various experimental conditions of acid concentration (0.5 to 1% m/v), hydrolysis time (9.5 to 111 min), and solids concentration (1.6 to 18%), and reported HMF contents ranging from 1 to 221 mg/L. The milder combined severity factor and lower solids concentration range in comparison with our study may explain the lower contents of HMF. HMF can cause damage to yeast cells as the presence of the hydroxymethyl group reduces both the hydrophobicity and membrane permeability, resulting in reduced rates of HMF conversion [18]. These same authors stated that in the presence of HMF, fermentative microorganisms redirect the energy of the cells to repair the damage, resulting in a reduction in intracellular levels of ATP and NADPH caused by enzymatic inhibition or by consumption/regeneration of cofactors.

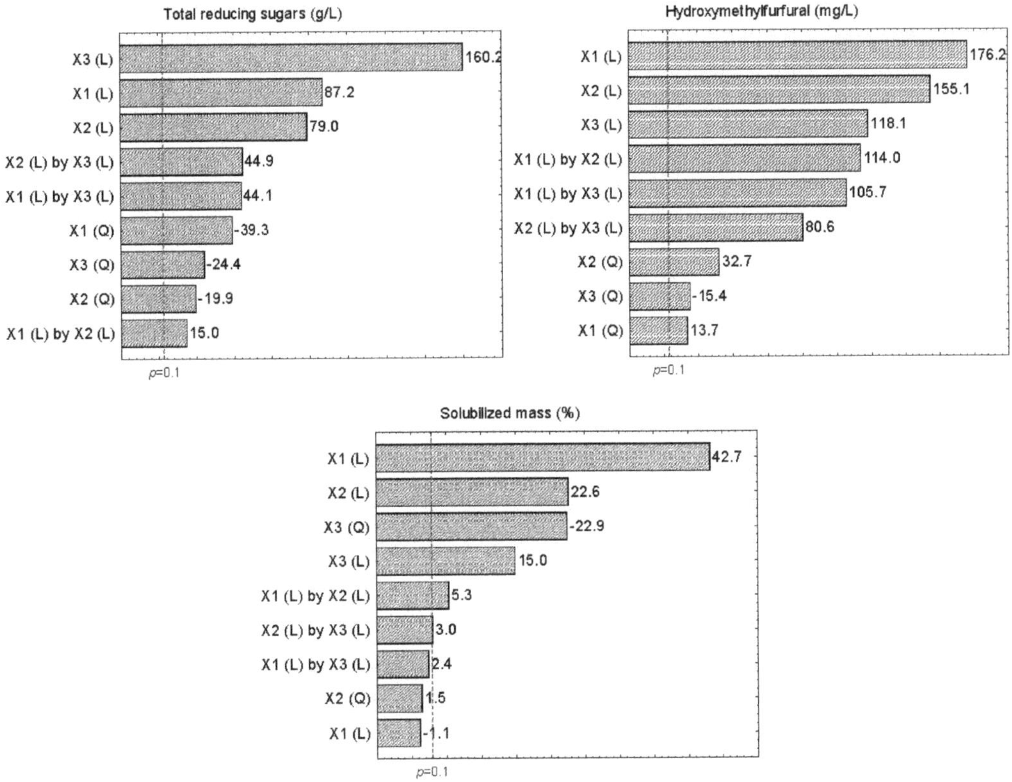

Figure 2. Pareto charts for the standardized effect estimates (absolute values) on the contents of total reducing sugars, hydroxymethylfurfural, and solubilized mass after the acid hydrolysis of dried açai seeds using H_2SO_4. X1 = Acid concentration (%, *w*/*v*); X2 = Hydrolysis time (min); X3 = Solids concentration (%, *w*/*v*); (L): linear effect; (Q) quadratic effect.

The Pareto charts for the contents of HMF (Figure 2) showed that acid concentration and hydrolysis time were the variables that most influenced ($p \leq 0.1$) its contents, with linear positive effects, followed by the percentage of solids, suggesting that the higher the concentration of H_2SO_4 and hydrolysis time, the higher the HMF contents in the liquor. These results were expected, as the use of high concentrations of acid, combined with high temperatures, promotes the dehydration of hexoses and consequently the production of furans [5,17].

Regarding the contents of solubilized mass of the solid fractions of the dried açai seeds after the acid hydrolysis with H_2SO_4, the lowest percentage (11.56%) was observed at 0% of diluted acid, and the mass solubilization increased in the runs with acid addition (1–5%) from 21.53 to 49.58% (Table 2). According to the Pareto charts (Figure 2), the positive linear effect of acid concentration was the variable that most affected ($p \leq 0.1$) the mass solubilization, followed by the negative quadratic effect of solids percentage and linear positive effect of hydrolysis time, suggesting that the solid fraction of the dried açai seeds was more soluble at the highest tested H_2SO_4 concentration and hydrolysis time. According to the literature, the acid pre-treatment of most lignocellulosic materials results in high solubilization percentages of hemicellulose and a partial digestion of cellulose leading the formation of monomer or oligomer sugars [15,16,19].

Table 2. Contents of sugars and microbiological inhibitors in the newly hydrolyzed liquor and after neutralization both produced after acid hydrolysis of dried açai seeds with H_2SO_4 at the optimal conditions.

	Hydrolyzed Liquor	
	Newly Hydrolyzed	**After Neutralization** *
Sugars		
Mannose (g/L)	51 ± 1^{a}	42.1 ± 0.1^{b}
Xylose (g/L)	1.9 ± 0.2^{a}	1.4 ± 0.2^{b}
Glucose (g/L)	1.3 ± 0.1^{a}	1.0 ± 0.1^{a}
Cellobiose (g/L)	0.9 ± 0.1^{a}	0.7 ± 0.1^{a}
Sum of the sugars (g/L)	55 ± 1^{a}	45.2 ± 0.4^{b}
Microbial inhibitors		
Hydroxymethylfurfural (mg/L)	338 ± 50^{a}	238 ± 20^{b}
Furfural (mg/L)	10 ± 1^{a}	9 ± 1^{a}
Acetic acid (g/L)	1.8 ± 0.1^{a}	1.76 ± 0.04^{a}

* Liquor hydrolyzed under the optimal acid hydrolysis conditions, followed by neutralization with K_2SO_4, autoclaved at 121 °C for 15 min and filtered, according to common practices used in fermentation processes. The values (mean ± standard deviation) with the same superscript letters in the same line are not statistically different ($p > 0.05$) (Student's *t*-test).

Figure 3 shows the behavior reflected in the distribution of residues, where the data from the present study are arranged randomly and not with a curved pattern for the solubilized mass (Figure 3c). Still, we no longer see the same behavior for TRS (Figure 3a) and HMF (Figure 3b), which indicates the need to adjust the mathematical model. In the calculated ANOVA (Supplementary Table S3), it was observed that there was a lack of fit for the data of this study; although the models studied were significant, they cannot be considered predictive; they can only be used as trend indicators, that is, only evaluated for the values of the range studied in the present work and cannot predict other values that may deviate from this range.

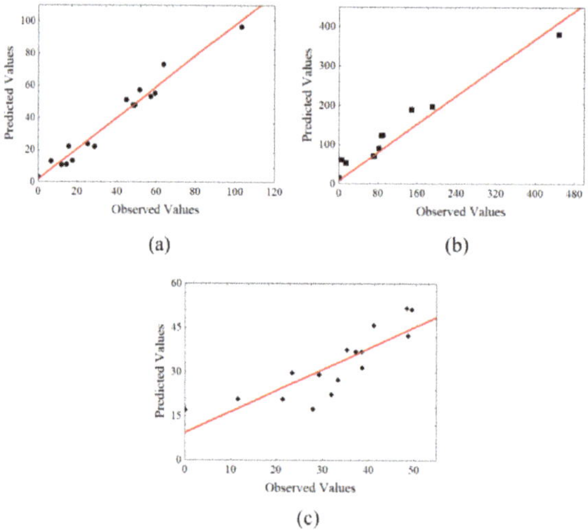

Figure 3. Graph of the observed and predicted values of the experimental design with central composite and axial levels of the studied parameters (**a**) TRS, (**b**) HMF, and (**c**) solubilized mass.

2.3. Optimal Conditions to Produce Liquor from Dried Açai Seeds by Acid Hydrolysis as Pre-Treatment Using H_2SO_4

After performing ANOVA ($p \leq 0.1$), the second order polynomial models proposed by the response surface methodology were not adequate for the prediction of TRS, HMF, and SM contents, since they showed great lack of fit; they were significant but non predictive, despite their high and moderate R^2 (0.96, 0.89, and 0.71, respectively) (Supplementary Table S3).

The regression models aim to establish a relationship that predicts the response corresponding to any independent variable value. For the test of the statistical significance of the regression, it was observed that the F distribution for TRS (20.11), with a 90% confidence interval, was greater than the F_{table} value (2.72). For HMF, the $F_{regression}$ value (6.92), with a 90% confidence interval, is greater than the F_{table} value (2.72). Regression (1.93) was more significant for soluble solids than the F_{table} (2.72), which rules out the null hypothesis that there is no linear relationship between the dependent and independent variables.

However, a lack of fit was observed in all experiments, which may be due to the failure to fully evaluate the response surface, as can be seen in Figure 4 (for constructing the contour graph). Therefore, the complementation of the results obtained in the regression analysis was considered satisfactory for continuing studies in the range of our experimental design.

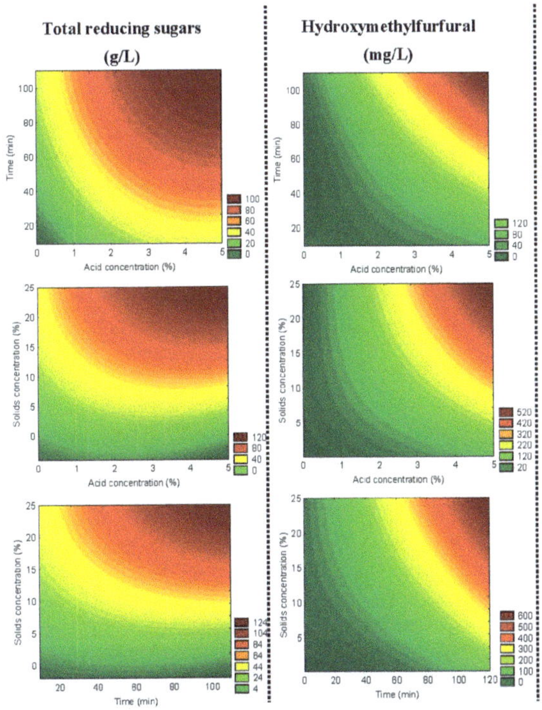

Figure 4. Contour plots to illustrate the contents of total reducing sugars (g/L), hydroxymethylfurfural (mg/L), and solubilized mass (%) for the determination of the optimal conditions for the acid hydrolysis of dried açai seeds using H_2SO_4.

Concerning the formation of HMF after carbohydrate hydrolysis, the highest levels tend to be formed at longer times and both higher acid and solid concentrations, as expected. However, according to Figure 4, no relevant increase in the HMF contents is expected to be found as long as the acid concentration is kept below 2%, even at longer hydrolysis time and higher solids concentration. Lastly, in relation to the solubilized mass, the highest solubilization of lignocellulosic material in dried açai seeds will be achieved by the combination of the highest acid concentrations and hydrolysis time.

Considering that the liquor with the highest TRS content (103.12 g/L) was obtained at the highest H_2SO_4 and longer time, which also showed the highest HMF level, we decided to use the desirability function to determine the optimal acid hydrolysis conditions with reduced CSF that minimize the formation of HMF. Therefore, after considering all the tendencies shown by the contour plots (Figure 4), the desirability function was set to predict the contents of TRS, HMF, and solubilized mass at 3.5% (w/v) acid concentration (m/v), 70 min of hydrolysis time, and 25% (w/v) solids concentration (Supplementary Figure S1). The correct selection of the acid hydrolysis conditions should be carefully and systematically further investigated to provide industrially feasible results to establish an economical viable process; the higher the acid concentration, the higher the probability of reactors' corrosion, increasing the maintenance price. Moreover, the need of high-power consumption due to long times as well as the need for additional costs for the neutralization of the acid liquor along with further purification steps must also be taken into account when designing any industrial project.

In order to validate the optimal acid hydrolysis conditions selected in our study for the dried açai seeds, we carried out the hydrolysis procedure (triplicate) and the produced liquor presented 82 $\pm$ 2 g/L of TRS, 265 $\pm$ 1 mg/L of HMF, and the solubilized mass was 41 $\pm$ 1%. The experimental values were close to the values predicted by the desirability function for TRS (94 g/L), HMF (273 mg/L), and solubilized mass (36%), with low relative standard deviation values (RSD = 9.6%, 2.1%, and 9.2%, respectively), demonstrating that the approached methodology is predictive and the conditions established for the acid hydrolysis procedure are reproductive.

2.4. Characterization of Sugars and Potential Microbial Inhibitors in the Newly Hydrolyzed Liquor and the Neutralized Liquor to Be Used as Substrate in Fermentation Processes

In our study, the newly hydrolyzed liquor obtained at the optimal acid hydrolysis conditions exhibited very low pH value (pH < 1.0) due to the use of 3.5% H_2SO_4, which characterizes it as a non-favorable substrate to be used for fermentation purposes, thus needing neutralization (pH close to 7.0). According to Table 2, the newly hydrolyzed liquor presented 55 g/L of sugars, mannose being the major monosaccharide (51 g/L) and accounting for about 93% of the detected sugars. After the neutralization of the liquor, the contents of mannose decreased by about 18% (42 g/L), and xylose by ≈26% (1.4 g/L), and were lower than those reported by Monteiro et al. [11] who found maximum values of 49.5 g/L for mannose in the liquor. For the contents of xylose, the values found hitherto varied from 1.9 to 1.4 (before and after neutralization) and were lower than the values observed by Oliveira et al. [6], which was 24.8 g/L, and by Monteiro et al. [11] with the maximum value of 3.26 g/L.

Several yeasts of the genus *Rhodotorula* spp., such as *R. glutinis*, *R. toruloides*, *R. graminis*, and *R. mucilaginosa*, have been investigated over the years as carotenoid producers by biotechnological means, and they were reported to be able to metabolize mannose and xylose as carbohydrate source [20–24], highlighting the high potential of açai seeds as low-cost substrates for the growth of selected microorganisms. As an example, Silva et al. [23] cultivated a strain of *R. glutinis* in a culture medium containing higher amounts of xylose (28 g/L) and compared to a glucose-containing medium (28 g/L, control). These authors observed that after 111 h of fermentation the maximum biomass concentration was almost the same for the two carbon sources (4.14 and 4.30 g/L for glucose and xylose, respectively). However, the specific growth rate for glucose (0.1 h) was higher than that observed for xylose (0.05 h). Another study carried out with *R. toruloides* tested a higher xylose concentration than our study (40 g/L) and the authors also reported that the growth rate in glucose-containing medium was higher (twice) than for xylose medium, the exponential phase with glucose lasting 72 h as compared to xylose (40 h) [24]. The high contents of mannose found demonstrates the great value of açaí seeds as a potential culture medium for similar yeasts, which was confirmed by Byrtusová et al. [25], who demonstrated the influence of mannose on yeast growth as similar to the influence of glucose, the most assimilable carbon source by yeasts, to perform growth.

Regarding the potential microbial inhibitors produced during the acid hydrolysis of carbohydrates in the hydrolyzed liquor of the dried açai seeds (Table 2), acetic acid was the major detected compound (1.8 g/L), followed by hydroxymethylfurfural (338 mg/L), and furfural (10 mg/L), and a significant decrease (≈29.5%) in the contents of hydroxymethylfurfural was observed after the liquor neutralization. Furfural and hydroxymethylfurfural are derived from pentoses and hexoses degradation, respectively, while acetic acid is formed from the acetic groups present in the hemicellulosic fraction generated during steam explosion; they have been reported as inhibitor for the growth of several microorganisms by prolongation of the lag phase during fermentation [5,17]. However, the inhibitory concentration ranges of each compound should be systematically investigated, particularly for each interested microorganism to establish the allowed limits in the culture media to enable the use of a great variety of substrates, especially those that are low-cost and derived from agroindustrial wastes.

Lin et al. [26] reported an inhibitory effect on *R. glutinis* when acetic acid concentration reached 12 g/L, and furfural at 800 mg/L also caused the inhibition of yeast growth for 12 h. In contrast, hydroxymethylfurfural at 200 mg/L had no inhibitory effect on yeast growth, but at 400 mg/L, an inhibitory effect on biomass was observed. In another study, Vajzovic et al. [27] reported that furfural at concentrations lower than 3 g/L showed no inhibitory effect on the growth of *R. mucilaginosa*, and that xylose metabolism was negatively affected only from this value onwards.

2.5. Scanning Electron Microscopy (SEM) of the Dried Açai Seeds before and after Acid Hydrolysis Treatment with H_2SO_4

Through the micrographs obtained by SEM (Figure 5), it was possible to observe that the biomass of dried açai seeds before the acid treatment (Figure 5A–C) showed no damage or cracks in its fiber structures, with small irregular protuberances due to the milling process of the fruit. However, after the acid treatment with diluted H_2SO_4 (Figure 5D–F), it was possible to observe the loss of mass density, cracks, and accentuated cell disruption, promoting partial structural disorganization, making the material more porous and amorphous. These modifications are the result of the acid-solubilized mass (mainly hemicellulose) and partial modification of cellulose crystallization. According to the literature, the cracks and ruptures observed in the micrographs of the hydrolyzed açai seeds may be due to the breaking of bonds between carbohydrates and lignin [28].

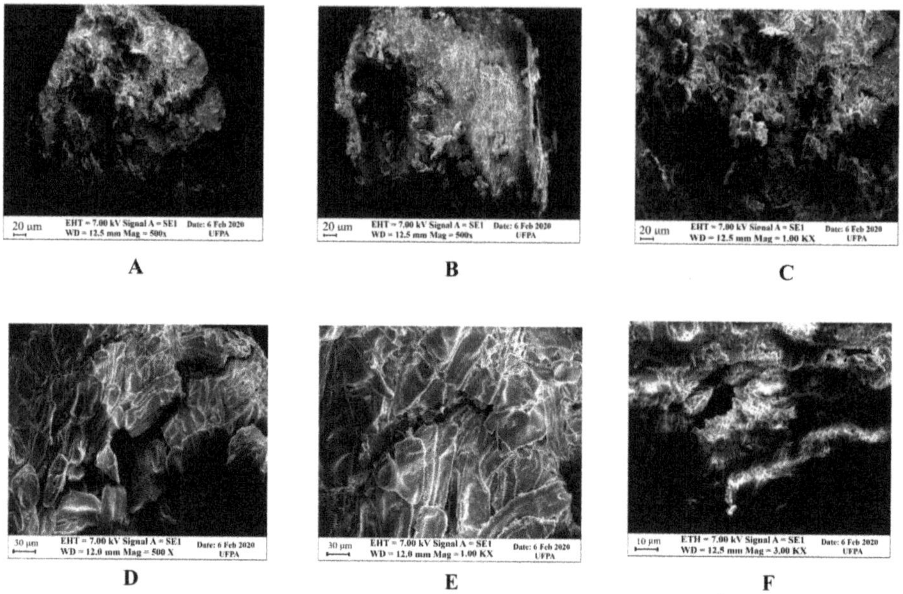

Figure 5. Micrographs of biomass of dried açai seeds before ((**A**) = 500× (**B**) = 500×; (**C**) = 1000×) and after the acid treatment ((**D**) = 500× (**E**) = 1000×; (**F**) = 3000×) with diluted H_2SO_4 at the optimal hydrolysis conditions (3.5% acid concentration (*w*/*v*), 25% solids concentration (*w*/*v*), during 70 min).

3. Materials and Methods

3.1. *Açai Seeds*

The açai seeds (*Euterpe oleracea* Mart.) (25 kg) were donated by a fruit pulp processing Company, whose fruits were grown in Igarapé-Miri city, Pará State, Brazil (latitude 1°58′37″ S, longitude 48°57′34″ W). The access to açai fruits was registered in the Brazilian National System for the Management of Genetic Heritage and Associated Traditional Knowledge (SisGen AF18EDC). After removing the pulp and rind to produce açai juice, the açai seeds (25 kg) were collected at the company and directed to the laboratory. The açai seeds were washed with running water to remove solid wastes and other impurities, followed by sanitization by immersion in a sodium hypochlorite solution (100 mg/L) for 10 min. The seeds were dried at 65 °C in an oven with forced air circulation for 48 h, until moisture was below 10%, as monitored by an infrared moisture analyzer, according to the National Renewable Energy Laboratory [12]. The dried açai seeds were ground in a knife mill, subjected to sieving at 20 mesh (0.841 mm), and stored under vacuum in polypropylene plastic bags at −20 °C until use.

3.2. *Characterization of Dried Açai Seeds*

The contents of total proteins (conversion factor 6.25 from total nitrogen), total lipids, total acidity (expressed in meq NaOH/100 g), ashes, and pH values were determined according to the Association of Official Analytical Chemists [29]. The total carbohydrate contents (%) were calculated by the difference between 100 and the sum of the percentages of moisture, proteins, lipids, and ashes. The contents of total extractives (secondary metabolite compounds of vegetables that may include waxes, fats, resins, tannins, gums, sugars, starches, and pigments) were determined by gravimetry according to Sluiter et al. [30] to eliminate the usual overestimation of structural carbohydrates. Briefly, 4 g of dried açai seeds were rinsed in reflux with ethanol (99%) until complete decolorization of the solvent, followed by washing with distilled water and drying in an oven until dryness. All the results, in triplicate, were expressed in g/100 g (%).

3.3. Determination of Structural Carbohydrates

The structural carbohydrates were determined in the dried açai seed samples, after removing the extractives, followed by acid hydrolysis with H_2SO_4 72%, according to Sluiter et al. [31]. The contents of structural carbohydrates were calculated after the determination of furfural, hydroxymethylfurfural, total lignin, monomeric sugars, and acetic acid.

3.3.1. Determination of Furfural and Hydroxymethylfurfural Content

The furfural and hydroxymethylfurfural (HMF) contents were determined in a HPLC (Agilent HPLC, model 1260 Infinity, Santa Clara, CA, USA), equipped with a quaternary pump (G1311C), a Rheodyne injection valve with a 20 µL loop, an oven (G1316A), and a diode array detector (DAD) (G1328C). For all the chromatographic analyses, samples and solvents were filtered using 0.22 and 0.45 µm membranes, respectively (Millipore, Billerica, MA, USA). The compounds were separated in a C_{18} column (Nova-Pak Waters, Milford, MA, USA) with an isocratic mobile phase of acetonitrile/water (1:8, v/v) with 1% acetic acid, at 30 °C, flow rate at 0.8 mL/min, and the chromatograms were monitored at 280 nm for 20 min [3]. The furfural and HMF contents were determined, in triplicate, by analytical curves (0.001 to 0.1 g/L, $R^2 > 0.99$), and the results are expressed in g/L.

3.3.2. Determination of Total Lignin

The content of soluble and insoluble lignin (Lignin of Klason) was determined according to Sluiter et al. [31] Since part of the insoluble material of the dried açai seeds is composed of minerals (ashes), which are not soluble in acid, the ashes contents were also determined to avoid the overestimation of insoluble lignin values. The content of soluble lignin was determined by spectrophotometry at 280 nm. All the results, in triplicate, are expressed in g/100 g (%).

3.3.3. Determination of Monomeric Sugars and Acetic Acid

The contents of cellobiose, glucose, mannose, xylose, and acetic acid in the hydrolysates were determined by HPLC (Thermo Fisher Scientific, SRVYER model, San Jose, CA, USA), coupled to a refractive index detector (SRVYER-RI), according to Martins et al. [3] The compounds were separated in an ion exchange column (Aminex HPX-87H) at 30 °C with a solution of 5 mM H_2SO_4 in deionized water as mobile phase at 0.6 mL/min (isocratic mode). The chromatographic separation was monitored by a refractive index detector (RID) at 35 °C for 20 min. The contents (g/L, $n = 3$) of each sugar and acetic acid were determined by external standardization with 6-point analytical curves ranging from 0.1 to 0.6 g/L ($r^2 > 0.99$) with D-(+)-cellobiose, D-glucose, D-mannose, D-xylose, and acetic acid standards.

3.3.4. Calculation of Structural Carbohydrates Contents

Cellobiose and glucose contents were converted into cellulose, using the conversion factors of 0.95 and 0.90, respectively, and mannose and xylose contents were converted into hemicellulose, using the conversion factor of 0.88. For the acetyl groups, the conversion factor from acetic acid to acetate of 0.717 was used, and for the calculation of the masses of furfural and HMF, the conversion factors of 1.375 and 1.286, respectively, were used, with furfural being converted to hemicellulose and HMF for cellulose. The conversion values are part of the known stoichiometry involved in the reaction, involving their respective molecular masses during the conversions, according to Sluiter et al. [32].

3.4. Determination of the Optimal Conditions for the Acid Hydrolysis of Dried Açai Seeds

The optimal conditions for the acid hydrolysis of the dried açai seeds as pre-treatment were determined, in a first moment, for two diluted acids (H_2SO_4 and H_3PO_4) using a central composite rotational design (CCRD) for each acid in order to select the most efficient acid in generating total reducing sugars. Both the CCRDs consisted of a 2^3 factorial design plus 6 axial points and 3 repetitions at the central point, totaling 17 experiments for each diluted acid, with three independent variables at two levels (−1, +1): acid concentration (1 to 4%, *w*/*v*), hydrolysis time (30 to 90 min), and solids concentration (5 to 20%, *w*/*v*) (Table 3), for which the selected ranges were based on the results obtained by previous studies [3,6].

Table 3. Range of the independent variables selected for the central composite rotational design (CCRD) during the determination of the optimal conditions for the acid hydrolysis of dried açai seeds using H_2SO_4 and H_3PO_4 as pre-treatment.

Independent Variable		Level				
		−1.68	**−1**	**0**	**+1**	**+1.68**
Acid concentration (%, *w*/*v*)	X_1	0	1	2.5	4	5
Hydrolysis time (min)	X_2	9.5	30	60	90	110
Solids concentration (%, *w*/*v*)	X_3	0.1	5	12.5	20	25

The response variables for the experimental designs using H_2SO_4 or H_3PO_4 were the content of total reducing sugars (TRS) and the combined severity factor of the acid treatment (CSF). The TRS contents were determined by spectrophotometry at 540 nm, according to the dinitro-3,5-salicylic acid (DNS) method [33], the contents (g/L, in triplicate) being obtained by external analytical curves with 10 points of D-glucose standard ranging from 0.1 to 1 g/L ($R^2 > 0.99$). The CSF was calculated according to Equation (1).

$$log(R_0) = log \cdot \left[t \cdot exp\left(\frac{T - T_{ref}}{14.75}\right)\right] - \text{pH} \quad (1)$$

where: t is the reaction time (min), T is the reaction temperature (°C), T_{ref} is the reference temperature (100 °C), pH is the dilute acid solution; 14.75 is an arbitrary constant (ω) based on the activation energy of the reaction when assuming pseudo-first order kinetics.

In a second step, after selecting the most efficient acid that produced liquors with the highest yields of TRS to continuous the investigation, we also determined the levels of hydroxymethylfurfural (HMF) and solubilized mass yields (SM) for the same CCRD. The HMF contents were determined by spectrophotometry at 284 and 336 nm, according to the methodology of AOAC [29], and the percentages of SM of the solid fraction of the hydrolyzed core were determined according Equation (2).

$$\text{SM}(\%) = \left(\frac{m_{final}}{m_{initial}} \times 100\right) - 100 \quad (2)$$

where $m_{initial}$ is the initial dry mass of lignocellulosic material; m_{final} is the final dry mass of lignocellulosic material.

In all the experiments of the acid hydrolysis, the ground dry açai seeds were weighed, transferred to borosilicate flasks, followed by the different acid and solid concentrations, and submitted to heating in an autoclave at 121 °C at different times, according to the selected CCRD experiment. After the acid hydrolysis step, the solid biomass was separated from the liquor by vacuum-filtration on qualitative filter paper (with mass previously known). The obtained liquor was directed for the TRS and HMF determinations by spectrophotometry during the CCRD experiments; and after establishing the optimal hydrolysis conditions, the newly hydrolyzed liquor was fully characterized by HPLC for the contents of cellobiose, glucose, xylose, arabinose, acetic acid, furfural, and HMF. In another set of experiments, the full characterization was also carried out for the newly hydrolyzed liquors produced at the optimal conditions and submitted to neutralization with K_2SO_4, autoclaved at 121 °C for 15 min, and filtered, according to common practices to be further used in fermentation processes. The solid biomasses obtained during vacuum-filtration were washed with distilled water until the washing water reached pH close to neutrality (pH 6.0–7.0). The biomass was dried in an oven at 105 °C with forced air circulation until constant mass for the SM (%) determination.

3.5. Scanning Electron Microscopy (SEM)

The ground dried açai seeds (in natura) and the dried biomass obtained after the pre-treatment with the selected acid, at the optimal hydrolysis conditions, were analyzed by SEM (Zeiss, EVO MA10, Germany) to observe the surface changes resulting from the pre-treatment. The samples were fixed with carbon tape on an aluminum support, then submitted to a 10 nm gold metallic coating in a metallizer. The operating conditions were electron beam current = 100 μA, constant accelerating voltage = 10 kV, and working distance = 8.5 mm.

3.6. Statistical Analysis

The results (mean ± standard deviation, $n = 3$) were analyzed by the Statistica 7.0 software (Statsoft Inc., Tulsa, OK, USA) through analysis of variance (ANOVA), Tukey's test, and Student's t-test ($p < 0.05$). All data obtained by the CCRD experiments were fitted to second order polynomial models to determine the optimal hydrolysis conditions by the response surface methodology, and Pareto charts were built at 10% of statistical significance ($p \leq 0.1$) to investigate the effects of acid hydrolysis on the studied variables. The adequacy to the second order models was determined by evaluating the coefficient determination (R^2), lack of fit, and Fisher test values (F-value) through ANOVA at 10% of statistical significance ($\alpha = 0.1$). The obtained mathematical models were used to generate response surfaces and contour plots to provide the optimal acid hydrolysis conditions to obtain a liquor with high contents of simple carbohydrates and low levels of potential microbial inhibitors.

4. Conclusions

Pre-treatment with diluted H_2SO_4 was more efficient than H_3PO_4. The optimal hydrolysis condition established for the açai seeds for fermentation processes provided a concentration of 55 g/L of sugars, consisting mainly of mannose (93%) and low levels of inhibitors. Thus, açai seeds can be seen as a promising byproduct with both high value and high potential to be used as a substrate in biotechnological processes, contributing to an environmental and socioeconomic strategy for the use of agroindustrial waste.

Supplementary Materials: The following supporting information can be downloaded at: https://www.mdpi.com/article/10.3390/molecules28186661/s1, Supplementary Table S1: Total reducing sugars (TRS) and combined severity factor (CSF) obtained during the central composite rotational design (DCCR) runs for the acid hydrolysis of açai seeds using phosphoric acid (H_3PO_4). Supplementary Table S2—t and p value at 90% confidence for the studied parameters in the experimental design. Supplementary Table S3—ANOVA (10% significance) for the adequacy of the quadratic response surface models to determine the contents of total reducing sugars (TRS), hydroxymethylfurfural (HMF) and solubilized mass (SM) after acid hydrolysis of dried açai seeds. Supplementary Figure S1—Profiles for the predicted values and desirability function used to predict the contents of total reducing sugars (TRS), hydroxymethylfurfural (HMF), and solubilized mass at the chosen optimal condition for the acid hydrolysis of dried açai seeds: with 3.5% H_2SO_4 + 25% solids during 70 min in autoclave.

Author Contributions: W.S.I., L.H.d.S.M. and R.R.d.A.: Data curation, Formal analysis, Investigation, Methodology, Validation, Visualization, Roles/Writing—original draft, review, and editing. J.A.R.d.O.: Roles/Writing—original draft, review, and editing. A.S.L.: Resources, Supervision, Visualization. R.C.C.: Conceptualization, Data curation, Funding acquisition, Investigation, Methodology, Project administration, Resources, Supervision, Visualization, Writing—original draft, review, and editing. All authors have read and agreed to the published version of the manuscript.

Funding: The authors thank Conselho Nacional de Desenvolvimento Científico e Tecnológico (CNPq, Brazil) for financial support (Project 403121/2016-7) and W.S.I. thanks Coordenação de Aperfeiçoamento de Pessoal de Nível Superior (CAPES, Brazil) for his Master's scholarship (code 001, Process 00214342204).

Data Availability Statement: All data generated or analyzed during this study are included in this published article and its Supplementary Information file.

Acknowledgments: Xingu Fruit Company for kindly donating the açaí seeds and Federal University of Pará (UFPA, Brazil) through PROPESP/UFPA.

Conflicts of Interest: The authors declare no conflict of interest.

Sample Availability: Not applicable.

References

1. EMBRAPA. Empresa Brasileira de Pesquisa Agropecuária. Available online: https://www.embrapa.br/busca-de-noticias/-/noticia/52276433/amazonia-em-foco-debate-desenvolvimento-rural-da-regiao?p_auth=XvNXOFir (accessed on 15 March 2023).
2. Almeida, A.V.C.; Melo, I.M.; Pinheiro, I.S.; Freitas, J.F.; Melo, A.C.S. Appreciation of acai core of a pulp producer from Ananindeua/PA: Proposal of reverse channel structure oriented by NPSW and reverse logistics. *GEPROS Gest. Prod. Oper. Sist.* **2017**, *3*, 59–83. [CrossRef]
3. Martins, L.H.S.; Rabelo, S.C.; Costa, A.C. Effects of the pretreatment method on high solids enzymatic hydrolysis and ethanol fermentation of the cellulosic fraction of sugarcane bagasse. *Bioresour. Technol.* **2015**, *191*, 312–321. [CrossRef]
4. Rocha, M.V.P.; Rodrigues, T.H.S.; Albuquerque, T.L.; Gonçalves, L.R.B.; Macedo, G.R. Evaluation of dilute acid pretreatment on cashew apple bagasse for ethanol and xylitol production. *Chem. Eng. J.* **2014**, *243*, 234–243. [CrossRef]
5. Maurya, D.P.; Singla, A.; Negi, S. An overview of key pretreatment processes for biological conversion of lignocellulosic biomass to bioethanol. *3 Biotech* **2015**, *5*, 597–609. [CrossRef]
6. Oliveira, J.A.R.; Komesu, A.; Filho, R.M. Hydrothermal pretreatment for enhancing enzymatic hydrolysis of seeds of açaí (*Euterpe oleracea*) and sugar recovery. *Chem. Eng. Trans.* **2014**, *37*, 787–792. [CrossRef]
7. Rambo, M.K.D.; Schmidt, F.L.; Ferreira, M.M.C. Analysis of the lignocellulosic components of biomass residues for biorefinery opportunities. *Talanta* **2015**, *144*, 696–703. [CrossRef]
8. Ferreira, D.S.; Gomes, A.L.; Silva, M.G.; Alves, A.B.; Agnol, W.H.D.; Ferrari, R.A.; Carvalho, P.R.N.; Pacheco, M.T.B. Antioxidant Capacity and Chemical Characterization of Açaí (*Euterpe oleracea* Mart.) Fruit Fractions. *Food Sci. Technol. (HRPUB)* **2016**, *4*, 95–102. [CrossRef]
9. Arruda, J.C.B.; Fonseca, L.A.B.; Pinto, L.C.P.; Pinheiro, H.C.O.; Monteiro, B.T.O.; Manno, M.C.; Lima, K.R.S.; Lima, A.R. Açaí seed bran in the feed of slow-growth broilers. *Acta Amazon.* **2018**, *48*, 298–303. [CrossRef]
10. Martins, M.A.; Mattoso, L.H.C.; Pessoa, J.D.C. Thermogravimetric evaluation of açaí fruit (*Euterpe oleracea* Mart.) agroindustry waste. *Rev. Bras. Frutic.* **2009**, *31*, 1150–1157. [CrossRef]
11. Monteiro, A.F.; Miguez, I.S.; Silva, J.P.R.B.; Silva, A.S. High concentration and yield production of mannose from açaí (*Euterpe oleracea* Mart.) seeds via mannanase-catalyzed hydrolysis. *Sci. Rep.* **2019**, *9*, 10939. [CrossRef]
12. NREL—National Renewable Energy Laboratory. *Biodiesel: Handling and Use Guide*; TP-540-43672; U.S. Department of Energy: Golden, CO, USA, 2009.
13. Ruales-Salcedo, Á.V.; Prado-Rubio, O.A.; Rojas González, A.F. Harnessing lignocellulosic waste for energy storage and generation for off-grid rural areas. *Clean Technol. Environ. Policy* **2018**, *20*, 1515–1526. [CrossRef]
14. Silva, A.S.; Teixeira, R.S.S.; Moutta, R.O.; Ferreira-Leitão, V.S.; de Barros, R.R.O.; Ferrara, M.A.; Bon, E.P.S. Sugarcane and Woody Biomass Pretreatments for Ethanol Production. In *Sustainable Degradation of Lignocellulosic Biomass—Techniques, Applications and Commercialization*; Chandel, A.K., da Silva, S.S., Eds.; IntechOpen: London, UK, 2013.
15. López-Linares, J.C.; Cara, C.; Moya, M.; Ruiz, E.; Castro, E.; Romero, I. Fermentable sugar production from rapeseed straw by dilute phosphoric acid pretreatment. *Ind. Crops Prod.* **2013**, *50*, 525–531. [CrossRef]
16. Nair, R.B.; Lundin, M.; Brandberg, T.; Lennartsson, P.R.; Taherzadeh, M.J. Dilute phosphoric acid pretreatment of wheat bran for enzymatic hydrolysis and subsequent ethanol production by edible fungi Neurospora intermedia. *Ind. Crops Prod.* **2015**, *69*, 314–323. [CrossRef]
17. Alvira, P.; Tomás-Pejó, E.; Ballesteros, M.; Negro, M.J. Pretreatment technologies for an efficient bioethanol production process based on enzymatic hydrolysis: A review. *Bioresour. Technol.* **2010**, *101*, 4851–4861. [CrossRef]
18. Sehnem, N.T.; Machado, A.S.; Leite, F.C.B.; Pita, W.B.; Morais Junior, M.A.; Ayub, M.A.Z. 5-Hydroxymethylfurfural induces ADH7 and ARI1 expression in tolerant industrial Saccharomyces cerevisiae strain P6H9 during bioethanol production. *Bioresour. Technol.* **2013**, *133*, 190–196. [CrossRef]
19. Kim, S.; Holtzapple, M.T. Lime pretreatment and enzymatic hydrolysis of corn stover. *Bioresour. Technol.* **2005**, *96*, 1994–2006. [CrossRef]
20. Hari, R.P.K.; Patel, T.; Martin, A.A. A new strain of *Rhodotorula rubra* isolated from yogurt. *J. Ind. Microbiol.* **1992**, *11*, 43–51. [CrossRef]
21. Muncner, D.; Augustin, J. Assimilation of benzoate by *Rhodotorula rubra, Rhodotorula glutinis* and *Rhodosporidium toruloides*, as affected by glucose or xylose. *World J. Microbiol. Biotechnol.* **1995**, *11*, 240–241. [CrossRef]

22. Xu, P.; Bura, R.; Doty, S.L. Genetic analysis of D-xylose metabolism by endophytic yeast strains of *Rhodotorula graminis* and *Rhodotorula mucilaginosa*. *Genet. Mol. Biol.* **2011**, *34*, 471–478. [CrossRef]
23. Silva, T.L.; Feijão, D.; Roseiro, J.C.; Reis, A. Monitoring *Rhodotorula glutinis* CCMI 145 physiological response and oil production growing on xylose and glucose using multi-parameter flow cytometry. *Bioresour. Technol.* **2011**, *102*, 2998–3006. [CrossRef]
24. Tiukova, A.I.; Brandenburg, J.; Blomqvist, J.; Sampels, S.; Mikkelsen, N.; Skaugen, M.; Arntzen, M.Ø.; Nielsen, J.; Sandgren, M.; Kerkhoven, E.J. Proteome analysis of xylose metabolism in *Rhodotorula toruloides* during lipid production. *Biotechnol. Biofuels* **2019**, *12*, 137. [CrossRef] [PubMed]
25. Byrtusová, D.; Szotkowski, M.; Kurowska, K.; Shapaval, V.; Márová, I. *Rhodotorula kratochvilovae* CCY 20-2-26—The Source of Multifunctional Metabolites. *Microorganisms* **2021**, *9*, 1280. [CrossRef] [PubMed]
26. Lin, Z.; Zhou, Y.; Wu, J.; Liu, H.; Zhang, J. Effect of multiple inhibitions in corncob hydrolysate on the lipid production by *Rhodotorula glutinis*. *Energy Fuels* **2017**, *31*, 12247–12255. [CrossRef]
27. Vajzovic, A.; Bura, R.; Kohlmeier, K.; Doty, S.L. Novel endophytic yeast *Rhodotorula mucilaginosa* strain PTD3 II: Production of xylitol and ethanol in the presence of inhibitors. *J. Ind. Microbiol. Biotechnol.* **2012**, *39*, 1453–1463. [CrossRef] [PubMed]
28. Oliveira, J.A.R.; Komesu, A.; Martins, L.H.S.; Maciel Filho, R. Evaluation of microstrutcture of açaí seeds biomass un-treated and treated with H_2SO_4 and NaOH by SEM, RDX and FTIR. *Chem. Eng. Trans.* **2016**, *50*, 379–384. [CrossRef]
29. AOAC—Association of Official Analytical Chemists. *Official Methods of Analysis*, 16th ed.; AOAC: Washington DC, USA, 1995.
30. Sluiter, A.; Ruiz, R.; Scarla, T.A.C.; Sluiter, J.; Templeton, D. *Determination of Extractives in Biomass*; Technical Report, NREL/TP-510-42619.7; National Renewable Energy Laboratory; U.S. Department of Energy: Golden, CO, USA, 2008.
31. Sluiter, A.; Hames, B.; Ruiz, R.; Scarlata, C.; Sluiter, J.; Templeton, D.; Crocker, D. *Determination of Structural Carbohydrates and Lignin in Biomass*; Technical Report, NREL/TP-510-42618; National Renewable Energy Laboratory; U.S. Department of Energy: Golden, CO, USA, 2011.
32. Sluiter, A.; Hames, B.; Ruiz, R.; Scarlata, C.; Sluiter, J.; Templeton, D. *Determination of Sugars, Byproducts, and Degradation Products in Liquid Fraction Process Samples*; Technical Report, NREL/TP-510-42623; National Renewable Energy Laboratory; U.S. Department of Energy: Golden, CO, USA, 2006.
33. Miller, G.L. Use of dinitrosalicylic acid reagent for determination of reducing sugar. *Anal. Chem.* **1959**, *3*, 426–428. [CrossRef]

molecules

MDPI

Article

Chemical and Energetic Characterization of the Wood of *Prosopis laevigata*: Chemical and Thermogravimetric Methods

Luis Fernando Pintor-Ibarra [1,*], José Juan Alvarado-Flores [1,*], José Guadalupe Rutiaga-Quiñones [1], Jorge Víctor Alcaraz-Vera [2], María Liliana Ávalos-Rodríguez [3] and Oswaldo Moreno-Anguiano [1]

[1] Facultad de Ingeniería en Tecnología de la Madera, Universidad Michoacana de San Nicolás de Hidalgo, Edif. D. Cd. Universitaria, Santiago Tapia No. 403, Centro, Morelia 58000, Mexico; jose.rutiaga@umich.mx (J.G.R.-Q.); 0234033j@umich.mx (O.M.-A.)

[2] Instituto de Investigaciones Económicas y Empresariales, Universidad Michoacana de San Nicolás de Hidalgo, Edif. D. Cd. Universitaria, Santiago Tapia No. 403, Centro, Morelia 58000, Mexico; jorge.alcaraz@umich.mx

[3] Centro de Investigaciones en Geografía Ambiental, Universidad Nacional Autónoma de México, Antigua Carretera a Pátzcuaro No. 8701, Col. Ex Hacienda de San José de la Huerta, Morelia 58190, Mexico; lic.ambientalista@gmail.com

* Correspondence: luis.pintor@umich.mx (L.F.P.-I.); jjalvarado@umich.mx (J.J.A.-F.)

Abstract: Diverse methodologies exist to determine the chemical composition, proximate analysis, and calorific value of biomass. Researchers select and apply a specific methodology according to the lignocellulosic material they study and the budgetary resources available. In this project, we determined the primary chemical constitution and proximate analysis of *Prosopis laevigata* (Humb. & Bonpl.) Jonhst wood using a traditional chemical method and a novel procedure based on the deconvolution of the DTG signal produced by TGA. The highest calorific value was verified using a calorimetric pump based on mathematical models. We also conducted elemental analysis and a microanalysis of ash, and applied Fourier transform infrared spectroscopic analysis (FT-IR). The means of the results obtained by the chemical method and TGA-DTG, respectively, were: hemicelluloses 7.36%–(8.72%), cellulose 48.28%–(46.08%), lignin 30.57%–(32.44%), extractables 13.53%–(12.72%), moisture 2.03%–(4.96%), ash 1.77%–(1.90%), volatile matter 75.16%–(74.14%), and fixed carbon 23.05%–(18.93%). The procedure with the calorimetric pump generated a calorific value above 20.16 MJ/kg. The range generated by the various models was 18.23–21.07 MJ/kg. The results of the elemental analysis were: carbon 46.4%, hydrogen 6.79%, oxygen 46.43%, nitrogen 0.3%, and sulfur 0.5%. The microanalysis of ash identified 18 elements. The most abundant ones were potassium > calcium > sodium. Based on the infrared spectrum (FT-IR) of *Prosopis laevigata* wood, we detected the following functional groups: OH, C-H, C=O, CH_2, CH_3, C-O-C, C-OH, and C4-OH. Our conclusion is that the TGA-DTG method made it possible to obtain results in less time with no need for the numerous reagents that chemical procedures require. The calorific value of *P. laevigata* wood is higher than the standards. Finally, according to our results, proximate analysis provides the best model for calculating calorific value.

Keywords: TGA-DTG; biomass; proximate analysis; calorific power; biofuels

1. Introduction

Worldwide, 44 species of the genus *Prosopis* have been registered. The predominant species in Mexico are *P. laevigata* and *P. glandulosa*. In this project, we studied the wood of *P. laevigata*, a species of the Fabaceae family that has a broad, natural distribution in arid and semiarid zones of the country. Estimates indicate that 50% of Mexico's national territory has these types of climates, concentrated in the states of Guanajuato, Hidalgo, Michoacán, Nuevo León, Oaxaca, Puebla, San Luis Potosí, and Veracruz [1,2]. *P. laevigata* is a species of great ecological and economic importance. Its wood is commercialized mainly in the manufacture of rustic furniture, artisanal crafts, flooring, railroad ties, moldings, doors, and windows, as well as charcoal production. The bark is used in tanning, the exudate known as gum arabic is used in cosmetic products, printing, and as a glue for paper, while the leaves and fruits have medicinal applications as antimicrobials and are utilized in cattle feed [3,4]. While information on the uses, diversity, and distribution of this wood is abundant, there are few studies of its chemical constituents and energy properties [5–7]. Given the heterogeneous nature of wood, it is especially important to determine the chemical and energy properties of its lignocellulosic biomass as a possible energy source. A renewable raw material, wood is neutral in terms of emissions and ensures a sustainable supply because the CO_2 it produces is equivalent to the amount of energy retained in the plants [8,9]. Today, bioenergy is the main source of renewable energy, representing around 10% of global energy consumption and 77% of all forms of renewable energy; that is, biofuels and hydroelectric, solar, eolic, and geothermic energy [10]. Determining the chemical and energy properties of this biomass can be achieved using various techniques that evaluate its principal chemical components, but proximate analysis of lignocellulosic materials using routine gravimetric procedures require considerable care and time [9,11]. Other methods include thermogravimetric analysis (TGA) and the deconvolution of the first derivative (DTG), which generate precise quantifications in less time, and can determine the primary constituents of biomass and proximate analysis [12,13]. Few researchers, however, have access to DTG because many laboratories lack the budgetary resources to acquire the equipment used to perform TGA. For this reason, they continue to use classic gravimetric methods for the chemical analysis of biomasses [9,14]. The objective of the present study is to contribute to the development of new methodologies that will improve our knowledge of the chemical and energy properties of *P. laevigata* wood.

2. Results and Discussion

2.1. Primary Constitution of the Wood: Chemical and TGA-DTG Methods

Diverse methodologies are available to determine the primary composition of biomass, but it is well known that each one has certain strengths and weaknesses [9,11]. This study found that statistically significant differences exist ($p \leq 0.05$) between the chemical and thermogravimetric (TGA-DTG) methods for the parameters of the primary constituents—hemicelluloses and lignin– and in a proximate analysis of moisture, volatile matter, and fixed carbon. In contrast, there were no significant differences for cellulose, extractables, and ash (Table 1). The differences encountered are related to the procedure used, since the chemical method makes its determinations individually, while thermogravimetry (TGA-DTG) obtains results through pyrolysis in a controlled system using mathematical calculations, such as multiple linear regression, where the fractions of the main chemical components can be optimized [15]. Independent thermal degradation by pyrolysis of the three principal chemical components of wood (cellulose, hemicelluloses, and lignin) can accurately predict the parameters of deconvolution of the DTG [13,16].

Table 1. Chemical composition and proximate analysis of *P. laevigata* wood by two methods.

Parameter	Chemical Analysis *		TGA-DTG Analysis *	
	Average (%)	SD (±)	Average (%)	SD (±)
Hemicelluloses	7.36	0.15	8.74	0.22
Cellulose	48.28	1.0	46.08	0.25
Lignin	30.57	0.74	32.44	0.40
Extractives	13.53	0.12	12.72	0.73
Moisture	2.03	0.14	4.96	0.11
Ashes	1.77	0.07	1.90	0.01
Volatile material	75.16	0.13	74.14	0.16
Fixed carbon	23.05	0.36	18.93	0.08

* Statistically significant difference ($p \leq 0.05$) between methods. SD: standard deviation.

Our review of other chemical studies of the duramen of *P. laevigata* wood using chemical methods found that published results vary. Carrillo et al. [5], for example, reported the following percentages: hemicelluloses 16%, cellulose 45.7%, lignin 31.4%, and extractables 16%. In contrast, based on a fiber analysis by the Van Soest method [7] published these yields: hemicelluloses 14.09%, cellulose 45.35%, lignin 22.14%, extractables 16.72%, and ash 1.70%. The results of our work are only comparable with respect to cellulose, lignin, and ash, according to these references (Table 1). It is important to emphasize that the relevant literature indicates that the chemical constituents of wood can vary by the material studied (duramen, sapwood, bark), the age of the tree, growth conditions, geographic location, altitude, climatic factors, the chemical solvents utilized, and experimental methodologies, among other factors [17,18].

Through proximate analysis, one can obtain an immediate profile of the quality of solid biofuels, so this is an important approach for commercialization and the design of combustion equipment [19], two parameters that affect the quality of biofuels. For example, high percentages of moisture and ash reduce the calorific value and efficiency of biomasses during combustion [20,21], while high percentages of volatile matter increase yields during the conversion of biomass into biodiesel by pyrolysis and favor the production of biogas and the calorific value in densified biofuels [22,23]. In this regard, high fixed carbon content improves the calorific value of densified biofuels and charcoal [24,25]. Our work did not find any statistically significant differences for ash content (Table 1). The statistically significant differences determined for the other parameters tested in the proximate analysis may be related to the procedures used in distinct methodologies; for example, the fact that in TGA pyrolysis is determined using a sample in a controlled system (Figure 1). However, they may also be attributable to the calculation method applied. In gravimetric methods, these are performed on the basis of biomass that is free of moisture using, for instance, the expression: % fixed carbon = [100 − (% ash + % volatile matter)], where adjusting for the moisture in the ash and volatile matter is performed previously. Meanwhile, the TGA method for calculating volatile matter uses the expression: % volatile matter = [100 − (% moisture + % ash + % fixed carbon)]. Based on these expressions, it is clear that the adaptation of the % of moisture influences results.

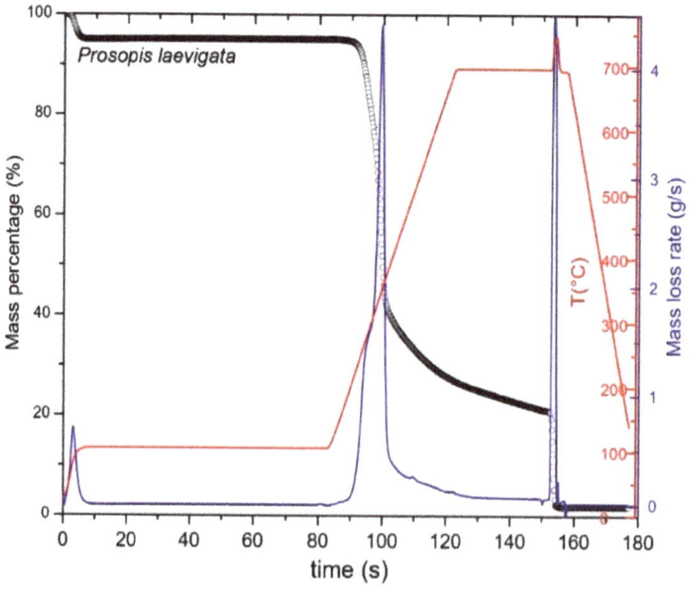

Figure 1. TGA and DTG curves of *P. laevigata* wood.

Because *P. laevigata* wood has an especially high amount of fixed carbon, it is a solid biofuel of good quality. The proximate analysis of the duramen of *P. laevigata* in a study using gravimetric methods obtained the following percentages: ash 1.70%, volatile matter 78.46%, and fixed carbon 19.83% [7]; while in their work, Martínez-Pérez et al. [6] reported 1.24% of ash in the duramen of this species. In comparative terms, only the percentage of ash obtained in these two reports is comparable to the references consulted (Table 1). In general, the results obtained with the two methods applied in our work are within the ranges reported for woods of latifoliate species: ash 0.49–4.51%, volatile matter 76.60–94.70%, and fixed carbon 9.32–21.27% [26,27].

2.2. Thermal Decomposition Process (TGA) and Deconvoluted (DTG)

In the first stage of the pyrolysis process of *P. laevigata* wood, conducted in the temperature range of 25–100 °C for a period of 80 min, a loss of moisture from the biomass occurred (Figure 1). Some studies have reported that this temperature causes the volatilization of the organic fraction of low molecular weight biomass, including some extractable substances [28].

In the second stage, when the temperature was increased from 100 to 700 °C for 30 min, we obtained the calculation of the fixed carbon content from the proximate analysis. This stage included the thermal decomposition of the hemicelluloses and cellulose. It is estimated that the loss of mass of these polymers occurs in a temperature range of 200–675 °C [29,30]. Another characteristic of this stage is a greater loss of the wood fraction. As Figure 1 shows, DTG indicates the degradation velocity of the mass (in blue). It is well known that wood is made up mainly of these polysaccharides, with estimated fractions as high as 70% [18,31]. Table 1 shows the percentages of these chemical constituents. Figure 2 presents the deconvolution of the DTG, where the orange color represents the primary component with lower thermal stability due to the presence of acetyl groups that correspond to the hemicelluloses. Other researchers have reported this tendency [12,15] in other lignocellulosic materials. The curve of greater size—in green—represents cellulose. This second polymer presents higher thermal resistance than the hemicelluloses due to its crystalline structure [32,33].

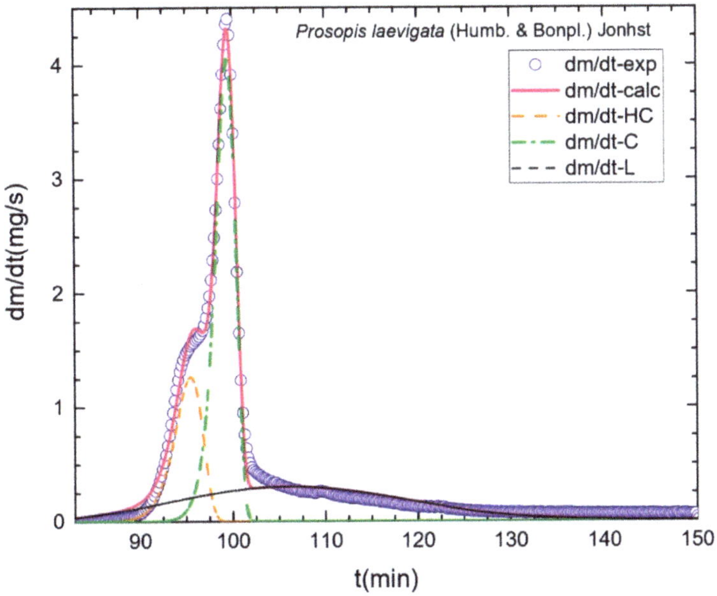

Figure 2. Deconvoluted DTG curve of *P. laevigata* wood.

This illustration includes a curve—in black—that corresponds to lignin, which began a reaction at low temperatures that persisted across a broad temperature range. This phenomenon is associated with the diverse functional groups in its structure, which vary in terms of thermal stability [34]. Reports in the literature include antecedents of this phenomenon of thermal degradation in some latifoliate species [12,32]. Figure 2 also shows the experimental graph—in blue—with calculations in pink. The similarity between them is evident, leading to the conclusion that they are correct. In the third stage (Figure 1), a peak was registered at an approximate time of 155 min. This corresponded to the influx of oxygen that was applied at 700 °C to eliminate all organic matter. Finally, the last fraction, consisting of the remaining residue, corresponds to the inorganic fraction made up of ash, which was obtained at the end of the cooling process from 700–25 °C in a time of 20 min.

2.3. Higher Heating Value (HHV)

The calorific value is the amount of energy per unit of mass or volume that is released in a complete combustion [35]. It is the most important parameter for characterizing materials used as fuels. The calorific value of wood depends on its chemical components. Some authors mention that lignin favors the calorific value of solid biofuels [36,37], though other studies have concluded that extractables increase calorific value, especially in species of *Pinus* spp. [38,39], while others report this same effect [40]. Our study did not identify any statistically significant differences (Tukey, $p \leq 0.05$) in the calorific value higher than—using a calorimetric pump—between the predictive mathematical models based on chemical composition and the proximate analysis (Table 2). We did, however, find that the model based on elemental analysis generated a lower value and presented statistically significant differences ($p \leq 0.05$) with respect to both the standard methodology and the other predictive models. These results are consistent with those reported in published studies that utilized the same models [26].

According to the results of the present study, the best model for predicting calorific value is proximate analysis, since there were no statistically significant differences between this model and the procedure that involved the use of the calorimetric pump. Moreover, this can be calculated by immediate analyses that do not require chemical reagents or large investments in expensive equipment. This concurs with reports by Rodríguez-Romero et al. [14]. Martínez-Pérez et al. [6], meanwhile, reported a lower calorific value of 18.31 MJ/kg in *P. laevigata* wood based on the approach using a calorimetric pump. They also obtained a larger interval (18.4–19.2 MJ/kg) than the one registered for latifoliate woods in UNE-EN ISO 17225-1 [41]. It is important to emphasize that the results for the caloric value obtained herein with the methodologies utilized are within the interval of 18.13–21.7 MJ/kg estimated for various species of latifoliate woods in Mexico [27,42]. According to international parameters (Table 3), the calorific value higher than of *P. laevigata* wood is above the quality parameter.

Table 2. Calculation of the higher calorific value (MJ/kg) of *P. laevigata* wood.

HHV using calorimetric pump [43]	20.16 (±0.25) a
HHV = 17.9017 + 0.07444 (L) + 0.0661 (E) [44]	21.07 (±0.06) a
HHV = 354.3 (FC) + 170.8 (VM) [45]	21.06 (±0.07) a
HHV = 0.335 (C) +1.423 (H) − 0.154 (O) − 0.145 (N) [46]	18.23 (±0.92) b

Different letters in the column indicate that there is a significant Tukey statistical difference ($p \leq 0.05$). Where: L = lignin, E = Extractives, FC = Fixed carbon, VM = Volatile material, C = carbono, H = Hydrogen, O = Oxygen, N = Nitrogen.

Table 3. Energetic properties, elemental analysis and microanalysis of ashes from *P. laevigata*.

Parameter	Magnitude	Calculated Value	Technical Parameters Limit [47]	Typical Variation for Hardwood Woods [41]
Energetic properties				
Moisture	%	7.36 (±0.15)	≤10	NS
Ashes	%	1.77 (±0.07)	≤1.5	0.2–1.0
Calorific value	MJ/kg	20.1 (±0.25)	16.3–19	18.4–19.2
Elementary analysis				
Carbon	%	46.4 (±0.42)	-	48–52
Hydrogen	%	6.79 (±0.15)	-	5.9–6.5
Oxygen	%	46.43 (±0.38)	-	41–45
Nitrogen	%	0.3 (±0.12)	≤0.5	< 0.1–0.5
Sulfur	%	0.05 (±0.01)	<0.03	<0.01–0.05
Ash microanalysis (parts per million, ppm)				
K	mg/kg	182,962.57	-	500–1500
Ca	mg/kg	122,616.47	-	800–20000
Na	mg/kg	11,042.69	-	10–200
Sr	mg/kg	3174.39	-	NS
S	mg/kg	1265.70	-	NS
Mg	mg/kg	972.01	-	100–400
P	mg/kg	649.19	-	50–200
Ba	mg/kg	106.76	-	NS
B	mg/kg	84.35	-	NS
Si	mg/kg	69.56	-	100–200
Fe	mg/kg	53.47	-	10–100
Cu	mg/kg	44.28	-	0.5–10
Al	mg/kg	35.98	-	<10–50
Li	mg/kg	32.40	-	NS
Mn	mg/kg	23.72	-	83
Zn	mg/kg	16.37	≤10.0	5–100
Ni	mg/kg	7.70	≤10.0	<0.1–10
Cr	mg/kg	4.48	≤10.0	0.2–10

NS = not specified.

2.4. Elemental Analysis

The elemental analysis of biomass is another important parameter since it can define the exact atomic elements that make up wood; namely, carbon, hydrogen, nitrogen, and sulfur. Criteria for the quality of biofuels registered in the literature include high percentages of C and O in biomass, as this favors calorific value [48,49]. This affirmation must be qualified; however, because Ali et al. [50] reported that lignocellulosic materials with high O content can corrode combustion equipment. Hydrogen, in contrast, has been found to contribute positively to calorific value, though its proportion in biomass is lower than that of C [48]. Studies also estimate that high amounts of N and S in biomass are the main causes of environmental impacts that can damage human health and lead to the formation of incrustations in combustion equipment [51]. Table 3 presents the results of our elemental analysis of *P. laevigata* wood, which determined a low proportion of C, while H and O were above the ranges reported in the literature for latifoliate woods: respectively, % C: 49.13–50.63, % H: 5.96–6.21, and % O: 43.62–44.49, compared to Rutiaga-Quiñones et al. [43]. Based on norm UNE-EN14961-1 [47] for solid biofuels, N complies with this parameter, but S is slightly above the limit and is comparable to the range estimated for woods of latifoliate species (Table 3).

2.5. Microanalysis of Ash

This analysis was performed to identify the inorganic fraction of the biomass, which consists of the minerals present in the soils where the trees grew. Although these exist in low proportions in wood, they can have negative effects on combustion. For example, Ca, K, P, and Mg can produce slag, corrosion, and fine particle emissions, and form incrustations in combustion equipment [52,53]. Na, Fe, and Si can cause problems like ash fusion, incrustations, and corrosion [54]. We identified 18 elements in our samples (Table 3). The most abundant ones were K >Ca >Na. Published studies have reported that Ca, K, P, and Mg are the most abundant chemical elements in wood ash [26,39]. This is consistent with our identifications. Sjöström [17] mentions that B, Cu, Mn, Si, and Zn can be found in lower proportions, and, in effect, we detected these elements. In relation to the elemental ranges of solid biofuels listed in norm UNE-EN 14961-1 [47], *P. laevigata* wood exceeds the $\leq$10.0 mg/kg of Zn. In contrast, Ni and Cr concentrations were below reported levels (Table 3). Other microanalyses of the wood ash of *P. laevigata* have also detected the presence of Ca, K, Mg, P, Si, and Al [6].

2.6. Fourier Transform Infrared Spectroscopy (FT-IR)

Figure 3 shows the infrared spectrum (FT-IR) of *P. laevigata* wood. We identified the OH group in the 3700–3000 cm^{-1} region. This functional group has also been reported for *P. juliflora* wood from our study area [55]. Published reports suggest that this functional group corresponds to the water molecules that constitute the biomass. However, it is important to note that significant broadening of this band indicates the presence of numerous hydrogen bonds. The existence of adsorbed water molecules or crystallization may be considered less important. This has been reported for other latifoliate woods at 3423 cm^{-1}. These stretching vibrations are related to the cellulose and hemicellulose polymers [56] in the absorption band of the 3000–2750 cm^{-1} region. We also identified the functional group CH and, as with other species of this genus, there was an absorption band in the 2980–2800 cm^{-1} region in *P. juliflora* wood [55]. According to the literature, absorption bands in this region are related to the CH group present in the structural components of wood (cellulose, hemicelluloses, lignin). The CH functional group has been found in other studies of latifoliate species, in the region of 2933 cm^{-1} [57,58]. The peak found between wave numbers 1800–1500 cm^{-1} is related to the C=O stretching vibrations of carboxylic acid and/or carbonyl (aldehydes, ketones, and esters) [59]. This result is comparable to other studies of latifoliate woods that report the presence of this functional group between 1740 and 1730 cm^{-1}. It is important to note that this functional group has been assigned to the hemicelluloses [58,60].

Turning to the carbonyl functional group, it was identified at an absorption of 1695 cm^{-1} in extracts isolated with acetone, and in aromatic substances between 1615 and 1520 cm^{-1} in extractable substances obtained with methanol and water in *P. kuntzei* wood. Other published reports affirm that these low molecular weight organic compounds are often present in derivatives of gum arabic, mainly in species of the genus *Prosopis* [61]. The position of the absorption bands between 1510 and 1260 cm^{-1} is associated with the functional groups CH_2 and CH_3, which correspond to the FT-IR spectrum of this wood. The CH_2 group has also been identified in *P. juliflora* wood [55]. These functional groups are associated with cellulose and lignin [58,62]. Other studies have detected them in latifoliate species in the region of 1464–1375 cm^{-1} [59,63]. Finally, the functional groups C-O-C, C-OH, and C4-OH were identified in vibrations at 1234, 1160, and 1028 cm^{-1}, respectively, and β-glucopyranose was found in the cellulose [64]. Similar values have been reported for other latifoliate species, including beech [65].

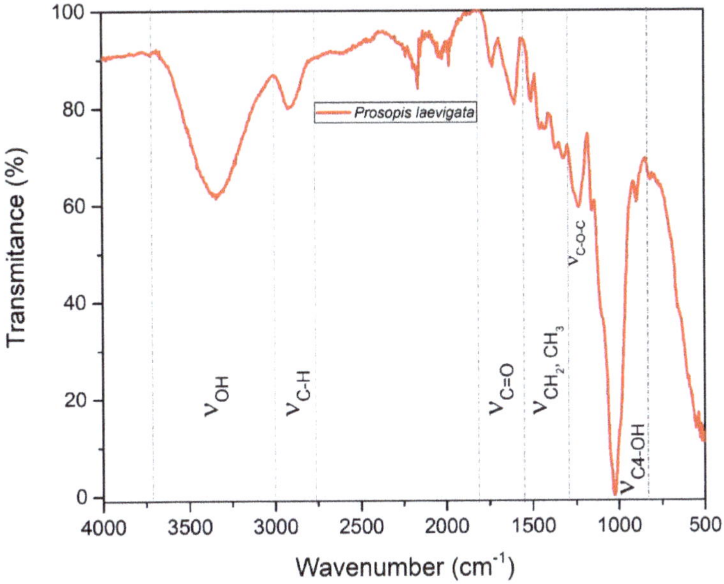

Figure 3. FT-IR infrared spectrum of *P. laevigata* wood.

3. Materials and Methods

3.1. Collection and Preparation of the Study Material

Samples of *P. laevigata* wood were collected in the Lake Cuitzeo Basin in the State of Michoacán, Mexico, between parallels 19°58′51″ N 101°7′15″ O and 19°58′21″ N 101°7′28″ O. They were taken from three individual trees of this species in 10-cm slices, cut at a height of 1.30 m (diameter at the height of the chest). The wood was separated from the bark, splintered manually, and left to dry in the shade until it reached a moisture equilibrium (approximately 12%). After drying, the splinters were ground in a conventional grinder, and then sieved in a RO-TAP device (Model RX-29, W.S. Tyler, Mentor, OH, USA) using standard meshes, as follows [66]: numbers 20, 40, and 60, and the #40 flour mesh (425 μm), in accordance with norm T 264 cm-97 [67]. The higher heating value (HHV) was utilized for the chemical, thermogravimetric (TGA), proximate analysis, elemental, and calorific analyses.

3.2. Primary Constitution and Proximate Analysis by Chemical Methods

3.2.1. Primary Analysis by the Chemical Method

Chemical characterization was conducted in an Ankom Model A200 fiber analyzer (ANKOM Technology, Macedon, NY, USA) following the methodology in Van Soest et al. [68]. Gravimetric calculations were performed to obtain the percentages of neutral detergent fiber (NDF), acid detergent fiber (ADF), insoluble lignin (IL), and ash, based on Equations (1)–(4), as reported by Torres et al. [69].

$$\%\text{Extractives} = 100 - \%\text{NDF} \quad (1)$$

$$\%\text{Hemicelluloses} = \%\text{NDF} - \%\text{ADF} \quad (2)$$

$$\%\text{Cellulose} = \%\text{ADF} - \%\text{IL} \quad (3)$$

$$\%Lignin = \%IL - \%Ash \quad (4)$$

3.2.2. Proximate Analysis by Gravimetric Methods

The percentage of moisture was determined using a thermobalance (Benetech Inc., model Gm640, Aurora, IL, USA), according to norm ASTM D4442-20 [70]. The percentage of ash was measured according to norm ASTM D1102-84 [71], while volatile matter content was verified based on norm ASTM E872-82 [72]. Fixed carbon was calculated as the following difference: % fixed carbon = [100 − (% ash + % volatile matter)], based on calculations reported previously by Ngangyo-Heya et al. [73].

3.3. Thermogravimetric (TGA) and Differential (DTG) Analyses

For the purposes of this study, a rapid methodology that integrated TGA-DTG analysis was applied using a PerkinElmer STA 6000 thermogravimetric analyzer (Pyris™ software, Version 11, PerkinElmer, Waltham, MA, USA), based on the parameters of the ASTM D5142-90 standard [74]. This analyzer has an alumina crucible that was filled with an amount of 30 ± 10 mg. The sample was distributed uniformly inside the crucible. The analysis was carried out in an inert atmosphere that required a flow of nitrogen gas (N_2) that was 99.99% pure. Before each experiment, the equipment was purged with N_2 for 15 min to eliminate, insofar as possible, the air inside the chamber where pyrolysis was conducted.

3.3.1. Proximate Analysis by TGA

The heating ramp was programmed in four stages. Stage (1): initial heating from 25 to 105 °C. This temperature was maintained for 80 min. The heating velocity was 30 °C/min, in a nitrogen atmosphere at a flow rate of 30 mL/min. Stage (2): temperature was increased to 700 °C at a rate of 15 °C/min, then maintained for 30 min under the same nitrogen flow. Stage (3): oxygen (O_2) was introduced at a flow rate of 30 mL/min. The temperature was kept at 700 °C for 5 min. Stage (4): a switch was made back to nitrogen. Cooling from 700 to 25 °C was performed at a velocity of 30 °C/min. In this final stage, the nitrogen flow was reduced to 15 mL/min.

3.3.2. Primary Constitution: Deconvolution of the DTG

Based on the algorithm developed and data from the TGA analysis, the fractions of the structural components of the wood (cellulose, hemicelluloses, lignin) were calculated utilizing the deconvolution of the curve of the TGA derivative that corresponded to the DTG analysis. The algorithm allowed us to determine the fractions of the structural components as a function of the loss of mass in the sample.

The graphs of the curves were created based on the specific algorithm written in the SciLab platform, which utilizes the ODE (ordinary differential equations) subroutine based on the Adams method (for non-rigid ODE problems). This approach integrates ordinary differential equations to describe the loss of weight associated with the degradation of hemicellulose, cellulose, and lignin, and the *Fminsearch* subroutine (based on Nelder and Mead's algorithm) to optimize the kinetic parameters and achieve the best adjustment of the experimental results [12]. Extractable substances were calculated as the following difference: % extractables = 100 − (% cellulose + % hemicelluloses + % lignin + % ash).

3.4. Calorific Value Higher Than (HHV)

The calorific value was determined in a calorimetric pump (LECO AC 600, LECCO Corporation, St. Joseph, MI, USA) following norm UNE-EN ISO 18125 [43]. The calorimeter was calibrated with benzoic acid before initiating the analysis. HHV was calculated using mathematical models according to the chemical composition (lignin and extractables), following White [44] and incorporating the results of the proximate analysis as a function of the fixed carbon and volatile matter. This procedure followed Cordero et al. [45] and employed the mathematical model that resulted from the elemental analysis [46].

3.5. Elemental Analysis

Carbon, hydrogen, nitrogen, and sulfur content were determined in a Perkin-Elmer, model 2400 CHNS-Or apparatus, using the modified Dumas method [75]. Sulfur content was calculated by the turbidimetric method with gum arabic. Oxygen content was calculated by difference, based on the reports by Ghetti et al. [76].

3.6. Microanalysis of Ash

Identification of the inorganic elements in the ash was conducted by means of inductively coupled plasma atomic emission spectroscopy (ICP-AES) (VARIAN 730-ES, Varian Inc., (Agilent), Mulgrave, Australia), under the operating conditions reported by Arcibar-Orozco et al. [77].

3.7. Fourier Transform Infrared Spectroscopy (FT-IR)

The functional groups of *P. laevigata* wood were obtained by FT-IR in a Perkin Elmer ATR model 400 spectrometer. The spectra were determined in 16 scans per sample in a range of 4000–500 cm^{-1} at a resolution of 4 cm^{-1}, according to the operating conditions reported by our research team in previous studies [15].

3.8. Statistical Analyses

The analyses of the primary constituents and the proximate analysis by the chemical method and thermogravimetry (TGA-DTG) were performed in triplicate. A student's t-test was used to compare the two treatments for each variable of the basic chemical composition and proximate analysis. For calorific value, an analysis of variance (ANOVA) ($p \leq 0.05$) was applied with a comparison of means using the Tukey test. All statistical analyses were performed with the R Studio program [78]. For the elemental analysis, we report the means and their standard deviation. The microanalysis of ash and spectroscopy (FT-IR) was carried out only once.

4. Conclusions

The novel TGA-DTG thermogravimetric method tested in this study can obtain results for the principal chemical components of wood and for proximate analysis in less time, once the algorithm has been obtained, with no need for numerous chemical reagents compared to traditional chemical procedures. We found statistically significant differences ($p \leq 0.05$) between these two methods in some parameters, but both approaches yielded results comparable to reports in the literature. The calorific value of *P. laevigata* wood is higher than international standards for solid biofuels. According to our results, the best model for calculating calorific value is the one obtained through proximate analysis, as this showed no statistically significant differences ($p \leq 0.05$) with studies that used a calorimetric pump. This method has two advantages: (i) it generates results immediately because no chemical reagents are required, and (ii) it does not require costly equipment. Regarding the elemental analysis, *P. laevigata* wood complies with the limit on nitrogen content for commercialization, while sulfur content was only slightly above the corresponding specification. The microanalysis of ash identified 18 elements, the most abundant ones being potassium >calcium >sodium. According to technical parameters, nickel and chrome concentrations comply with international allowances. Finally, the FT-IR analysis detected eight functional groups: OH, C-H, C=O, CH_2, CH_3, C-O-C, C-OH, and C4-OH.

Author Contributions: Conceptualization, J.J.A.-F., J.V.A.-V., L.F.P.-I. and O.M.-A.; methodology, J.J.A.-F., J.V.A.-V., L.F.P.-I. and J.G.R.-Q.; software, L.F.P.-I. and J.J.A.-F.; validation, J.J.A.-F., J.V.A.-V., M.L.Á.-R. and L.F.P.-I.; formal analysis, L.F.P.-I., J.J.A.-F. and J.V.A.-V.; investigation, J.J.A.-F., L.F.P.-I., J.V.A.-V., M.L.Á.-R. and J.G.R.-Q.; resources, J.J.A.-F., J.V.A.-V., M.L.Á.-R., L.F.P.-I., J.G.R.-Q. and O.M.-A.; data curation, J.J.A.-F., J.V.A.-V., M.L.Á.-R. and O.M.-A.; writing—original draft preparation,

L.F.P.-I. and J.J.A.-F.; writing—review and editing, J.J.A.-F., J.V.A.-V., M.L.Á.-R., L.F.P.-I., J.G.R.-Q. and O.M.-A.; visualization, L.F.P.-I. and J.J.A.-F.; supervision, J.J.A.-F., L.F.P.-I., J.V.A.-V., M.L.Á.-R., J.G.R.-Q.; project administration, L.F.P.-I., J.J.A.-F., J.G.R.-Q. and J.V.A.-V.; funding acquisition, J.J.A.-F., L.F.P.-I., J.V.A.-V., M.L.Á.-R. and O.M.-A. All authors have read and agreed to the published version of the manuscript.

Funding: The authors thank the Coordination of Scientific Investigation of the Universidad Michoacana de San Nicolás de Hidalgo (UMSNH) for financing project JGRQ-CIC-UMSNH-2023. The author L.F.P.-I. adds his thanks to CONAHCYT 1621400 scholarship number for the funding provided during his studies in the Doctoral program in the Science and Technology of Wood, also at the UMSNH.

Institutional Review Board Statement: Not applicable.

Informed Consent Statement: Not applicable.

Data Availability Statement: All data supporting the reported results are available on request from the corresponding author.

Acknowledgments: The authors also thank the Gaspar-Arévalo, Vega-Izquierdo, and Pintor-González families for donating the samples, and José Alberto Soria González for his technical support.

Conflicts of Interest: The authors declare no conflicts of interest.

References

1. Garatuza-Payan, J.; Argentel-Martinez, L.; Yepez, E.A.; Arredondo, T. Initial response of phenology and yield components of wheat (*Triticum durum* L.; CIRNO C2008) under experimental warming field conditions in the Yaqui Valley. *Peer J.* **2018**, *6*, e5064. [CrossRef] [PubMed]
2. Romero, A.P.; Muñoz, E.J.; Laguna, R.R.; Zárate, R.R. Distribución potencial de *Prosopis laevigata* (Humb. et Bonpl. ex Willd.) MC Johnst. en el estado de Hidalgo, México. *Rev. Mex. Cienc. For.* **2021**, *12*, 71–87. [CrossRef]
3. Saucedo, J.C.R.; Caciano, R.T.; Nuñez, L.M.V.; Pérez, G.S.; Serna, R.R. *Importancia de las Poblaciones de Mezquite en el Norte-Centro de México*; CENID-RASPA, Instituto Nacional de Investigaciones Forestales Agrícolas y Pecuarias: Durango, México, 2011; pp. 1–21.
4. Rodríguez-Sauceda, E.N.; Argentel-Martínez, L.; Morales-Coronado, D. Water regime and gas exchange of *Prosopis laevigata* (Humb. & Bonpl. ex Willd.) M. C. Johnst. in two semi-arid ecosystems in southern Sonora. *Rev. Chapingo Ser. Cienc. For. Y Ambiente* **2019**, *25*, 107–121. [CrossRef]
5. Carrillo, A.; Mayer, I.; Koch, G.; Hapla, F. Wood anatomical characteristics and chemical composition of *Prosopis laevigata* grown in the Northeast of Mexico. *IAWA J.* **2008**, *29*, 25–34. [CrossRef]
6. Martínez-Pérez, R.; Pedraza-Bucio, F.E.; Orihuela-Equihua, R.; López-Albarrán, P.; Rutiaga-Quiñones, J.G. Calorific value and inorganic material of ten Mexican wood species. *Wood Res.* **2015**, *60*, 281–292.
7. Salazar-Herrera, F.; Pintor-Ibarra, L.F.; Musule, R.; Nava-Berumen, C.A.; Alvarado-Flores, J.J.; González-Ortega, N.; Rutiaga-Quiñones, J.G. Chemical and Energetic Properties of Seven Species of the Fabaceae Family. *South-East Eur. For.* **2023**, *14*, 215–224. [CrossRef]
8. Quiñones-Reveles, M.A.; Ruiz-García, V.M.; Ramos-Vargas, S.; Vargas-Larreta, B.; Masera-Cerutti, O.; Ngangyo-Heya, M.; Carrillo-Parra, A. Assessment of pellets from three forest species: From raw material to end use. *Forests* **2021**, *12*, 447. [CrossRef]
9. Espinosa-Negrín, A.M.; López-González, L.M.; Casdelo-Gutiérrez, N.L. Pretratamientos aplicados a biomasas lignocelulósicas: Una revisión de los principales métodos analíticos utilizados para su evaluación. *Rev. Cuba. Quím.* **2022**, *34*, 87–110.
10. Masera, O.; Sacramento Rivero, J.C. Promoting a sustainable energy transition in Mexico: The role of solid biofuels. *BioEnergy Res.* **2022**, *15*, 1691–1693. [CrossRef]
11. Segura, F.; Echeverri, R.; Patiño Ll, A.C.; Mejía, A.I. Descripción y discusión acerca de los métodos de análisis de fibray of the valor nutricional de forrajes y alimentos para animales. *Vitae* **2007**, *14*, 72–81.
12. Saldarriaga, J.F.; Aguado, R.; Pablos, A.; Amutio, M.; Olazar, M.; Bilbao, J. Fast characterization of biomass fuels by thermogravimetric analysis (TGA). *Fuel* **2015**, *140*, 744–751. [CrossRef]
13. Rego, F.; Dias, A.P.S.; Casquilho, M.; Rosa, F.C.; Rodrigues, A. Fast determination of lignocellulosic composition of poplar biomass by thermogravimetry. *Biomass Bioenergy* **2019**, *122*, 375–380. [CrossRef]
14. Rodríguez-Romero, L.A.; Gutiérrez-Antonio, C.; García-Trejo, J.F.; Feregrino-Pérez, A.A. Estudio comparativo de modelos matemáticos para predecir el poder calorífico de residuos agrícolas mexicanos. *TecnoLógicas* **2022**, *25*, e2142. [CrossRef]
15. Flores, J.J.A.; Vera, J.V.A.; Rodríguez, M.L.Á.; Quiñones, J.G.R.; Espino, J.V.; Martínez, S.J.G.; Ríos, E.T.; Aguado, R.Z. Kinetic, thermodynamic, FT-IR, and primary constitution analysis of *Sargassum* spp. from Mexico: Potential for hydrogen generation. *Int. J. Hydrogen Energy* **2022**, *47*, 30107–30127. [CrossRef]
16. Kim, H.; Yu, S.; Kim, M.; Ryu, C. Progressive deconvolution of biomass thermogram to derive lignocellulosic composition and pyrolysis kinetics for parallel reaction model. *Energy* **2022**, *254*, 124446. [CrossRef]
17. Sjöström, E. *Wood Chemistry: Fundamentals and Applications*; Academic Press, Inc.: London, UK, 1981; pp. 87–102.
18. Fengel, D.; Wegener, G. *Wood Chemistry, Ultrastructure, Reactions*; Walter de Gruyter: Berlin, Germany, 1984; pp. 66–222.

19. Pintor-Ibarra, L.F.; Méndez-Zetina, F.D.; Rutiaga-Quiñones, J.G.; Alvarado-Flores, J.J. Capítulo 5: Caracterización proximal de los biocombustibles sólidos. In *Aplicaciones Energéticas de la Biomasa: Perspectivas para la Caracterización Local de Biocombustibles Sólidos*; Universidad Intercultural Indígena de Michoacán: Pátzcuaro, México, 2023; pp. 87–116.
20. Pintor-Ibarra, L.F.; Carrillo-Parra, A.; Herrera-Bucio, R.; López-Albarrán, P.; Rutiaga-Quiñones, J.G. Physical and chemical properties of timber byproducts from *Pinus leiophylla*, *P. montezumae* and *P. pseudostrobus* for a bioenergetic use. *Wood Res.* **2017**, *62*, 849–861.
21. Martínez-Gómez, O.; Pintor-Ibarra, L.F.; Rutiaga-Quiñones, J.G.; Corona-Terán, J. Chemical Composition and Energy Evaluation of *Abies* spp. and *Pinus* spp. Sawdust Collected as a Byproduct of the Primary Wood Sawing. *South-East Eur. For.* **2022**, *13*, 89–96. [CrossRef]
22. Carrillo-Parra, A.; Ngangyo-Heya, M.; Colín-Urieta, S.; Foroughbakhch Pournavab, R.; Rutiaga-Quiñones, J.G.; Correa-Méndez, F. Physical, mechanical and energy characterization of wood pellets obtained from three common tropical species. *Peer J.* **2018**, *6*, e5504. [CrossRef]
23. López-Sosa, L.B.; Alvarado-Flores, J.J.; Corral-Huacuz, J.C.; Aguilera-Mandujano, A.; Rodríguez-Martínez, R.E.; Guevara-Martínez, S.J.; Alcaraz-Vera, J.V.; Rutiaga-Quiñones, J.G.; Zárate-Medina, J.; Ávalos-Rodríguez, M.L.; et al. A prospective study of the exploitation of pelagic *Sargassum* spp. as a solid biofuel energy source. *Appl. Sci.* **2020**, *10*, 8706. [CrossRef]
24. Cruz-Montelongo, C.D.L.; Herrera-Gamboa, J.; Ortiz-Sánchez, I.A.; Ríos-Saucedo, J.C.; Rosales-Serna, R.; Carrillo-Parra, A. Caracterización energética of the carbon vegetal producido en el Norte-Centro de México. *Madera Bosques* **2020**, *26*, e2621971. [CrossRef]
25. Ríos-Saucedo, J.C.; Rosales-Serna, R.; Jiménez-Ocampo, R.; Domínguez-Martínez, P.A.; Carrillo-Parra, A.; Valenzuela-Nuñez, L.M. Calidad de pélets a partir de biomasa de ocho especies dendroenergéticas de crecimiento rápido. *Agrociencia* **2021**, *55*, 557–568.
26. Rutiaga-Quiñones, J.G.; Pintor-Ibarra, L.F.; Orihuela-Equihua, R.; González-Ortega, N.; Ramírez-Ramírez, M.A.; Carrillo-Avila, N.; Carrillo-Parra, A.; Navarrete-García, M.A.; Ruiz-Aquino, F.; Rangel-Méndez, J.R.; et al. Characterization of Mexican waste biomass relative to energy generation. *BioResources* **2020**, *15*, 8529–8553. [CrossRef]
27. Ruiz-Aquino, F.; Jiménez-Mendoza, M.E.; Santiago-García, W.; Suárez-Mota, M.E.; Aquino-Vásquez, C.; Rutiaga-Quiñones, J.G. Energy Properties of 22 Timber Species from Oaxaca, Mexico. *South-East Eur. For.* **2022**, *13*, 107–113. [CrossRef]
28. Ong, H.C.; Chen, W.H.; Singh, Y.; Gan, Y.Y.; Chen, C.Y.; Show, P.L. A state-of-the-art review on thermochemical conversion of biomass for biofuel production: A TG-FTIR approach. *Energy Convers. Manag.* **2020**, *209*, 112634. [CrossRef]
29. Ma, Z.; Chen, D.; Gu, J.; Bao, B.; Zhang, Q. Determination of pyrolysis characteristics and kinetics of palm kernel shell using TGA–FTIR and model-free integral methods. *Energy Convers. Manag.* **2015**, *89*, 251–259. [CrossRef]
30. Granados, D.A.; Ruiz, R.A.; Vega, L.Y.; Chejne, F. Study of reactivity reduction in sugarcane bagasse as consequence of a torrefaction process. *Energy* **2017**, *139*, 818–827. [CrossRef]
31. Zhang, C.; Ho, S.H.; Chen, W.H.; Xie, Y.; Liu, Z.; Chang, J.S. Torrefaction performance and energy usage of biomass wastes and their correlations with torrefaction severity index. *Appl. Energy* **2018**, *220*, 598–604. [CrossRef]
32. Yang, H.; Yan, R.; Chen, H.; Zheng, C.; Lee, D.H.; Liang, D.T. In-depth investigation of biomass pyrolysis based on three major components: Hemicellulose, cellulose and lignin. *Energy Fuels* **2006**, *20*, 388–393. [CrossRef]
33. Sebio-Puñal, T.; Naya, S.; López-Beceiro, J.; Tarrío-Saavedra, J.; Artiaga, R. Thermogravimetric analysis of wood, holocellulose, and lignin from five wood species. *J. Therm. Anal. Calorim.* **2012**, *109*, 1163–1167. [CrossRef]
34. Brebu, M.; Vasile, C. Thermal degradation of lignin—A review. *Cellul. Chem. Technol.* **2010**, *44*, 353.
35. *UNE-EN ISO 16559*; Biocombustibles Sólidos, Terminología, Definiciones y Descripciones. Asociación Española de Normalización y Certificación (AENOR): Madrid, Spain, 2015.
36. Moya, R.; Tenorio, C. Fuelwood characteristics and its relation with extractives and chemical properties of ten fast-growth species in Costa Rica. *Biomass Bioenergy* **2013**, *56*, 14–21. [CrossRef]
37. Varfolomeev, M.A.; Grachev, A.N.; Makarov, A.A.; Zabelkin, S.A.; Emel'yanenko, V.N.; Musin, T.R.; Gerasimov, A.V.; Nurgaliev, D.K. Thermal analysis and calorimetric study of the combustion of hydrolytic wood lignin and products of its pyrolysis. *Chem. Technol. Fuels Oils* **2015**, *51*, 140–145. [CrossRef]
38. Uceda Castillo, M.E. Determinación of the poder calorífico de 20 especies forestales of the Amazonía Peruana. *Rev. For. Perú* **1984**, *12*, 1–15.
39. Ngangyo-Heya, M.; Foroughbahchk-Pournavab, R.; Carrillo-Parra, A.; Rutiaga-Quiñones, J.G.; Zelinski, V.; Pintor-Ibarra, L.F. Calorific value and chemical composition of five Semi-Arid Mexican tree species. *Forests* **2016**, *7*, 58. [CrossRef]
40. Demirbaş, A.; Demirbaş, A.H. Estimating the calorific values of lignocellulosic fuels. *Energy Explor. Exploit.* **2004**, *22*, 135–143. [CrossRef]
41. *UNE-EN ISO 17225-1*; Biocombustibles Sólidos. Especificaciones y Clases de Combustibles. Parte 1: Requisitos Generales. Asociación Española de Normalización y Certificación (AENOR): Madrid, Spain, 2014.
42. Gutiérrez-Acosta, J.M.; Orihuela-Equihua, R.; Pintor-Ibarra, L.F.; González-Ortega, N.; Hernández-Solís, J.J.; Ruíz-Aquino, F.; Navarrete-García, M.A.; Rutiaga-Quiñones, J.G. On the basic chemical composition of selected biomass types from four regions of Mexico, for bioenergetic purposes. *BioResources* **2021**, *16*, 5694–5705. [CrossRef]
43. *UNE-EN ISO 18125*; Biocombustibles Sólidos. Determinación of the Poder Calorífico. Asociación Española de Normalización y Certificación (AENOR): Madrid, Spain, 2018.
44. White, R.H. Effect of lignin content and extractives on the higher heating value of Wood. *Wood Fiber Sci.* **1987**, *19*, 446–452.

45. Cordero, T.; Márquez, F.; Rodríguez-Mirasol, J.; Rodríguez, J.J. Predicting heating values of lignocellulosics and carbonaceous materials from proximate analysis. *Fuel* **2001**, *80*, 1567–1571. [CrossRef]
46. Demirbaş, A. Calculation of higher heating values of biomass fuels. *Fuel* **1997**, *76*, 431–434. [CrossRef]
47. *UNE-EN 14961-1*; Especificaciones y Clases de Combustibles. Parte 1: Requisitos Generales. Asociación Española de Normalización y Certificación (AENOR): Madrid, Spain, 2011.
48. García, R.; Pizarro, C.; Lavín, A.G.; Bueno, J.L. Characterization of Spanish biomass wastes for energy use. *Bioresour. Technol.* **2012**, *103*, 249–258. [CrossRef]
49. Palamanit, A.; Khongphakdi, P.; Tirawanichakul, Y.; Phusunti, N. Investigation of yields and qualities of pyrolysis products obtained from oil palm biomass using an agitated bed pyrolysis reactor. *Biofuel Res. J.* **2019**, *6*, 1065–1079. [CrossRef]
50. Ali, L.; Ahmed Baloch, K.; Palamanit, A.; Raza, S.A.; Laohaprapanon, S.; Techato, K. Physicochemical Characterisation and the Prospects of Biofuel Production from Rubberwood Sawdust and Sewage Sludge. *Sustainability* **2021**, *13*, 5942. [CrossRef]
51. Obernberger, I.; Brunner, T.; Bärnthaler, G. Chemical properties of solid biofuels-significance and impact. *Biomass Bioenergy* **2006**, *30*, 973–982. [CrossRef]
52. Obernberger, I.; Thek, G. *The Pellet Handbook*, 1st ed.; Earthscan: London, UK; Washington, DC, USA, 2010.
53. Telmo, C.; Luousada, K.; Moreira, N. Proximate analysis, backwards stepwise regression between gross caloric value, ultimate and chemical analysis of wood. *Bioresour. Technol.* **2010**, *101*, 3808–3815. [CrossRef]
54. Obernberger, I.; Thek, G. Physical characterisation and chemical composition of densified biomass fuels with regard to their combustion behaviour. *Biomass Bioenergy* **2004**, *27*, 653–669. [CrossRef]
55. Saravanapriya, G.; Shreelavaniya, R. Characterization of *Prosopis juliflora* and its products as a potential feedstock for energy generation. *Pharma Innov.* **2023**, *12*, 1673–1678.
56. Michell, A.; Higgins, H. *Infrared Spectroscopy in Australian Forest Products Research*; CSIRO Forestry and Forest Products: Melbourne, Australia, 2002; p. 60.
57. Carrillo, F.; Colom, X.; Sunol, J.J.; Saurina, J. Structural FTIR analysis and thermal characterisation of lyocell and viscose-type fibres. *Eur. Polym. J.* **2004**, *40*, 2229–2234. [CrossRef]
58. Bodirlau, R.; Teaca, C.A. Fourier transform infrared spectroscopy and thermal analysis of lignocellulose fillers treated with organic anhydrides. *Rom. J. Phys.* **2009**, *54*, 93–104.
59. Lopez-Velazquez, M.A.; Santes, V.; Balmaseda, J.; Torres-Garcia, E. Pyrolysis of orange waste: A thermo-kinetic study. *J. Anal. Appl. Pyrolysis* **2013**, *99*, 170–177. [CrossRef]
60. Pucetaite, M. *Archaeological Wood from the Swedish Warship Vasa Studied by Infrared Microscopy*; Lund University Press: Lund, Sweden, 2012.
61. Scholz, G.; Windeisen, E.; Liebner, F.; Bäucker, E.; Bues, C.T. Wood anatomical features and chemical composition of *Prosopis kuntzei* from the Paraguayan Chaco. *IAWA J.* **2010**, *31*, 39–52. [CrossRef]
62. Ona, T.; Ito, K.; Shibata, M.; Ootake, Y.; Ohshima, J.; Yokota, S.; Yoshizawa, N.; Sonoda, T. In Situ Determination of Proportion of Cell Types in Wood by Fourier Transform Raman Spectroscopy. *Anal. Biochem.* **1999**, *268*, 43–48. [CrossRef]
63. Moosavi Nejad, S.M.; Madhoushi, M.; Rasouli, D.; Vakili, M. Nondestructive evaluation of wood chemical compounds used in Gorgan historical building via FT-IR spectroscopy. *J. Wood For. Sci. Technol.* **2017**, *23*, 313–328. [CrossRef]
64. Moosavinejad, S.M.; Madhoushi, M.; Vakili, M.; Rasouli, D. Evaluation of degradation in chemical compounds of wood in historical buildings using FT-IR and FT-Raman vibrational spectroscopy. *Maderas Cienc. Tecnol.* **2019**, *21*, 381–392. [CrossRef]
65. Yamauchi, S.; Iijima, Y.; Doi, S. Spectrochemical characterization by FT-Raman spectroscopy of wood heat-treated at low temperatures: Japanese larch and beech. *J. Wood Sci.* **2005**, *51*, 498–506. [CrossRef]
66. *ASTM E11-20*; Standard Specification for Woven Wire Test Sieve Cloth and Test Sieves. ASTM International: West Conshohocken, PA, USA, 2020.
67. *TAPPI T264 cm-97*; Preparation of Wood for Chemical Analysis. TAPPI Press: Atlanta, GA, USA, 2000.
68. Van Soest, P.V.; Robertson, J.B.; Lewis, B.A. Methods for dietary fiber, neutral detergent fiber, and nonstarch polysaccharides in relation to animal nutrition. *J. Dairy Sci.* **1991**, *74*, 3583–3597. [CrossRef]
69. Jaramillo, D.T.; Vélez, S.P.M.; Díaz, J.C.Q. Evaluación de pretratamientos químicos sobre materiales lignocelulósicos. *Ingeniare Rev. Chil. Ing.* **2017**, *25*, 733–743. [CrossRef]
70. *ASTM D4442-20*; Standard Test Methods for Direct Moisture Content Measurement of Wood and Wood-Based Materials. ASTM International: West Conshohocken, PA, USA, 2020.
71. *ASTM D1102-84*; Standard Test Method for Ash in Wood. ASTM International: West Conshohocken, PA, USA, 2013.
72. *ASTM E872-82*; Standard Test Method for Volatile Matter in the Analysis of Particulate Wood Fuels. ASTM International: West Conshohocken, PA, USA, 2013.
73. Heya, M.N.; Hernández, A.L.R.; Pournavab, R.F.; Pintor, L.F.I.; Díaz-Jiménez, L.; Heya, M.S.; Cruz, L.R.S.; Parra, A.C. Physicochemical Characteristics of Biofuel Briquettes Made from Pecan (*Carya illinoensis*) Pericarp Wastes of Different Particle Sizes. *Molecules* **2022**, *27*, 1035. [CrossRef]
74. *ASTM D5142-90*; Standard Test Methods for Proximate Analysis of the Analysis Sample of Coal and Coke by Instrumental Procedures. ASTM International: West Conshohocken, PA, USA, 1998.
75. Rotz, L.; Giazzi, G. *Characterization of Pharmaceutical Products by the Thermo Scientific FLASH 2000 Elemental Analyzer*; Thermo Fischer Scientific: Milan, Italy, 2012.

76. Ghetti, P.; Ricca, L.; Angelini, L. Thermal analysis of biomass and corresponding pyrolysis products. *Fuel* **1996**, *75*, 565–573. [CrossRef]
77. Arcibar-Orozco, J.A.; Josue, D.B.; Hurtado, J.C.R.; Méndez, J.R.R. Influence of iron content, surface area and charge distribution in the arsenicremoval by activated carbons. *Chem. Eng. J.* **2014**, *249*, 201–209. [CrossRef]
78. R Core Team. *A Language and Environment for Statistical Computing;* R Foundation for Statistical Computing: Vienna, Austria, 2020.

molecules

MDPI

Article

Pyrolysis and Physicochemical, Thermokinetic and Thermodynamic Analyses of *Ceiba aesculifolia* (Kunth) Britt and Baker Waste to Evaluate Its Bioenergy Potential

José Juan Alvarado Flores [1,*], Luis Fernando Pintor Ibarra [1], Fernando Daniel Mendez Zetina [1], José Guadalupe Rutiaga Quiñones [1], Jorge Víctor Alcaraz Vera [2] and María Liliana Ávalos Rodríguez [3,*]

1 Facultad de Ingeniería en Tecnología de la Madera, Universidad Michoacana de San Nicolás de Hidalgo, Edif. D. Cd. Universitaria, Santiago Tapia No. 403, Centro, Morelia 58000, Mexico; luis.pintor@umich.mx (L.F.P.I.); 1614346b@umich.mx (F.D.M.Z.); jose.rutiaga@umich.mx (J.G.R.Q.)

2 Instituto de Investigaciones Económicas y Empresariales, Universidad Michoacana de San Nicolás de Hidalgo, Cd. Universitaria, Santiago Tapia No. 403, Centro, Morelia 58000, Mexico; jorge.alcaraz@umich.mx

3 Centro de Investigaciones en Geografía Ambiental, Universidad Nacional Autónoma de México, Antigua Carretera a Pátzcuaro No. 8701, Col. Ex Hacienda de San José de la Huerta, Morelia 58190, Mexico

* Correspondence: jjalvarado@umich.mx (J.J.A.F); lrodriguez@ciga.unam.mx (M.L.Á.R.)

Abstract: *Ceiba aesculifolia* is an important species in Mexico that generates significant amounts of biomass waste during its exploitation, which can be utilized to produce energy. This study presents the characterization of this waste based on chemical (proximal and elemental) and thermal analyses (TGA-DTG) at different heating rates (β = 10–30 °C/min (283–303 K/min)) in the presence of nitrogen and in a temperature range of 25–900 °C. Kinetic parameters were calculated and analyzed as well. Activation energy (*Ea*) and the pre-exponential factor (*A*) were determined using the Friedman (132.03 kJ/mol, 8.11E + 10 s^{-1}), FWO (121.65 kJ/mol, 4.30E + 09), KAS (118.14 kJ/mol, 2.41E + 09), and Kissinger (155.85 kJ/mol, 3.47E + 11) kinetic methods. Variation in the reaction order, *n* (0.3937–0.6141), was obtained by Avrami's theory. We also calculated the thermodynamic parameters (ΔH, ΔG, ΔS) for each kinetic method applied. The results for *Ea*, *A*, *n*, ΔH, ΔG, and ΔS show that this biomass waste is apt for use in pyrolysis. Moreover, the moisture (<10%), ash (<2%), volatile material (>80%), and HHV (>19%) contents of *C. aesculifolia* allowed us to predict acceptable performance in generating energy and fuels. Finally, infrared spectroscopy analysis (FT-IR) allowed us to identify important functional groups, including one that belongs to the family of the aliphatic hydrocarbons.

Keywords: pyrolysis; physiochemical properties; TGA-DTG; iso-conversional kinetic methods; thermodynamic analyses and bioenergy potential

1. Introduction

Today, one of the most attractive options worldwide for developing energies based on alternative sources is biomass. Mexico is no exception in this regard. Every day in western Mexico, thousands of tons (1,018,407.17 m^3) of biomass is produced as lignocellulosic waste from the exploitation of forest species like pine, oak, mahogany, parotta (elephant-ear tree), and ceiba, among others [1]. This waste consists mainly of sawdust, branches, leaves, bark, and stumps. Unfortunately, it is rarely used for purposes other than as fuel for cooking or drying wood, or in ovens to cook adobe bricks [2]. However, research into the energy potential of biomass waste, focused on utilizing it to produce fuels that could eventually replace fossil fuels, has expanded around the world [3]. Biomass waste from *Ceiba aesculifolia* (Kunth) Britt and Baker (hereinafter, *C. aesculifolia*, known locally as pochote) is an option that may help achieve this goal. This tree, which grows in western Mexico, including the state of Michoacán, is notable for its natural ability of constant regeneration. It is well known that lignocellulosic biomass—terrestrial and marine—can be thermally transformed to generate energy. Currently, thermochemical processes (pyrolysis, gasification, combustion) are viable options for converting biomass into products with high added value, such as bio-oil, bio-char, and combustible gas (Figure 1), all of which can be used to generate electrical energy by means of fuel cells [4,5].

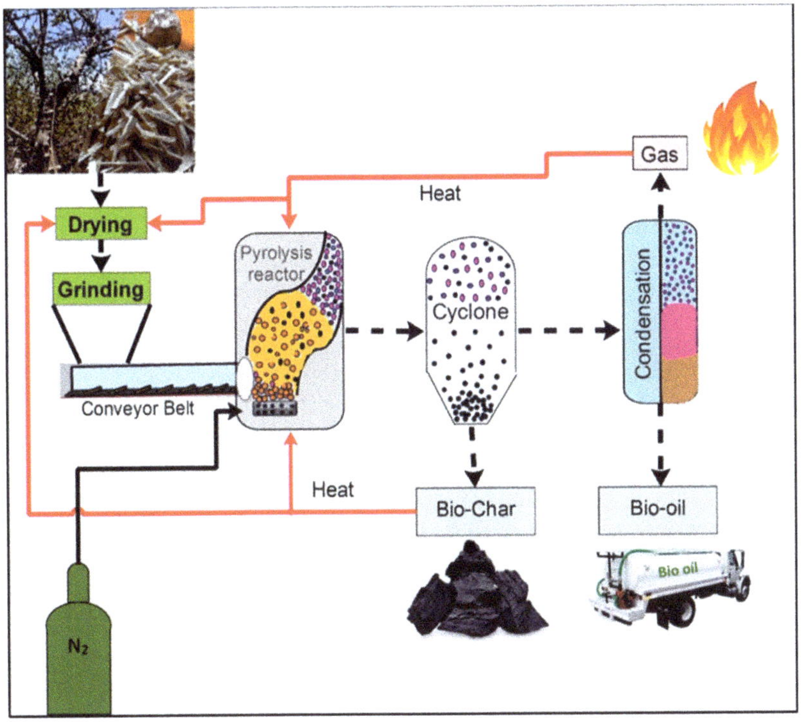

Figure 1. Generation of high-value-added products from biomass.

Of the processes mentioned above, pyrolysis is one of the most widely used forms of thermochemical transformation of biomass waste not only into lignocellulosic biomass, but also into animal biomass waste, such as manure [6], and even in materials such as microplastics [7]. This thermal process, also called destructive distillation, is carried out in an inert atmosphere (e.g., nitrogen and argon), in contrast to the process that requires oxygen (complete combustion). Pyrolysis of biomass can be used to obtain several substances of importance for industry, such as oil, coal, hydrocarbon chains (e.g., methane), aromatic compounds, and hydrogen (H_2), known as syngas, which is usually obtained industrially through petroleum refining processes. Pyrolysis generally requires temperatures of 400–600 °C and a pressure of 1–5 bar [8]. Given the efficiency of hydrogen, the possibility of obtaining this gas from biomass, and its potential for generating clean, renewable, environmentally friendly electrical energy, numerous researchers have taken up the challenge of optimizing processes to obtain it. Published studies show good average yields of 55, 40, 48, 50, 48, and 50% H_2 in biomass derived from nogal and cotton, and waste from tea, olive, beech, and fir production, respectively [9]. Significantly, studies of biomass include marine residues like algae [10]. At the time of our study, however, there were no reports on the biomass waste of *C. aesculifolia*.

It is well known that, thanks to its physicochemical properties and lignocellulosic composition, biomass is an excellent option as an energy source. One key aspect in optimizing yields consists of analyzing the kinetic parameters of the thermochemical conversional process utilized (in the present case, pyrolysis). This study documents the chemical, energy, and thermal characterization of this material and discusses the handling of the data obtained from thermogravimetric analyses and their application in well-known mathematical methods that make it possible to optimize the energy extracted from this biomass. We applied three iso-conversional methods—Friedman, Flynn–Wall–Ozawa (FWO), and Kissinger–Akahira–Sunose (KAS)—and Kissinger's non-iso-conversional type to obtain precise values for three kinetic parameters—apparent energy activation, Ea, the pre-exponential factor, A, and the reaction order, n—as well as the Gibbs free energy change as a thermodynamic parameter, and changes in enthalpy and entropy. In addition to the fact that this thermokinetic study is the first to be carried out for *C. aesculifolia*, we believe that it is necessary to make better energy use of the enormous amounts of biomass waste such as sawdust (more than 2000 kg) and shavings (more than 5000 kg) that are generated per week in some communities in Mexico [11].

2. Results and Discussion

2.1. Determination of Basic Density

The basic density of wood is a physical characteristic that has a crucial role in processes of biomass combustion [12,13]. Studies of this factor show that an increase in basic density leads to an increase in calorific value; that is, a larger amount of energy stored per unit of volume. According to the literature [14], *C. aesculifolia* wood is classified as "heavy", with a basic density of 0.70 g/cm^3. For the wood of the group of leafy trees, basic density varies from 0.475 to 0.814 g/cm^3 [15], so *C. aesculifolia* is within this range.

2.2. Basic Chemical Composition

2.2.1. Holocellulose, Cellulose, and Hemicellulose

Holocellulose constitutes the total fraction of the polysaccharides present in wood, including cellulose (CE) and hemicelluloses (HMs) [16,17]. In this study, the sum of CE and HM resulted in a total of 77.74%, a percentage above those reported for other hardwoods in Mexico, which tend to fall between 63.8 and 73.58% [18,19]. It is important to clarify that the total content of the basic composition shown in Figure 2, practically 100%, is only the sum of EC, HMs, lignin, and extractables.

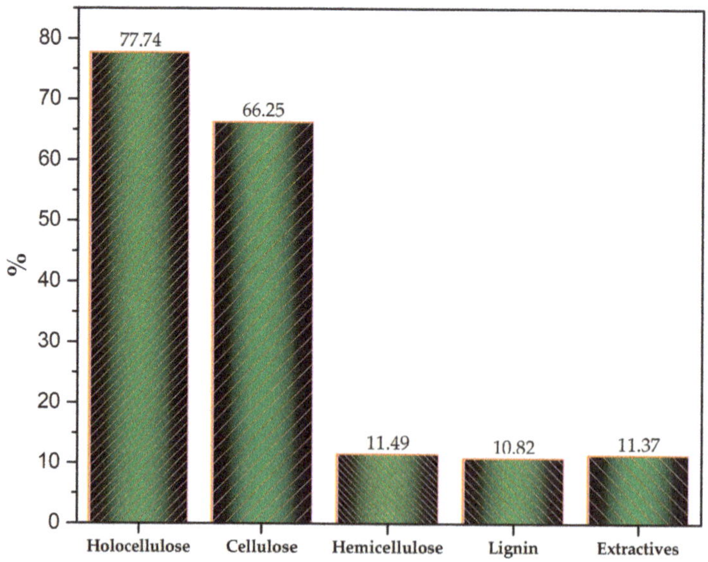

Figure 2. Basic chemical composition of *C. aesculifolia* wood.

The cellulose content of *C. aesculifolia* wood (see Figure 2) exceeds the 41.29% reported for *Ceiba pentandra* [20]; in fact, the value obtained herein is above the range of 41.32–58.83% registered for Mexican hardwoods [21]. This suggests its potential for use in producing bioethanol, although the CE is found mainly in crystalline form, which can impede its depolymerization for fermentation due to problems of accessibility and the consumption of chemical reagents [22]. This high percentage (66.25%) means that *C. aesculifolia* wood can be effective in obtaining CE destined for producing paper and its derivatives, a process that is expected to give high yields [23]. The percentage of hemicelluloses (11.49%) can be considered within the range (11.75–24.38%) found in five hardwood species native to Mexico that have characteristics similar to those of the biomass used in this study [24]. It is important to mention that the HMs present in amorphous regions are more easily hydrolyzed using enzymes than the crystalline forms, so they could be useful in fermentation processes [25].

2.2.2. Lignin

Lignin is the second-most-abundant amorphous biopolymer in nature. Its principal derivatives are coniferyl, coumaryl, and sinapyl alcohol. Lignin represents 15–30% of the composition of wood [26]. We found that *C. aesculifolia* contains 10.82% lignin (Figure 2), although other authors have recorded percentages from 14.86 to 36.57% in Mexican hardwoods [19]. Clearly, the value found in our work is below this range—a fact that could favor bioethanol production, since lignocellulosic materials with high percentages of lignin can impede fermentation treatments [22]. High lignin content in wood, in contrast, can improve its quality as a solid biofuel, because this compound increases the higher heating value [27,28].

2.2.3. Extractives

In general, extractives are organic substances of low molecular mass that are not part of the wood cell wall. They are also commonly called accessory substances or extraneous components, are formed from the secondary metabolism of plants, are removed by water and organic solvents (water, alcohol, acetone, benzene, ether, and mixtures between solvents), and contain various compounds such as waxes, fats, aromatic compounds (phenolics), terpenes, alcohols, esters, tannins, flavonoids, and carbohydrates [29]. Although extractives do not change the structure of wood, and they contribute only a low percentage of mass, they significantly impact its properties and transformation processes [30]. Recent reports on some Mexican hardwoods affirm that the content of extractive components ranges from 8.2 to 30.7% [19]. Our results show a percentage (11.37%) of extractives in *C. aesculifolia* within this range (Figure 2). We should mention that, according to the chemical and/or thermal transformation, the presence of extractives in lignocellulosic materials can have disadvantages, such as a greater risk of corrosion and an increase in the consumption of reagents in chemical transformation equipment. However, they may offer an advantage by increasing the higher heating value [27].

2.3. *Proximal Analysis and the Higher Heating Value*

Proximal analysis is widely used to evaluate biofuels because it provides initial information on the possible qualities of a biomass as a fuel source. This method involves measuring such parameters as the moisture, ash, fixed carbon, and volatile compounds present in biomass samples [31]. Table 1 shows the results. It should be clarified that, in the case of gravimetric analysis, moisture is omitted because the calculations are made on a dry weight basis, i.e., 100% should be considered in relation to the percentages of volatile materials, fixed carbon, and ash.

Table 1. Average results of proximate analysis and superior calorific value of *C. aesculifolia*.

Moisture (%)	VM (%) [1]	FC (%) [2]	Ash (%)	HHV (MJ/kg) [3]
9.37 ± 0.2	83.83 ± 2	14.33 ± 2.1	1.84 ± 0.1	19.12 ± 0.2

[1] VM: volatile material, [2] FC: fixed carbon, and [3] HHV: higher heating value.

2.3.1. Moisture Content

As Table 1 shows, the moisture content detected in *C. aesculifolia* was 9.37%, a figure within the range mentioned by other authors [32]. The marked variability of 6–75% in the initial moisture levels of our biomass samples was due primarily to the fact that some samples were taken from recently processed wood, while others were drawn from material that had been exposed to the environment. It is important to mention that high moisture content in biomass causes a weight increase that generates higher transport costs and greater energy consumption during the chipping process and can affect the quality of the splinters obtained [23]. High percentages of moisture in samples also require implementing drying procedures, although these raise costs and prolong drying times before the material can be used as biofuel [33].

It is worth mentioning that we have observed that humidity varies slightly even on the same day at different times. For that reason, the chemical analysis was performed with anhydrous wood, and the humidity of 9.37% corresponds to the humidity of the sample in the proximal analyses determined by gravimetric methods. It should also be mentioned that the proximal analyses were carried out on the basis of anhydrous weight; that is, the humidity of the flour was adjusted in the calculations. Due to the two procedures—chemical and TGA (Section 2.6) methods—there are some differences in the results, such as the humidity of 9.37% and 6.65% respectively.

2.3.2. Volatile Materials

The proportion of volatile materials in lignocellulosic biomass can vary from 76 to 86% by weight [34]. *C. aesculifolia* had a volatile material content within this range, at approximately 84% (Table 1). This volatile fraction was converted into gas [35], which was also within the reported limits of 82.6–84.9% for biomass collected in diverse regions of Mexico [18]. In analyses of volatile materials, a high content is an indicator that favors ignition even at relatively low temperatures, as various authors have noted [36]. However, high levels can also increase the speed of the combustion process, which may be disadvantageous in the case of solid biofuels [37], although they can be beneficial for biogas production [38]. As can be seen, *C. aesculifolia* is rich in volatiles, which, when thermally degraded, can be implemented for the generation of biogas such as hydrogen from a reforming of the so-called synthesis gas. This product would be a significant bioresource for such important purposes as the generation of electrical energy from the use of fuel cells (for example, solid oxides) [39].

2.3.3. Fixed Carbon

Fixed carbon is generated by the combustion of biomass through the oxidation of solid material, which lasts considerably longer than that of gases in fuels [34]. *C. aesculifolia* had a fixed carbon value of 14.33% (Table 1), placing it in the highest range according to the figures of 14.1–14.3% reported for diverse hardwood species [14]. This factor is deemed favorable for bioenergy processes [40], although it is important to note that low fixed carbon values are not considered to be disadvantageous and, in fact, can be appropriate for biogas and biodiesel production [41].

2.3.4. Ash

Ash includes chemically bonded, salt-based compounds. Lignocellulosic materials acquire ash during growth in their natural environment. Individual species may have distinct components present in variable amounts [42]. The ash content in lignocellulosic materials ranges from 0.94 to 17.89% [43]. For the biomass under study, we found a percentage of 1.84% (Table 1). It is important to point out that some of the compounds that make up this ash are volatilized and incorporated into the gaseous phase, but that the residual part can cause serious problems in equipment (for example, corrosion and obstruction of the gaseous flow in combustion boilers). According to the aforementioned range, the ash content of *C. aesculifolia* (1.84%) is low, leading us to believe that these would not be common problems. Although low, this level would impede the fabrication of products like pellets and briquettes, since current norms stipulate a maximum ash content of 0.70 for the former and 0.5% for the latter [44].

2.3.5. Higher Heating Value (HHV)

Another fundamental characteristic of biofuels is the higher heating value, which is determined by their chemical composition. It is widely held that elevated HHV values indicate good capacity as a fuel, while low values suggest the opposite [45]. Some researchers hold that the gross HHV of diverse raw materials ranges from 19.8 to 20.7 MJ/kg [34]. As Table 1 shows, the HHV value for *C. aesculifolia* wood (19.12 MJ/kg) is within this range. Similar reports exist for some hardwood species with high commercial value, such as the leaves of Ebenopis ebano [27]. This HHV result can be considered appropriate for the generation of high-value-added bioenergy products such as methane and hydrogen. This affirmation is based on the results observed in research on lignocellulosic biomasses that present an approximate HHV value of 19.12 MJ/kg, where gases such as bio-methane [46] and hydrogen [47] have been obtained from pyrolysis. However, due to the amount of fixed carbon obtained (14.33%), it is possible not to consider the production of materials such as briquettes, although this value may also represent an advantage in reducing CO_2 emissions [48].

2.4. Microanalysis of Ash

The results of the microanalysis of ash (see Table 2) identified 18 elements, with a marked presence of potassium (K), followed by calcium (Ca), phosphorus (P), sodium (Na), and magnesium (Mg).

Table 2. Microanalysis of *C. aesculifolia* ashes.

Element	ppm
K	257,523.69
Ca	35,917.89
P	15,824.95
Na	2373.19
Mg	1329.53
S	1198.21
Sr	292.95
Si	164.05
Ba	153.00
Fe	112.42
Li	91.78
B	69.67
Cu	41.73
Al	38.88
Mn	35.36
Zn	26.46
Ni	0.90
Cr	<0.05

Note: ppm = parts per million. The elements are represented by their symbols.

Of these five elements, the most common one (K) is characteristic of diverse woods and their derivatives, including Persian lime (leaves), orange trees' branches, and lemon peel. In the case of the leaves of Persian lime, phosphorus and magnesium have similar values, and the value for sodium approaches those obtained for *Pinus* spp. [18]. The predominant presence of these elements is significant because they have the potential to cause complications during combustion processes by affecting the fusion point of ash, causing the formation of slag, corroding equipment, emitting fine particles, and causing incrustations that can damage ovens and boilers [49]. However, some authors stress that elements like Ca and Mg may favor combustion processes by raising the fusion point of ash, thus reducing the amount inside the combustion equipment and contributing to environmental safety by reducing the toxicity of the residues dispersed Into the air [50]. In contrast to these two elements, sodium, iron, and silicon interfere in the fusion of ash and lead to the formation of incrustations. Here, we should mention that sulfur ($\approx$1200 ppm) can cause serious corrosion problems due to the formation of iron (II) chloride [46]. *C. aesculifolia* also had trace amounts of Cu, Ni, and Zn, which, even in small quantities, can have a direct negative impact on the environment. Actions have been undertaken to apply them to the benefit of the environment [51].

2.5. Elemental Analysis

This analysis was utilized primarily to determine five elements: carbon, hydrogen, oxygen, nitrogen, and sulfur. The first three are important because they supply the highest amounts of energy in the biomass. However, nitrogen and sulfur, even in small amounts, are also very significant, for when derived from processes like pyrolysis they indicate the possible formation of gases and/or acids (Nox, Sox, NH_3) that are harmful to the environment when expelled into the atmosphere. Table 3 shows the results of this analysis, which are within the ranges cited in recent publications. Of these five elements, we would emphasize the low sulfur content of *C. aesculifolia* (a value of just 0.08), although the nitrogen content is high. In thermal processes like pyrolysis, the atmospheric emissions of Sox types and nitrogenous compounds must be reduced to a minimum [52].

Table 3. Elemental analysis of *C. aesculifolia* wood.

Element	This Research (%)	Literature (%) [42,53]
C	45.60	40–46
H	6.90	4–6
O	47.20	45–50
N	0.30	0.11–0.32
S	0.08	0.00–0.96

2.6. Thermal Properties of *C. aesculifolia*: TGA-DTG

The thermal behavior generated by the mass degradation of the lignocellulosic material under study was characterized using a thermogravimetric analyzer. In Figure 3, the two vertical axes show the loss of mass (left axis) at a certain velocity (right axis). The latter represents the first derivative (DTG) of the thermogravimetric curve (TGA) throughout the temperature range. The thermogram reveals a behavior similar to that of diverse lignocellulosic materials [54].

Three regions at distinct heating speeds are discernible in Figure 3 (β = 10–30 °C/min; 283–303 K/min). The first, located at temperatures between 300 and 415 K, can be identified with the loss of surface water from the biomass. As the graph advances, it presents the greatest loss of mass ($\approx$60 %) in region 2 (415–680 K). For lignocellulosic materials, this zone is distinguished by the elimination of volatile materials, especially the two main polysaccharides: hemicellulose and cellulose. This is normally the area where the chemical reactions that decompose these compounds and the greatest generation of gases (methane, hydrogen, nitrogen, carbon monoxide and dioxide) occur, so it is called the active stage of pyrolysis. Finally, in the range of 680–1100 K, the degradation of lignin—the main cementing compound in these materials—occurs. Region 3 also reveals the formation of fixed carbon. In this stage, both curves (TGA and DTG) are practically flat.

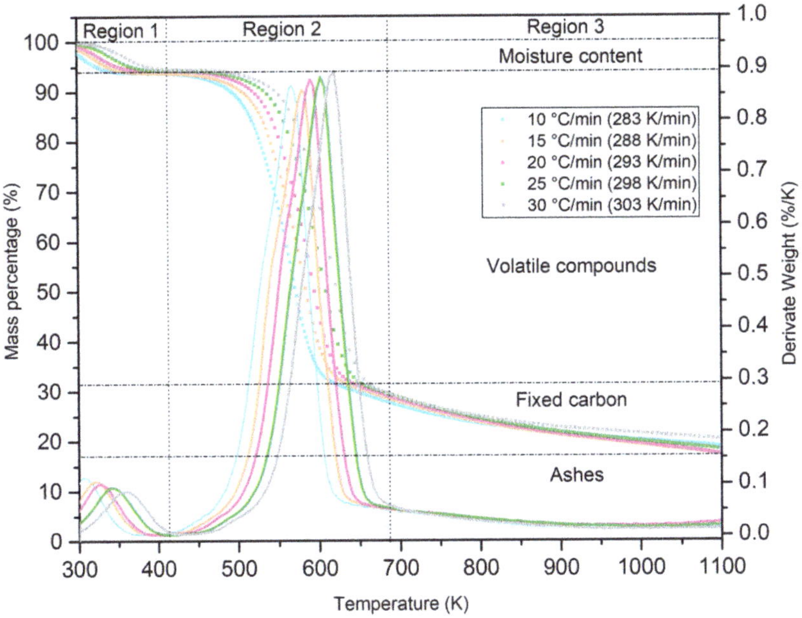

Figure 3. TGA-DTG test of *C. aesculifolia* for five heating rates.

In addition to the mass degradation curves (TGA), Figure 3 displays the characteristic peaks of the derivative of TGA—that is, the DTG that represents the ratio of change in weight loss with respect to temperature throughout the thermal process. Like TGA, DTG provides valuable information on key changes in the biomass, such as the maximum reaction temperatures (peaks) and decomposition velocities at any given point [55].

Parallel to TGA, the two main peaks in DTG represent the decomposition velocities of water, polysaccharides (hemicellulose and cellulose), and lignin. It is interesting to note that although this biomass is made up of these three compounds, as well as water, only two principal peaks encompass them. As mentioned, the polysaccharides degrade at distinct temperatures. Reports generally indicate that, due to heterogeneity and the size of the fiber in hardwoods [56], the decomposition of hemicellulose and cellulose occurs between 573 and 673 K [57], while the range for lignin is wider: 300–1200 K [58]. In this study, this summarizes the thermal behavior of *C. aesculifolia* wood. Another aspect observable in Figure 3 is the left-to-right shift of the TGA-DTG curves as the heating rate (10–30 °C/min; 283–303 K/min) and temperature increase. The explanation of this phenomenon is based on heat transfer inside the biomass, which is inversely proportional to the heating rate [59].

2.7. Kinetic Analysis of C. aesculifolia *Wood: Ea and A*

Based on the data obtained from the thermogravimetric analysis and the methods of chemical kinetics proposed by Friedman, Flynn–Wall–Ozawa, Kissinger–Akahira–Sunose, and Kissinger, it was possible to determine the principal kinetic parameters—that is, activation energy (Ea) and the pre-exponential factor (A). The reaction order, meanwhile, was determined using Avrami's method, as explained in the materials and methods section. These methods are extremely important for improving our understanding of the pyrolytic process of *C. aesculifolia* wood in an inert atmosphere. Since each one uses distinct aspects to determine the reaction mechanism, associated kinetics, and regression techniques according to the handling of the data obtained from the thermal process, distinct interpretations may be reached for each result, as occurs with Ea. Because each mathematical method applied in reaction kinetics studies represents diverse mechanisms, the values for the kinetic parameters (Ea and A) will show certain differences [60]. Figure 4a–d and Table 4 show the results for these parameters according to the four methods used to describe interrelations among the degree of advance, temperature, and velocity. They also show the calculation of the correlation coefficient, R^2 (Table 4). This is particularly important because, according to the value obtained—preferably between 0.90 and 1.0—any graph of the kinetic methods proposed is considered acceptable. In this study, the graph of *C. aesculifolia* presented low R^2 values (<0.80) when the degree of conversion (α) was >0.70. These values were not included in the analysis. It is important to mention that some research related to kinetic studies (Friedman, FWO, and KAS) of biomass have reported similar behavior, in fact with a lower degree of progress (α = 0.65). This situation can be attributed to the degree of complexity in the generation of carbon in the pyrolysis process [61]. Recent research has had to adjust its results to values where the degree of advance is also 0.70 due to poor fitting of the data—for example, in Trapa natans peel biomass [62]. However, this does not mean that the results are not valid for the study in question.

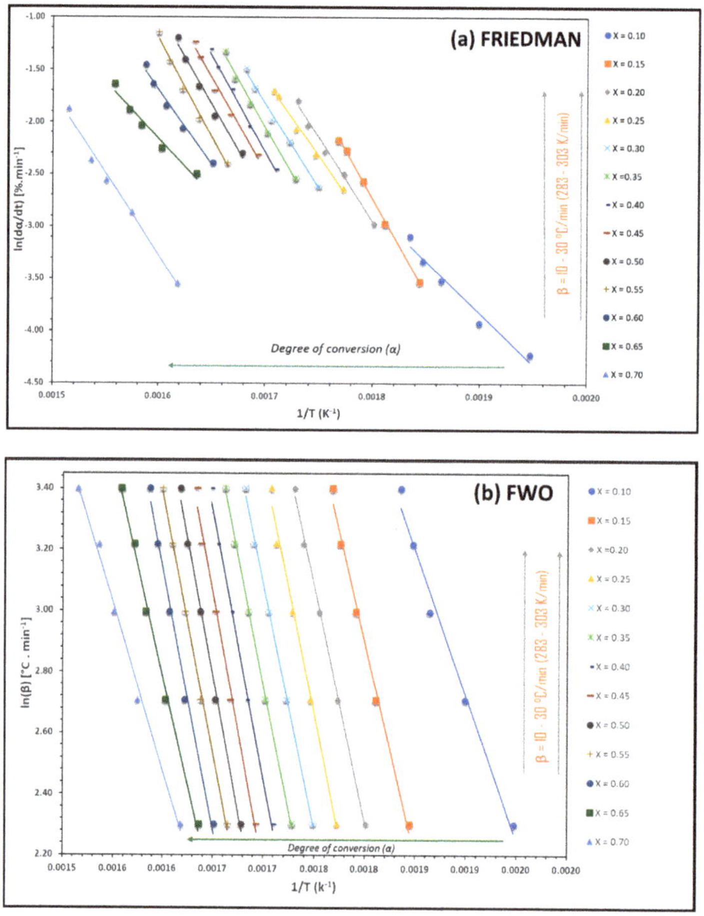

Figure 4. *Cont.*

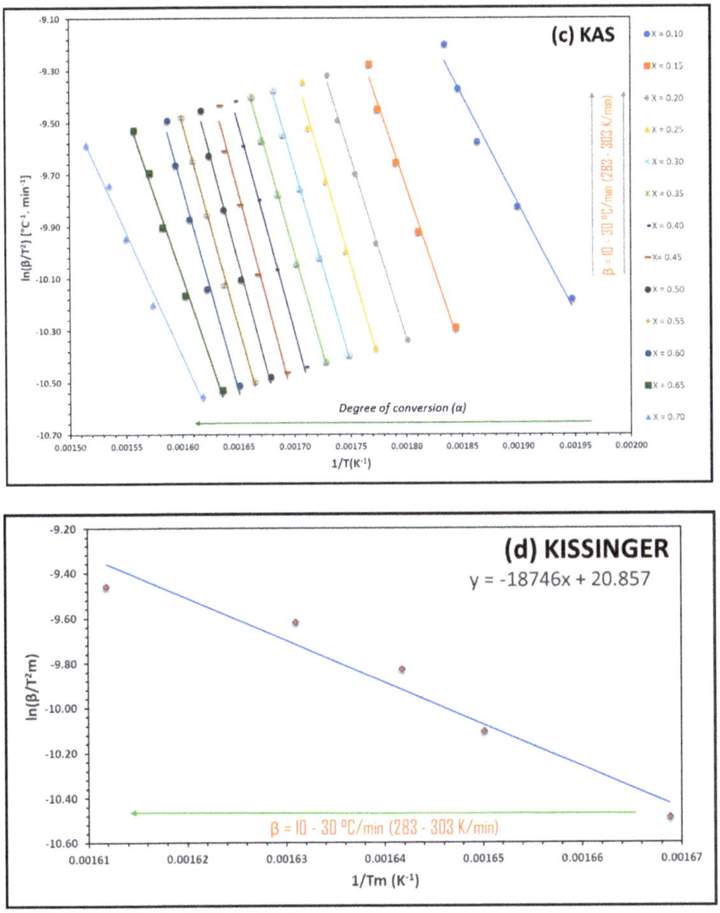

Figure 4. Regression lines to apparent activation energy proposed by Friedman (**a**), Flynn–Wall–Ozawa (FWO) (**b**), Kissinger–Akahira–Sunose (KAS) (**c**), and Kissinger (**d**) method free plots at the different heating rates for *C. aesculifolia* biomass.

It is important to emphasize that the average R^2 value for the four methods was >0.95, a figure practically equal to those of the FWO (0.9938) and KAS (0.9865) methods. We should clarify that Kissinger's method does not use the degree of conversion as its basis for analysis, although it must also fulfill the condition that R^2 is >0.90 in order to achieve more approximate values [63]. According to Table 4, the average Ea of the Friedman, FWO, and KAS methods was 132.03, 121.65, and 118.14 kJ/mol, respectively, while the figure for Kissinger's method was 155.85 kJ/mol. The first three results are quite similar, with KAS having the lowest value and Kissinger's method the highest. These differences are due to the mathematical bases employed in each method. Friedman's method is differential, based on the rate of change in the degree of advance with respect to time ($d\alpha/dt$), while FWO and KAS are based on the heating rate (β), and both are integral methods. Thus, we must keep in mind that these kinetic methods are complementary and should not be understood as being in competition [64]. Also important, however, is that Kissinger's method reports only a value for Ea. Since it is not possible to appreciate the degree of advance of the pyrolytic process, the complexity of that process is not observed clearly. From the phenomenological point of view, one of the variables used by these methods (Friedman, FWO, KAS, and Kissinger) is temperature. Such differences in activation energy can also be explained as a function of increasing temperature and advancing degree of advancement, causing a reduction in the molecular mobility of the biomass structure, resulting in an increase in activation energy. As the process continues, the activation energy is reduced as the kinetics progresses [65]. It can be observed that the value of the activation energy in the FWO method is slightly higher compared to the KAS method; this may be related to the behavior of the bonds in the biomass structure, the weakest being the ones that are first randomly broken by scission effects [66].

Table 4. Kinetic parameters for pyrolysis of *C. aesculifolia* biomass.

Degree of Conversion (α)	Equation ($y = mx + b$)	R^2	Ea (kJ.mol^{-1})	A (s^{-1}), (β = 15 °C/min)
		Friedman method		
0.10	y = −9849.9x + 14.886	0.9675	81.89	7.68E + 04
0.15	y = −18044x + 29.732	0.9981	150.02	1.05E + 11
0.20	y = −15786x + 25.46	0.9892	131.24	2.21E + 09
0.25	y = −14761x + 23.488	0.9894	122.72	3.81E + 08
0.30	y = −16454x + 26.142	0.9919	136.80	6.93E + 09
0.35	y = −17888x + 28.348	0.9921	148.72	8.03E + 10
0.40	y = −18415x + 29.008	0.9946	153.10	1.97E + 11
0.45	y = −17748x + 27.696	0.9821	147.56	6.33E + 10
0.50	y = −17833x + 27.585	0.9812	148.26	7.31E + 10
0.55	y = −18989x + 29.174	0.9887	157.87	5.25E + 11
0.60	y = −14395x + 21.333	0.9813	119.68	2.03E + 08
0.65	y = −10742x + 15.022	0.9657	89.31	3.65E + 05
0.70	y = −15536x + 21.579	0.9878	129.17	1.44E + 09
Average		**0.9854**	**132.03**	**8.11E + 10**
		Flynn–Wall–Ozawa method		
0.10	y = −9463.7x + 20.702	0.9869	74.79	1.71E + 04
0.15	y = −13993x + 28.077	0.9926	110.59	3.08E + 07
0.20	y = −15222x + 29.708	0.9979	120.30	2.31E + 08
0.25	y = −16296x + 31.168	0.9910	128.79	1.33E + 09
0.30	y = −16226x + 30.666	0.9981	128.23	1.19E + 09
0.35	y = −16540x + 30.871	0.9985	130.72	1.98E + 09
0.40	y = −17355x + 31.951	0.9941	137.16	7.46E + 09
0.45	y = −17915x + 32.609	0.9897	141.58	1.85E + 10
0.50	y = −17793x + 32.141	0.9942	140.62	1.52E + 10
0.55	y = −17108x + 30.759	0.9975	135.21	4.99E + 09
0.60	y = −17094x + 30.485	0.9915	135.09	4.88E + 09
0.65	y = −14206x + 25.514	0.9944	112.27	4.37E + 07
0.70	y = −10898x + 19.913	0.9929	86.13	1.87E + 05
Average		**0.9938**	**121.65**	**4.30E + 09**
		Kissinger–Akahira–Sunose method		
0.10	y = −8406.1x + 6.1598	0.9836	69.89	6.05E + 03
0.15	y = −12885x + 13.443	0.9914	107.13	1.50E + 07
0.20	y = −14089x + 15.029	0.9975	117.14	1.20E + 08
0.25	y = −15146x + 16.459	0.9897	125.92	7.37E + 08
0.30	y = −15060x + 15.929	0.9978	125.21	6.36E + 08
0.35	y = −15360x + 16.110	0.9983	127.70	1.06E + 09
0.40	y = −16163x + 17.171	0.9933	134.38	4.21E + 09
0.45	y = −16713x + 17.811	0.9883	138.95	1.08E + 10
0.50	y = −16579x + 17.324	0.9934	137.84	8.60E + 09
0.55	y = −15883x + 15.923	0.9972	132.05	2.61E + 09
0.60	y = −15859x + 15.633	0.9902	131.85	2.50E + 09
0.65	y = −12954x + 10.635	0.9934	107.70	1.69E + 07
0.70	y = −9622.7x + 4.996	0.9911	80.00	5.16E + 04
Average		**0.9865**	**118.14**	**2.41E + 09**
		Kissinger method		
α not involved	y = −18746x + 20.857	**0.9506**	**155.85**	**3.47E + 11**

Table 4 also shows that the maximum values for Ea (Friedman, KAS, FWO) are within the range of the degree of advance (α)—that is, between 0.40 and 0.55. This means that, based on this value (0.55), the amount of mass available to participate in the thermal process will be lower. These maximum values can also be interpreted as a function of the greater decomposition of the biomass that, according to the TGA-DTG graph (Figure 3), corresponds to the elimination of the HM and CE that make up the wood of *C. aesculifolia*.

Another significant aspect is the variability in Ea over the range considered for the methods applied: 81.89–153.10, 74.79–141.78, and 69.89–138.95 kJ/mol for the Friedman, FWO, and KAS methods, respectively. These results mean that the pyrolytic process is complex, occurs in multiple stages, and presents diverse types of chemical reactions across the temperature range [67]. In general, reports in the literature affirm that the range of Ea for lignocellulosic biomass made up of woody tissue lies between 60 and 170 kJ/mol [68]. Our results are within this range. Other biomasses present similar values for Ea: nutshell (≈136 kJ/mol) [69], Indian almond (≈133 kJ/mol) [70], Brazil nut (≈137 kJ/mol) [71], products of figs (≈160 kJ/mol) [72], fistula cane (Cassia fistula L.) (≈137 kJ/mol), peach palm (≈112 kJ/mol) [73], Manilkara zapota seeds (≈132 kJ/mol), Delonix regia (≈143 kJ/mol), and Cascabela thevetia (≈152 kJ/mol) [74].

Another key parameter in kinetic analyses of thermogravimetric processes is the reaction order (n). The variation in this parameter for *C. aesculifolia* is shown in Figure 5 and Table 5. This was calculated using Avrami's equation, as indicated in the Materials and Methods section. Calculating the reaction order requires considering the heating rate (β = 10–30 °C/min; 283–303 K/min) and the temperature range of 500–650 K for the most important event of pyrolysis (region 2)—that is, the TGA-DTG analysis. Clearly, according to the lines in Figure 5, the value of the lineal correlation coefficient (R^2) lies between 0.95 and 0.99, with an average of 0.9903, which is acceptable for calculating the reaction order.

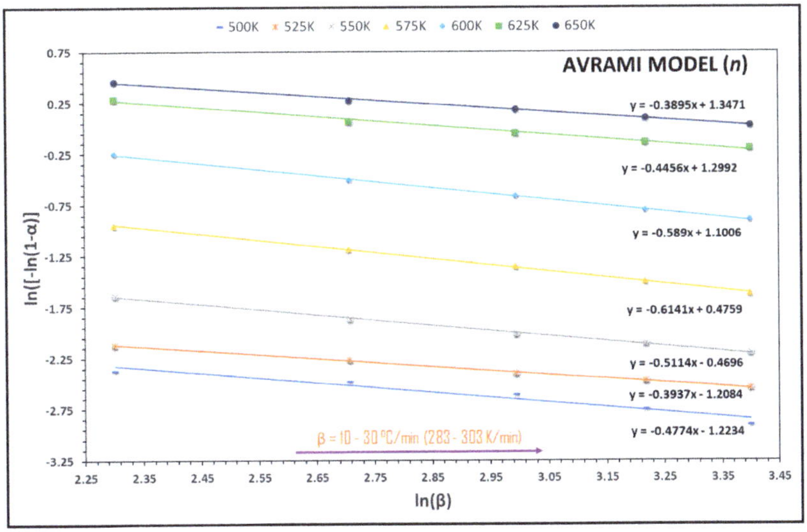

Figure 5. Regression lines to reaction order (n) proposed by Avrami's theory.

As Table 5 shows, the reaction order initially increased from 0.3937 to 0.6141, before decreasing to 0.3895, with an average value of 0.4887. This result is similar to those obtained by authors who have analyzed pyrolysis using agricultural waste and applying Avrami's equation, as in the cases of corn straw and rice husk, for which average reaction orders of 0.365 and 0.539 were reported, respectively. It is true that this value can change from one type of biomass to another. These differences may be related to the composition of ash content, since this exerts a significant effect on the alkaline metals present and, in turn, their impact on the thermal process [75].

Table 5. Reaction order (n) deduced from Avrami's theory for *C. aesculifolia*.

T (K)	Slope (m) of the Equation	Reaction Order ($n = -m$)	Correlation Coefficient (R^2)
500	−0.4774	0.4774	0.9531
525	−0.3937	0.3937	0.9986
550	−0.5114	0.5114	0.9976
575	−0.6141	0.6141	0.9992
600	−0.5890	0.5890	0.9986
625	−0.4456	0.4456	0.9889
650	−0.3895	0.3895	0.9963
Average		**0.4887**	**0.9903**

2.8. *Thermodynamic Analysis of C. aesculifolia: A, ΔH, ΔG, and ΔS*

The first parameter obtained in thermodynamic analyses is the pre-exponential factor, calculated using the Ea values throughout the range of the degree of advance (α = 0.10–0.70) and the temperature at the maximum peak of the DTG (Figure 3). Table 4 shows the pre-exponential factor (A) obtained at a low heating rate (β = 15 °C/min). This reflects the fact that the frequency of molecular collisions increases proportionally to the heating rate [76]. In this regard, applying a low value of β reduces the degree of interaction among the constituents of the biomass in pyrolysis. Recent research (on cattle manure and pistachio shells) has also opted for the calculation of the pre-exponential factor at even lower heating rates (10 °C/min) [77,78]. It is worth mentioning that, with respect to different regions of mass degradation, more accurate results can be obtained by performing thermogravimetric analysis at low heating rates, even as low as 1 °C/min [79]. In this regard, and based on the information reported in the aforementioned investigations, thermodynamic analysis of *C. aesculifolia* is performed at a low heating rate. In this study, A varied in the range of 10^4–10^{11} s^{-1}, defining the occurrence of both reactions and complex structures in this process due to the thermal conversional of the biomass [60]. This variability in A may be conditioned by other factors, such as particle size, the presence of a catalyzer, and even the diverse heating rate applied. It is indicative of the collisions that occur among the particles during the thermal process. Here, the number of collisions increased with the increase in this value. The average values shown in Table 4 for the KAS, FWO, Friedman, and Kissinger methods are 2.41E + 09, 4.30E + 09, 8.11E + 10, and 3.47E + 11 s^{-1}, respectively.

Similar results have been reported where A varied from 10^7 to 10^{12} s^{-1}—for example, with red pepper, rice husk, and bran [80]. Other reports affirm that, depending on whether the value of A is <10^9 s^{-1} or $\geq 10^9$ s^{-1}, either collisions or superficial reactions and the formation of a simple compound will occur, respectively. However, if this value lies between 10^{10} and 10^{12} s^{-1}, there is a possibility of activating rotations of certain compounds that were passive at first [81]. This may cause the size of the complex formed to increase (unimolecular reaction) or remain unchanged (monomolecular reaction) in relation to the interaction with its neighbors [82]. In this case, regardless of the method applied, the constant values of the pre-exponential factor (10^{10}–10^{11} s^{-1}) represent the zone where the main loss of mass occurs—that is, degradation of HM and CE when the degree of advance (α) lies between 0.40 and 0.55 throughout the pyrolytic process. The highest value for A (3.47E + 11 s^{-1}) was calculated by Kissinger's method, perhaps reflecting the fact that the rotation of the active–reactive compound pair remained unchanged during the thermal process [83]. However, in light of the disparity with the results of the Friedman, FWO, and KAS methods, it is likely that Kissinger's method presented this result because of the method's mathematical basis, as explained previously.

In addition to the pre-exponential factor, and as part of the thermodynamic study of the pyrolysis process of *C. aesculifolia*, Table 6 shows the variation in three other key parameters that make it possible to analyze energy behavior in terms of spontaneity, conservation, equilibrium, and quality [84], namely, analyses of enthalpy (ΔH, kJ/mol), Gibbs free energy (ΔG, kJ/mol), and entropy (ΔS, J/mol. K). As mentioned above, these parameters were calculated at a low heating rate. Table 6 displays the average values for these thermodynamic properties (ΔH, ΔG, and ΔS) according to the Friedman (127.12, 174.89, −78.83), FWO (116.75, 175.30, −96.61), KAS (113.23, 175.46, −102.68), and Kissinger (150.95, 173.97, −37.98) kinetic methods.

Table 6. Thermodynamic properties of *C. aesculifolia*.

	Friedman			FWO			KAS		
α	ΔH	ΔG	ΔS	ΔH	ΔG	ΔS	ΔH	ΔG	ΔS
0.10	77.51	177.21	−164.51	70.41	177.67	−176.98	65.51	178.01	−185.63
0.15	145.43	174.16	−47.41	106.00	175.69	−115.01	102.53	175.85	−120.99
0.20	126.56	174.83	−79.66	115.61	175.27	−98.45	112.45	175.40	−103.89
0.25	117.96	175.17	−94.40	124.03	174.93	−84.00	121.16	175.04	−88.91
0.30	131.97	174.62	−70.38	123.41	174.95	−85.05	120.38	175.07	−90.24
0.35	143.84	174.20	−50.11	125.83	174.85	−80.89	122.82	174.97	−86.06
0.40	148.16	174.06	−42.73	132.22	174.61	−69.95	129.44	174.71	−74.71
0.45	142.57	174.24	−52.26	136.60	174.45	−62.46	133.96	174.54	−66.96
0.50	143.23	174.22	−51.13	135.59	174.48	−64.19	132.81	174.58	−68.93
0.55	152.80	173.90	−34.82	130.13	174.68	−73.52	126.97	174.80	−78.92
0.60	114.55	175.30	−100.23	129.97	174.69	−73.79	126.73	174.81	−79.34
0.65	84.12	176.77	−152.89	107.08	175.62	−113.09	102.51	175.83	−120.98
0.70	123.89	174.91	−84.20	80.85	176.95	−158.59	74.72	177.33	−169.31
Average	**127.12**	**174.89**	**−78.83**	**116.75**	**175.30**	**−96.61**	**113.23**	**175.46**	**−102.68**
Kissinger		ΔH			ΔG			ΔS	
		150.95			**173.97**			**−37.98**	

ΔH and ΔG are given in kJ/mol, and ΔS in J/mol. K.

In the process of lignocellulosic biomass pyrolysis, i.e., the generation of compounds such as solids, liquids, and gases, the enthalpy can be defined as the energy that will be required by the biomass for the generation of these products [85]. Table 6 shows that, regardless of the mathematical method applied, enthalpy begins with low values, increases proportionally with the reaction order up to high values, and then decreases. This behavior revealed that the biomass contained compounds that require greater energy to achieve their total transformation. The positive value of enthalpy is another important factor, for it indicates that the thermal degradation process of *C. aesculifolia* involves an endothermic reaction [86]. Recently, similar enthalpy values have been obtained from the biomass of Manilkara zapota seeds, with reports of an average value of 137 kJ/mol [75]. Other studies, one with almond husk and another with acorn pericarp [87], have reported values close to the ones found in our work. Here, we would emphasize that the difference between activation energy (Table 4) and enthalpy (Table 6) is <5 kJ/mol, so we can affirm that the results for the degradation of *C. aesculifolia*, the formation of activated compounds, and the conversion into other compounds will all be favorable [88]. This means that the energy barrier to carry out the pyrolysis process is low, and the formation of pyrolytic products will be favorable.

In lignocellulosic biomass pyrolysis processes, the possibility of generating activated complexes is defined according to the total energy available in the thermal system, i.e., Gibbs free energy (ΔG) [82]. As Table 6 shows, this parameter is positive throughout the range of the degree of advance (α = 0.10–0.70), indicating that the process does not develop directly or automatically but, rather, is oxidative in nature and requires external energy to achieve the pyrolysis reaction [89]. This need for additional energy input may be a disadvantage; however, it is important to highlight that, in pyrolytic biomass processes, the results of entropy (negative) and ΔG > ΔH (Table 6) imply consideration that a small amount of energy is surplus as input to the thermal system [90]. It is also clear that an approximate value of 175 kJ/mol is maintained in all three iso-conversional methods (Friedman, FWO, KAS), as well as for Kissinger's non-iso-conversional method. Recent studies of corncobs and pine wood report approximate ΔG values of 173 kJ/mol [91] and 180 kJ/mol [86], respectively. These values are relatively high for lignocellulosic biomass, which means that pyrolysis of *C. aesculifolia* consumes a large amount of energy.

Theoretically, the entropy (ΔS) of a system represents its molecular disorder and randomness [76]; however, in a pyrolytic system, this function may represent the level of order or disorder of the carbon layers formed in the thermal process [82]. Results for this factor are shown in Table 6, where negative values indicate that the system will undergo only relatively small physicochemical changes; that is, the level of disorder in the thermal process of *C. aesculifolia* is relatively low compared to some of its degraded products [92]. The highest ΔS value was found for the KAS method (−102.68 J/mol. K), while the lowest was found for Kissinger's method (−37.98 J/mol. K). Studies of biomass waste (red pepper) have reported such negative values ranging from −100 J/mol. K to almost −250 J/mol. K over a wide conversion range. [93]. This variability in entropy may be related to the mathematical method applied and to the precise components of the biomass, which can interfere with the thermal process, such as the presence of certain inorganic and alkaline earth compounds, or even the production method of the biomass [94]. However, it has also been determined that such negative entropy values (i.e., reduction in the randomness of the system) as those reported in this study for *C. aesculifolia* can be generated due to physicochemical aging processes, which can lead to a higher thermodynamic equilibrium [95].

2.9. Fourier-Transform Infrared Analysis (FT-IR)

The FT-IR spectra of raw *C. aesculifolia* biomass are portrayed in Figure 6. There, we can identify the characteristic signals that correspond to the various functional groups that constitute the structure of wood, including polysaccharides, lignin, and other low-weight molecular compounds. The absorption band visible at 3354 cm^{-1} corresponds to the stretching vibrations of hydroxyl groups (OH) that are interlaced intermolecularly. These vibrations are related to the main constituents of wood, including cellulose, hemicellulose, lignin, and some proteins [75]. In another study, this functional group was reported for the bark of *C. pentandra* at 3418 cm^{-1} [96]. There are reports in the literature that this functional group is associated with the constitutional water present in the cell walls of wood [97]. Another functional group identified was CH, as has occurred with other species of this genus, like the bark of C. pentandra, where a similar signal to that of *C. aesculifolia* was observed at 2935 cm^{-1} [96]. According to the literature, the signal at 2911 cm^{-1} is associated with symmetric and non-symmetric vibrations, suggesting the presence of groups of aliphatic chains (CH_2, CH_3) derived from the elemental structure of the raw material, such as cellulose, hemicellulose, and lignin [98]. The CH functional group has been found in other studies of hardwood species in the range of 2938–2933 cm^{-1} [99]. Likewise, for species of this genus, there are reports of a vibration in the region around 2918 cm^{-1} in the cellulose of the fruit of *C. pentandra* [100]. Published reports also affirm that the bands centered around 2911 cm^{-1} have been associated with extractable substances located on the surface of the primary cell wall. A study of *C. aesculifolia* seeds found these kinds of substances at 2918 cm^{-1}, while another found them in C. speciosa fruit (also a member of this genus). In that case, the signals at 2900 cm^{-1} were attributed to lipids [101,102].

The signal at 1734 cm^{-1} is associated with stretching vibrations of the C=O type, such as the aldehyde, ketone, and ester functional groups present in cellulose, hemicellulose, and lignin [103]. This finding is similar to those of a study on *C. aesculifolia* seeds, which reported a signal at 1736 cm^{-1} [102], and to those of reports on hardwoods that found this functional group between 1740 and 1730 cm^{-1} [104]. In the absorption band centered at 1594 cm^{-1}, there were stretching vibrations characteristic of double bonds (C=C) in the aromatic skeleton of lignin [105]. A study of *C. aesculifolia* seeds reported this group (C=C) at 1594 cm^{-1}, a result consistent with the value reported here [102]. The functional groups CH_2 and CH_3 have been found in studies of hardwood species at around 1464–1375 cm^{-1} [104]. The signal at 1238 cm^{-1} is also linked to those groups, associated mainly with the CE and lignin in *C. aesculifolia* [106]. This signal has also been associated with the stretching of the C-O bonds in the xylene and syringyl ring of lignin and hemicellulose [107]. A study of cellulose using *C. pentandra* fruit identified C-O at 1249 cm^{-1} [99].

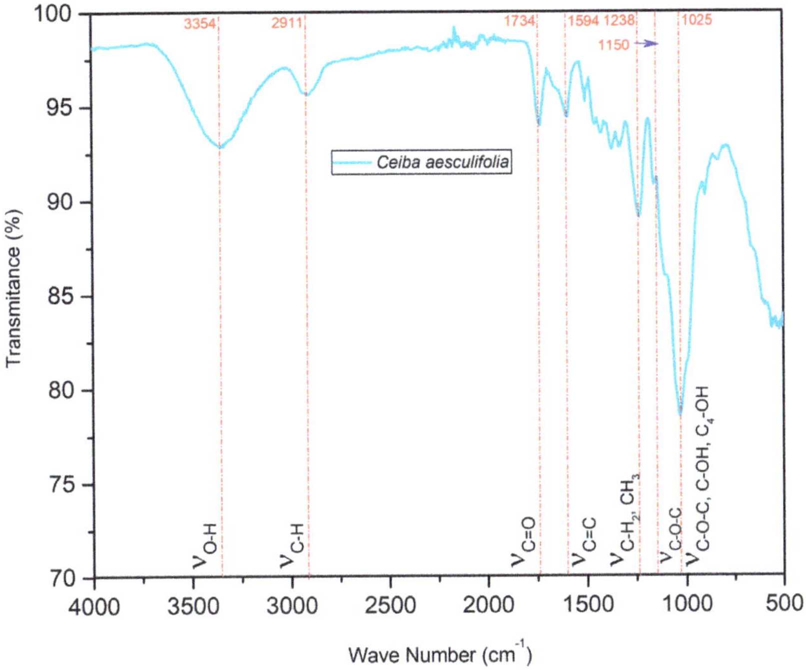

Figure 6. FT-IR spectra of *C. aesculifolia*.

The peaks at 1150 cm^{-1} are assigned to the stretching vibration of the C-O-C bridge of the characteristic esters of cellulose and hemicellulose. One study of *C. aesculifolia* seeds reported this bond at 1157 cm^{-1} [100]. The functional groups C-O-C, C-OH, and C_4-OH were identified in the absorption band at 1025 cm^{-1}, where mainly β-glucopyranose was located in the cellulose [101]. Similar values have been obtained for other hardwood species, including Prosopis laevigata [101]. Some authors affirm that the signal centered at 1025 cm^{-1} corresponds to the symmetric stretching of the C-OH groups in lignin, cellulose, and hemicellulose [108].

There are reports that vibrations around 930 cm^{-1} correspond to glycosidic bonds in cellulose and hemicellulose [109]. A study of cellulose in *C. pentandra* fruit identified this at 900 cm^{-1} [100]. Likewise, the vibration around 725 cm^{-1} can be attributed to the balance of the CH_2 group in crystalline cellulose I [110]. A study of *C. aesculifolia* seeds located this at 710 cm^{-1} [102]. These peaks are characteristic of native cotton [111]. Finally, the signal found with values near 600 cm^{-1} was thought to be due to the deformation of OH groups [112]. A study of cellulose in *C. pentandra* fruit identified this at 614 cm^{-1} [100].

3. Materials and Methods

3.1. Collection Area and Preparation of the Study Materials

As Figure 7 shows, the *C. aesculifolia* waste analyzed was collected in the Lake Cuitzeo basin in the state of Michoacán, Mexico (coordinates in Table 7). Three trees were selected, and 10 cm wide slices were cut at a height of 1.30 m (DAP: diameter at the height of the chest). We then separated the wood and bark of the slices. The wood was splintered manually and left to dry in the shade until it reached moisture equilibrium (approximately 12%). After drying, the splinters were pulverized in a mill (Model K20F, series 236, Micron S.A. de C.V., Mexico City). Finally, the pulverized material was sifted in a Ro-Tap machine (Model RX-29, W.S. Tyler, Mentor, OH, USA), utilizing meshes of 20, 40, and 60. The material from flour mesh 40 (425 μm) was used for the chemical, energy, and thermal characterizations (TGA-DTG) and thermodynamic analysis.

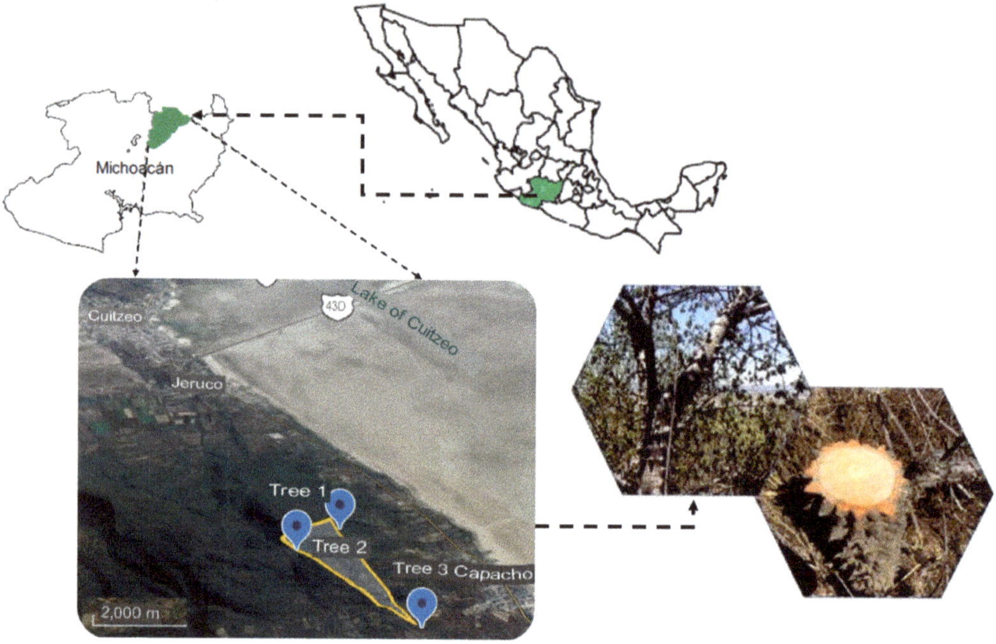

Figure 7. Geographic location of *C. aesculifolia*.

Table 7. Characteristics of *C. aesculifolia* trees sampled.

Trees	Geographic Location	DBH (cm) [1]	Tree Height (m)	SLHA (m) [2]
1	19°58′37″N, 101°5′42″O	16	5.30	1850
2	19°58′13″N, 101°5′38″O	23	8.50	1940
3	19°58′18″N, 101°6′3″O	17	5.40	1920

[1] Diameter at breast height. [2] Height above sea level.

3.2. Determination of Basic Density

Three pieces of *C. aesculifolia* wood were selected from each tree, following the recommendation to take samples that are free of knots and in good condition with a moisture content of 12%. The samples were cut into rectangles to facilitate measuring. A Vernier caliper was used to precisely measure the length, width, and thickness of each sample in order to calculate their volume [113]. Mass was calculated on an Ohaus PA214 scale at a precision of 0.0001 g, and density was determined (g/cm^3).

3.3. Basic Chemical Composition

The chemical composition was verified according to well-known methodologies [114]. The main polymeric components of the wood (CE, HM, lignin) were determined by fiber analysis using Van Soest's gravimetric method with α-amylase. Extractable substances were calculated by subtracting the percentages of CE, HM, lignin, and ash at 100% application, as well as a correction factor for ash as in earlier studies [115]. The equipment utilized was an ANKOM fiber analyzer (model AMKON200, ANKOM Technology, Macedon, NY, USA).

3.4. Proximal Analysis

The moisture content of each dry sample was determined in triplicate based on the UNE-EN ISO18134-1 standard [116]. The percentage of ash was calculated according to the UNE-EN 14775 standard [117], and the volatile material content was determined in accordance with ASTM E 872-82 [118]. As Equation (1) shows, fixed carbon [119] was calculated by the differences of the means. Average values and standard deviations are reported.

$$\%Fixed\ carbon = 100 - (\%volatile\ matter + \%ash) \tag{1}$$

3.5. Microanalysis of Ash

This analysis was conducted by inductively coupled plasma optical emission spectrophotometry (VARIAN 730-ES, Varian Inc., (Agilent), Mulgrave, Australia), following well-established procedures [120].

3.6. Elemental Analysis

This analysis encompassed determining the primary components of the biomass, such as carbon, hydrogen, oxygen, nitrogen, and sulfur (CHONS). Calculations of C, H, N, and S were carried out in triplicate using a COSTECH elemental analyzer (Model 4010; COSTECH International S.P.A., Milan, Italy), in strict adherence to the UNE-CEN/TS 15104 EX standard [121]. Oxygen content was calculated from the difference in each analysis performed.

3.7. Higher Heating Value (HHV)

The higher heating value was determined in triplicate according to the UNE-EN 14918 standard [122], using a semiautomatic calorimeter (LECO AC600, LECCO Corporation, St. Joseph, MO, USA). The calorimetric pump was calibrated with benzoic acid. Mean values and standard deviations are reported.

3.8. Fuel Value Index

It is necessary to evaluate lignocellulosic material destined for use as fuel in terms of its suitability for bioenergy production. In this regard, it is important to mention that the fuel value index is a measure used to categorize the biomass by taking its heating value and density as positive parameters, and ash as a negative one. This analysis was performed in triplicate. Average values and standard deviations are reported. This parameter was also determined using well-established methods according to Equation (2) [123].

$$Fuel\ value\ index = \frac{(Higher\ heating\ value,\ MJ/kg)(basic\ density,\ g/cm^3)}{ash\ content\ (\%)} \tag{2}$$

3.9. Thermogravimetric (TGA-DTG), Thermodynamic, and FT-IR Analyses

3.9.1. TGA-DTG Analysis

The thermal analysis (TGA-DTG) of *C. aesculifolia* wood (Kunth) Britt and Baker was carried out with an STA 6000 device (Perkin Elmer Inc., Wellesley, MA, USA). Pyrolytic conditions were maintained using ultrahigh-purity nitrogen gas (99.99% pure) acquired from INFRA, Mexico. This grade of purity was required to minimize possible mass transfer.

The nitrogen flow was kept at 30 mL/min. Approximately 30 ± 5 mg of the sample was placed uniformly in an aluminum crucible for each assay, heated through a programmed range of 25–900 °C, and kept at this temperature for 10 min, before being allowed to cool at a rate of 30 °C/min. Figure 8 depicts this process. It is important to note that the kinetic and thermodynamic analyses of *C. aesculifolia* waste required considering as many as five heating rates (β = 10–30 °C/min; 283–303 K/min). To minimize error and achieve greater precision, all tests were performed in triplicate. The heating ramps were also carried out in triplicate when the conversional difference was >5%, or when the data contained excessive background noise. OriginPro Graphing and Analysis software 2016 was used to graph the data.

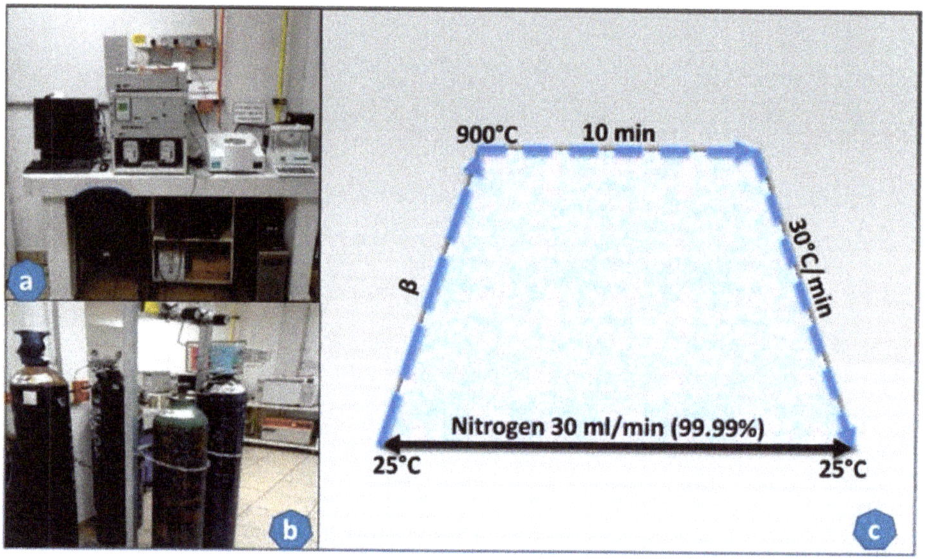

Figure 8. Equipment for thermogravimetric analysis (**a**), nitrogen tanks used in the experiment (**b**) and heating ramp used (**c**).

3.9.2. Theory of Kinetic Analysis

According to the iso-conversion principle, the methods used for kinetic analyses are based on a constant conversional value, such that the reaction velocity is a function of temperature [121,122]. The pyrolysis process of solid biomass can be simplified as a one-step global process with a first-order reaction, as depicted in Figure 9.

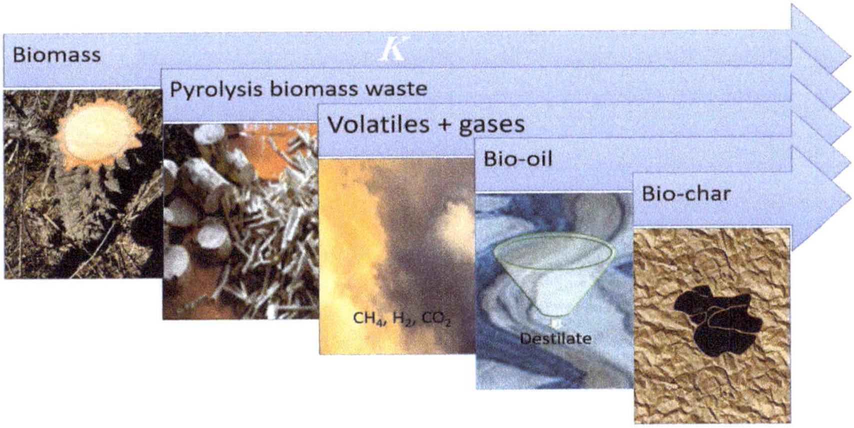

Figure 9. General biomass pyrolysis process.

According to Figure 9, volatiles are considered to be indicative of the gas generated, while the variable *k* refers to the velocity constant of the pyrolysis process. The conversion degree in each fraction, denoted as α, in any temperature range, is defined as the mass conversion in the pyrolytic reaction. It can be defined as in Equation (3):

$$\alpha = (m_i - m_t)/\left(m_i - m_f\right) \tag{3}$$

where m_i, m_t, and m_f represent the initial, instant, and final mass of the sample, respectively.

Parallel to this, the constant of velocity (*k*) is directly related to the temperature (*T*) of the entire pyrolytic process. As Equation (4) shows, Arrhenius's ratio makes it possible to analyze this relation:

$$k(T) = Aexp^{\left(\frac{-E_a}{RT}\right)} \tag{4}$$

This equation includes several variables of vital importance for analyses of thermogravimetric processes. The indicators *A*, *Ea*, *R*, and *T* represent the pre-exponential factor (s^{-1}), the apparent activation energy (kJ/mol), the universal constant of ideal gases (8.314 J/Kmol), and the absolute temperature (Kelvin), respectively.

Equation (4) can be considered together with a reaction method (*fα*) in order to determine the ratio of change in the degree of mass conversion with respect to time (*dα/dt*), as shown here:

$$\frac{d\alpha}{dt} = K(T) \times f(\alpha) \tag{5}$$

By substituting the value of *k(T)* from Equation (4) into Equation (5), we obtained the fundamental expression of the analytical methods applied to study the kinetic parameters in relation to the results of the thermogravimetric analysis:

$$\frac{d\alpha}{dt} = Aexp^{\left(\frac{-E_A}{RT}\right) \times f(\alpha)} \tag{6}$$

Non-isothermal processes generally consider a heating rate (*dT/dt*), denoted as β. Upon including this variable in Equation (6) and simplifying it, we obtained

$$\frac{d\alpha}{dt} = \frac{A}{\beta} \cdot exp^{\left(\frac{-E_A}{RT}\right) \times f(\alpha)} \tag{7}$$

The solution of this expression can be approached using methods like those of Doyle [123], Agrawal, Gorbatchev, and Frank-Kamenetskii [124], which permit calculating the principal kinetic parameters involved in the thermal process. The resulting equation is

$$g(\alpha) = \int_0^{\alpha} \frac{d\alpha}{f(\alpha)} = \frac{A}{\beta} \int_{T_0}^{T} exp^{\left(\frac{-E_a}{RT}\right)} dT \tag{8}$$

Determination of the Kinetic Parameters

Since a complex physicochemical interaction of mass–heat transfer occurs during pyrolysis, it is necessary to optimize the energy and/or reduce contaminants in the industrial equipment used in this process of thermally transforming lignocellulosic material like the biomass analyzed. This means that a study of the kinetics involved must be conducted first to analyze the main parameters: activation energy, the pre-exponential factor, and the reaction order. Figure 10 shows some of the mathematical methods most often employed to calculate the first two parameters—the same ones that were applied in this study.

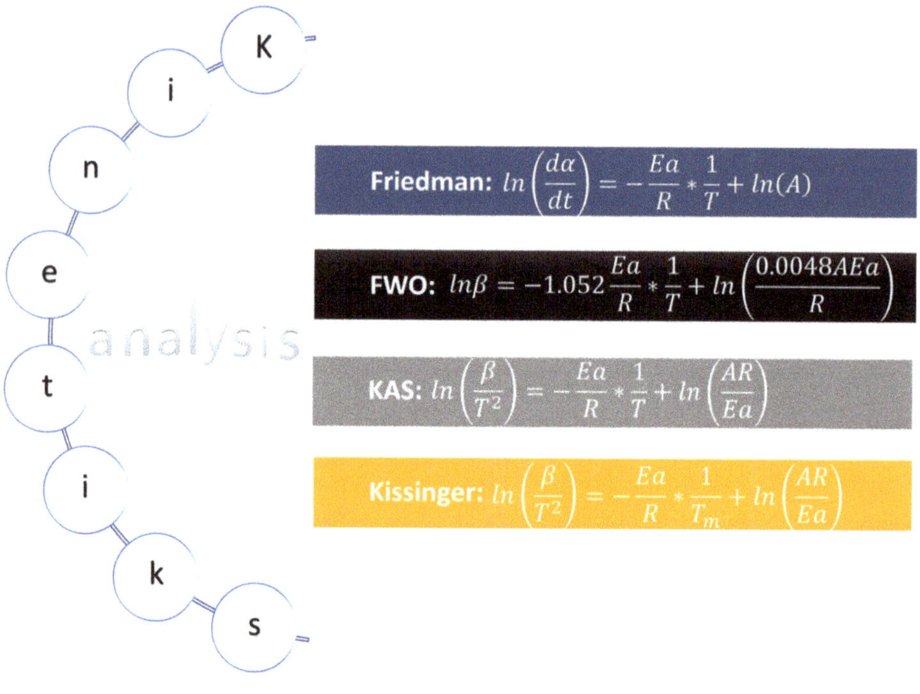

Figure 10. Mathematical method to determine the kinetic parameters of *C. aesculifolia*.

To calculate the reaction order, we began with Avrami's equation [124]. The ratio required for this parameter is outlined in the following procedure:

Avrami's method: $\alpha = 1 - e^{\left(\frac{-k(T)}{\beta^n}\right)}$

Transposing and applying logarithms:

$$-ln(1-\alpha) = -\left[\left(\frac{-k(T)}{\beta^n}\right)\right]$$
$$-ln(1-\alpha) = -\left[\frac{-k(T)}{\beta^n}\right]$$
$$-ln(1-\alpha) = \left[\frac{k(T)}{\beta^n}\right]$$

Applying logarithms again: $ln[-ln(1-\alpha)] = ln\left(\frac{k(T)}{\beta^n}\right)$

Applying logarithmic properties: $ln[-ln(1-\alpha)] = lnk(T) - ln\beta^n$

Substituting $k(T)$ according to Equation (3): $ln[-ln(1-\alpha)] = ln\left[Ae^{\left(-\frac{Ea}{RT}\right)}\right] - ln\beta^n$

Applying logarithmic properties:

$$ln[-ln(1-\alpha)] = lnA + \left(-\frac{Ea}{RT}\right) - ln\beta^n$$
$$ln[-ln(1-\alpha)] = lnA - \frac{Ea}{RT} - ln\beta^n$$

Equation that calculates the reaction order: $ln[-ln(1-\alpha)] = lnA - \frac{Ea}{RT} - nln\beta$

The graph is obtained in analogy to the equation $y = mx + b$, where

$$y = ln[-ln(1-\alpha)]$$
$$m = -n$$
$$x = ln\beta$$
$$b = lnA - \frac{Ea}{RT}$$

Hence, the value for the reaction order (n) is given by the slope (m) calculated on the y vs. x graph.

3.9.3. Determination of the Thermodynamic Parameters

Once Ea was determined by the Friedman, FWO, KAS, and Kissinger methods, we were able to calculate the three thermodynamic parameters enthalpy (ΔH), Gibbs free energy (ΔG), and entropy (ΔS), at a low heating rate of 15 °C/min. In this stage, we also calculated the pre-exponential factor (A) considering, first, the Ea value obtained and, second, the heating rate. Figure 11 shows the equations utilized.

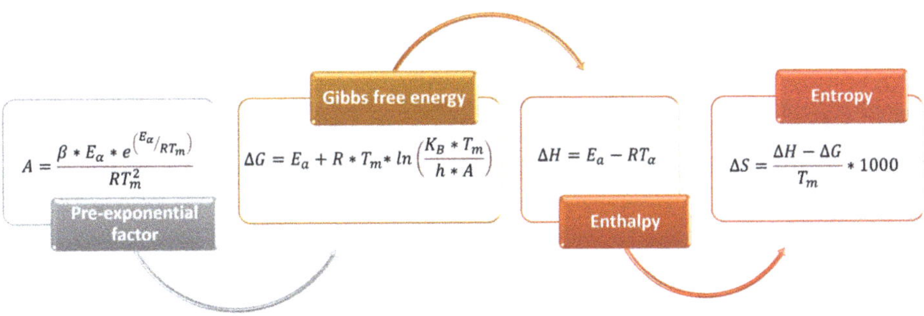

Figure 11. Equations for the determination of thermodynamic parameters, where E_a is the activation energy, R is the ideal gas constant, T_α is the temperature according to the degree of conversion, T_m is the temperature at the maximum peak of the DTG, K_B is the Boltzmann constant (1.381E-23 J/K), h is Planck's constant (6.626E-34 J.s), and β is the heating rate (15 °C/min).

3.9.4. Fourier-Transform Infrared Analysis (FT-IR)

The FT-IR spectrum was obtained from the raw biomass. The functional groups of *C. aesculifolia* waste were analyzed by FT-IR in a Perkin Elmer ATR model 400 spectrometer. Spectra were obtained in 16 scans per sample in a range of 4000–500 cm^{-1}, at a resolution of 4 cm^{-1}. The samples selected were mixed with potassium bromide (KBr) and then dried in an oven for 24 h.

4. Conclusions

The waste of *Ceiba aesculifolia* wood constitutes an abundant renewable energy source. This study was designed to determine the thermochemical characterization of this biomass during pyrolysis. The results for moisture (9.37%), ash (1.84%), volatile material content (83.83%), and HHV (19.12%) demonstrate that it is feasible to consider this biomass an excellent candidate with high energy potential. The TGA analysis indicated that the greatest degradation, around 60%, occurred in the temperature range of 415–680 K. This is the first study to conduct a kinetic analysis of the thermogravimetric process of *C. aesculifolia* waste with a view to determining the kinetic triplet using three iso-conversional methods and one non-iso-conversional approach. The average values estimated for activation energy by the Friedman, FWO, KAS, and Kissinger methods were 132.03, 121.65, 118.14, and 155.85 kJ/mol, respectively. The Friedman, FWO, and KAS methods showed variations throughout pyrolysis, but Kissinger's method showed no changes. According to Avrami's equation, the reaction order presented a variation of 0.3937–0.6141 across the entire temperature range (500–650 K). It is important to emphasize that a high correlation coefficient was obtained (R^2 = approximately 1) in all of the graphs of the mathematical methods applied, a result that is deemed preferable in kinetic analyses. The value for the pre-exponential factor varied from 10^3 to 10^{11} s^{-1}, indicating that pyrolysis of *C. aesculifolia* waste transpires at a high reaction velocity. The thermodynamic analysis showed that the pyrolysis was endothermic, since the difference between the apparent activation energy and enthalpy was <5 kJ/mol. We conclude that this process is not totally spontaneous, since the Gibbs free energy and entropy were positive and negative, respectively. The principal vibrations detected by FT-IR were O-H, C-H, C=O, C=C, CH_2, CH_3, C-O-C, C-OH, and C_4-OH. Based on these findings, we can affirm that it is possible to optimize the times, temperatures, and heating rates of *C. aesculifolia* waste during pyrolysis to generate bioenergy.

Author Contributions: Conceptualization, J.J.A.F., J.V.A.V., L.F.P.I., and F.D.M.Z.; methodology, J.J.A.F., J.V.A.V., L.F.P.I., F.D.M.Z., and J.G.R.Q.; software, L.F.P.I., and J.J.A.F.; validation, J.J.A.F., J.V.A.V., M.L.Á.R., and L.F.P.I.; formal analysis, L.F.P.I., J.J.A.F., and J.V.A.V.; investigation, J.J.A.F., L.F.P.I., J.V.A.V., M.L.Á.R., and J.G.R.Q.; resources, J.J.A.F., J.V.A.V., M.L.Á.R., L.F.P.I., J.G.R.Q., and F.D.M.Z.; data curation, J.J.A.F., J.V.A.V., M.L.Á.R., and F.D.M.Z.; writing—original draft preparation, L.F.P.I., J.J.A.F., and F.D.M.Z.; writing—review and editing, J.J.A.F., J.V.A.V., M.L.Á.R., L.F.P.I., J.G.R.Q., and F.D.M.Z.; visualization, L.F.P.I., and J.J.A.F.; supervision, J.J.A.F., L.F.P.I., J.V.A.V., M.L.Á.R., and J.G.R.Q.; project administration, L.F.P.I., J.J.A.F., J.G.R.Q., and J.V.A.V.; funding acquisition, J.J.A.F., L.F.P.I., J.V.A.V., M.L.Á.R., and F.D.M.Z. All authors have read and agreed to the published version of the manuscript.

Funding: The authors thank the Coordination of Scientific Investigation of the Universidad Michoacana de San Nicolás de Hidalgo (UMSNH) for financing project JJAF-CIC-UMSNH-2023-2024. The author F.D.M.-Z. adds his thanks to the CONAHCYT scholarship for the funding provided during his studies in the master's degree program in the Science and Technology of Wood, also at the UMSNH.

Institutional Review Board Statement: Not applicable.

Informed Consent Statement: Not applicable.

Data Availability Statement: All data supporting the reported results are available upon request from the corresponding author.

Conflicts of Interest: The authors declare no conflicts of interest.

References

1. Food and Agriculture Organization of the United Nations (FAO). FAOSTAT Forestry Production and Trade. 2022. Available online: https://www.fao.org/faostat/es/#data/FO/visualize (accessed on 1 December 2023).
2. Jiménez-Mendoza, M.E.; Ruiz-Aquino, F.; Aquino-Vásquez, C.; Santiago-García, W.; Santiago-Juárez, W.; Rutiaga-Quiñones, J.G. Use of firewood in a community in the Southern Sierra of Oaxaca, Mexico. *Rev. Mex. Cienc. For.* **2023**, *14*, 22–49. [CrossRef]
3. González, J.R.Q.; González, L.E.Q. Perspectivas del potencial energético de la biomasa en el marco global y latinoamericano. *Gestión Y Ambiente* **2015**, *18*, 179–188.
4. Shamsi, M.; Obaid, A.; Farokhi, S.; Bayat, A. A novel process simulation model for hydrogen production via reforming of biomass gasification tar. *Int. J. Hydrogen Energy* **2022**, *47*, 772–781. [CrossRef]
5. Li, B.; Mbeugang, C.; Liu, D.; Zhang, S.; Wang, S.; Wang, Q.; Xu, Z.; Hu, X. Simulation of sorption enhanced staged gasification of biomass for hydrogen production in the presence of calcium oxide. *Int. J. Hydrogen Energy* **2020**, *45*, 26855–26864. [CrossRef]
6. Jha, S.; Nanda, S.; Acharya, B.; Dalai, A. A review of thermochemical conversion of waste biomass to biofuels. *Energies* **2022**, *15*, 6352. [CrossRef]

7. Ding, Z.; Li, W.; Chen, Z.; Wang, L.; Huang, S.; Evrendilek, F.; Yang, C.; Cai, H.; Zhong, S.; Yang, Z.; et al. Microplastics as emerging contaminants in textile dyeing sludge: Their impacts on co-combustion/pyrolysis products, residual metals, and temperature dependency of emissions. *J. Hazard. Mater.* **2024**, *466*, 133465. [CrossRef]
8. Demirbaş, A. Biomass resource facilities and biomass conversion processing for fuels and chemicals. *Energy Convers Manag.* **2001**, *42*, 1357–1378. [CrossRef]
9. Demirbas, A.; Arin, G. Hydrogen from biomass via pyrolysis: Relationships between yield of hydrogen and temperature. *Energy Sources* **2004**, *26*, 1061–1069. [CrossRef]
10. Duman, G.; Uddin, M.; Yanik, J. Hydrogen production from algal biomass via steam gasification. *Bioresour. Technol.* **2014**, *166*, 24–30. [CrossRef]
11. Morales-Máximo, M.; García, C.A.; Pintor-Ibarra, L.F.; Alvarado-Flores, J.J.; Velázquez-Martí, B.; Rutiaga-Quiñones, J.G. Evaluation and characterization of timber residues of *Pinus* spp. as an energy resource for the production of solid biofuels in an indigenous community in Mexico. *Forests* **2021**, *12*, 977. [CrossRef]
12. Demirbas, T.; Demirbas, C. Fuel properties of wood species. *Energy Sources Part A Recovery Util. Environ. Eff.* **2009**, *31*, 1464–1472. [CrossRef]
13. Kumar, R.; Pandey, K.K.; Chandrashekar, N.; Mohan, S. Effect of tree-age on calorific value and other fuel properties of *Eucalyptus hybrid*. *J. For. Res.* **2010**, *21*, 514–516. [CrossRef]
14. Lincoln, W.A. *World Woods in Colour*; Linden Pub.: Egestorf, Germany, 1986; pp. 13–14.
15. Ruiz-Aquino, F.; Jiménez-Mendoza, M.E.; Santiago-García, W.; Suárez-Mota, M.E.; Aquino-Vásquez, C.; Rutiaga-Quiñones, J.G. Energy properties of 22 timber species from Oaxaca, Mexico. *S. E. Eur. For. SEEFOR* **2022**, *13*, 107–113. [CrossRef]
16. Sjöström, E. *Wood Chemistry: Fundamentals and Applications*; Academic Press, Inc.: London, UK, 1981; pp. 87–102.
17. Fengel, D.; Wegener, G. (Eds.) *Wood: Chemistry, Ultrastructure, Reactions*; Walter de Gruyter: Berlin, Germany, 2011.
18. Ruiz-Aquino, F.; Bautista, L.L.; Bautista, A.E.L.; Santiago-García, W.; Pintor-Ibarra, L.F.; Rutiaga-Quiñones, J.G. Anatomical characterization, physical, and chemical properties of wood of *Quercus macdougallii* Martínez, endemic species of the Sierra Juárez of Oaxaca, Mexico. *BioResources* **2020**, *15*, 5975–5998. [CrossRef]
19. Gutiérrez-Acosta, J.M.; Orihuela-Equihua, R.; Pintor-Ibarra, L.F.; González-Ortega, N.; Hernández-Solís, J.J.; Ruíz-Aquino, F.; Rutiaga-Quiñones, J.G. On the basic chemical composition of selected biomass types from four regions of Mexico, for bioenergetic purposes. *BioResources* **2021**, *16*, 5694–5705. [CrossRef]
20. Honorato-Salazar, J.A.; Colotl-Hernández, G.; Apolinar-Hidalgo, F.; Aburto, J. Principales componentes químicos de la madera de *Ceiba pentandra, Hevea brasiliensis* y *Ochroma pyramidale*. *Madera Y Bosques* **2015**, *21*, 131–146. [CrossRef]
21. Salazar-Herrera, F.; Pintor-Ibarra, L.F.; Musule, R.; Nava-Berumen, C.A.; Alvarado-Flores, J.J.; González-Ortega, N.; Rutiaga-Quiñones, J.G. Chemical and energetic properties of seven species of the Fabaceae family. *S. E. Eur. For. SEEFOR* **2023**, *14*, 215–224. [CrossRef]
22. Kruyeniski, J. Influencia del Pretratamiento de Residuos Foresto Industriales Sobre la Producción de Bioetanol. Ph.D. Thesis, Presentada a la Facultad de Ciencias Exactas, Químicas y Naturales de la Universidad Nacional de Misiones, Posadas, Argentina, 2018.
23. Pintor-Ibarra, L.F.; Rivera-Prado, J.J.; Ngangyo-Heya, M.; Rutiaga-Quiñones, J.G. Evaluation of the chemical components of *Eichhornia crassipes* as an alternative raw material for pulp and paper. *BioResources* **2018**, *13*, 2800–2813. [CrossRef]
24. Ruiz-Aquino, F.; Ruiz-Ángel, S.; Feria-Reyes, R.; Santiago-García, W.; Suárez-Mota, M.E.; Rutiaga-Quiñones, J.G. Wood chemical composition of five tree species from Oaxaca, Mexico. *BioResources* **2019**, *14*, 9826–9839. [CrossRef]
25. Díaz-Oviedo, A.F.; Ramón-Valencia, B.A.; Moreno-Contreras, G.G. Caracterización físico-química de la cáscara de mazorca de cacao como posible uso en la elaboración de tableros aglomerados. *Rev. Investig. Desarro. E Innovación* **2022**, *12*, 97–106. [CrossRef]
26. Acosta, R.A.; Arango, J.A.; Prieto, A. *Una Revisión de Los Tratamientos Térmicos Aplicados a la Madera*; CIBEN: Lisboa, Portugal, 2017; pp. 23–26.
27. Ngangyo-Heya, M.; Foroughbahchk-Pournavab, R.; Carrillo-Parra, A.; Rutiaga-Quiñones, J.G.; Zelinski, V.; Pintor-Ibarra, L.F. Calorific value and chemical composition of five semi-arid Mexican tree species. *Forests* **2016**, *7*, 58. [CrossRef]
28. Pintor-Ibarra, L.F.; Carrillo-Parra, A.; Herrera-Bucio, R.; López-Albarrán, P.; Rutiaga-Quiñones, J.G. Physical and chemical properties of timber byproducts from *Pinus leiophylla, P. montezumae* and *P. pseudostrobus* for a bioenergetic use. *Wood Res.* **2017**, *62*, 849–861.
29. Sjostrom, E. *Wood Chemistry: Fundamentals and Applications*; Elsevier: Amsterdam, The Netherlands, 2013.
30. Ávila, L.E.; Herrera, M.A. Efecto de los extraíbles en tres propiedades físicas de la madera de *Enterolobium cyclocarpum* procedente de Michoacán, México. *Bosque* **2012**, *33*, 227–232. [CrossRef]
31. Inegbedion, F. Estimation of the moisture content, volatile matter, ash content, fixed carbon and calorific values of saw dust briquettes. *MANAS J. Eng.* **2022**, *10*, 17–20. [CrossRef]
32. Rutiaga-Quiñones, J.G.; Pintor-Ibarra, L.F.; Orihuela-Equihua, R.; Gonzalez-Ortega, N.; Ramírez-Ramírez, M.A.; Carrillo-Parra, A.; Lujan-Alvarez, C. Characterization of Mexican waste biomass relative to energy generation. *BioResources* **2020**, *15*, 8529–8553. [CrossRef]
33. Morales-Máximo, C.N.; López-Sosa, L.B.; Rutiaga-Quiñones, J.G.; Corral-Huacuz, J.C.; Aguilera-Mandujano, A.; Pintor-Ibarra, L.F.; Morales-Máximo, M. Characterization of agricultural residues of Zea mays for their application as solid biofuel: Case study in San Francisco Pichátaro, Michoacán, México. *Energies* **2022**, *15*, 6870. [CrossRef]

34. Obernberger, I.; Thek, G. Physical characterisation and chemical composition of densified biomass fuels with regard to their combustion behaviour. *Biomass Bioenergy* **2004**, *27*, 653–669. [CrossRef]
35. Velázquez-Martí, B. *Aprovechamiento de la Biomasa para uso Energético*, 2nd ed.; Editorial Reverté: Barcelona, Spain, 2018; pp. 25–30.
36. Garcia-Maraver, A.; Rodriguez, M.L.; Serrano-Bernardo, F.; Díaz, L.F.; Zamorano, M. Factors affecting the quality of pellets made from residual biomass of olive trees. *Fuel Process. Technol.* **2015**, *129*, 1–7. [CrossRef]
37. Holt, G.A.; Blodgett, T.L.; Nakayama, F.S. Physical and combustion characteristics of pellet fuel from cotton gin by-products produced by select processing treatments. *Ind. Crops Prod.* **2006**, *24*, 204–213. [CrossRef]
38. Ríos-Saucedo, J.C.; Rosales-Serna, R.; Jiménez-Ocampo, R.; Domínguez-Martínez, P.A.; Carrillo-Parra, A.; Valenzuela-Nuñez, L.M. Calidad de pélets a partir de biomasa de ocho especies dendroenergéticas de crecimiento rápido. *Agrociencia* **2021**, *55*, 557–568. [CrossRef]
39. David, E.; Kopac, J.; Armeanu, A.; Niculescu, V.; Sandru, C.; Badescu, V. Biomass-alternative renewable energy source and its conversion for hydrogen rich gas production. In *E3S Web of Conferences*; EDP Sciences: Les Ulis, France, 2019; Volume 122, p. 01001.
40. Tenorio, C.; Moya, R.; Tomazello-Filho, M.; Valaert, J. Quality of pellets made from agricultural and forestry crops in Costa Rican tropical climates. *BioResources* **2015**, *10*, 482–498. [CrossRef]
41. Bandara, W.; Kowshayini, P. Evaluation of the performances of biomass briquettes produced with invasive *Eichhornia crassipes* (water hyacinth), wood residues, and cow dung for small and medium scale industries. *J. Fundam. Renew. Energy Appl.* **2018**, *8*, 1. [CrossRef]
42. Smołka-Danielowska, D.; Jabłońska, M. Chemical and mineral composition of ashes from wood biomass combustion in domestic wood-fired furnaces. *Int. J. Environ. Sci. Technol.* **2022**, *19*, 5359–5372. [CrossRef]
43. Onokwai, A.O.; Ajisegiri, E.S.A.; Okokpujie, I.P.; Ibikunle, R.A.; Oki, M.; Dirisu, J.O. Characterization of lignocellulose biomass based on proximate, ultimate, structural composition, and thermal analysis. *Mater. Today Proc.* **2022**, *65*, 2156–2162. [CrossRef]
44. Obernberger, I.; Thek, G. *The Pellet Handbook*, 1st ed.; Earthscan: London, UK; Washington, DC, USA, 2010.
45. Villarroel Bastidas, J.; Macías Vera, J. Biocombustible sólido a partir de residuos que generan los procesos agroindustriales del sector El Empalme. *Ing. E Innov.* **2021**, 8. [CrossRef]
46. Görling, M.; Larsson, M.; Alvfors, P. Bio-methane via fast pyrolysis of biomass. *Appl. Energy* **2013**, *112*, 440–447. [CrossRef]
47. Demirbas, A. Hydrogen-rich gaseous products from tea waste by pyrolysis. *Energy Sources* **2001**, *23*, 739–746. [CrossRef]
48. Mishra, R.; Mohanty, K. Pyrolysis kinetics and thermal behavior of waste sawdust biomass using thermogravimetric analysis. *Bioresour. Technol.* **2018**, *251*, 63–74. [CrossRef] [PubMed]
49. Telmo, C.; Louzada, K.; Moreira, N. Proximate analysis, backwards stepwise regression between gross caloric value, ultimate and chemical analysis of wood. *Bioresour. Technol.* **2010**, *101*, 3808–3815. [CrossRef]
50. De la Cruz Montelongo, C.; Herrera Gamboa, J.; Ortiz Sánchez, I.A.; Ríos Saucedo, J.C.; Rosales Serna, R.; Carrillo-Parra, A. Caracterización energética del carbón vegetal producido en el Norte-Centro de México. *Madera Y Bosques* **2020**, *26*, e2621971. [CrossRef]
51. Ugwu, E.I.; Agunwamba, J.C. A review on the applicability of activated carbon derived from plant biomass in adsorption of chromium, copper, and zinc from industrial wastewater. *Environ. Monit. Assess.* **2020**, *192*, 240. [CrossRef] [PubMed]
52. Tripathi, M.; Sahu, J.N.; Ganesan, P. Effect of process parameters on production of biochar from biomass waste through pyrolysis: A review. *Renew. Sustain. Energy Rev.* **2016**, *55*, 467–481. [CrossRef]
53. Mbougueng, P.; Sachindra, N.; Nodem, F.; Ngoune, L. Characterization of volatile compounds of liquid smoke flavourings from some tropical hardwoods. *Sci. Afr.* **2020**, *8*, e00443. [CrossRef]
54. Díez, D.; Urueña, A.; Piñero, R.; Barrio, A.; Tamminen, T. Determination of hemicellulose, cellulose, and lignin content in different types of biomasses by thermogravimetric analysis and pseudocomponent kinetic model (TGA-PKM method). *Processes* **2020**, *8*, 1048. [CrossRef]
55. Subramanian, S.; Ragula, U.B.R. Pyrolysis kinetics of *Hibiscus rosa sinensis* and *Nerium oleander*. *Biofuels* **2018**, *11*, 903–917. [CrossRef]
56. Stelte, W.; Sanadi, A.R. Preparation and characterization of cellulose nanofibers from two commercial hardwood and softwood pulps. *Ind. Eng. Chem. Res.* **2009**, *48*, 11211–11219. [CrossRef]
57. Toscano, G.; Duca, D.; Rossini, G.; Mengarelli, C.; Pizzi, A. Identification of different woody biomass for energy purpose by means of soft independent modeling of class analogy applied to thermogravimetric analysis. *Energy* **2015**, *83*, 351–357. [CrossRef]
58. Pasangulapati, V.; Ramachandriya, K.D.; Kumar, A.; Wilkins, M.R.; Jones, C.L.; Huhnke, R.L. Effects of cellulose, hemicellulose and lignin on thermochemical conversion characteristics of the selected biomass. *Bioresour. Technol.* **2012**, *114*, 663–669. [CrossRef]
59. Ceylan, S.; Topçu, Y. Pyrolysis kinetics of hazelnut husk using thermogravimetric analysis. *Bioresour. Technol.* **2014**, *156*, 182–188. [CrossRef]
60. Manić, N.; Janković, B.; Dodevski, V. Model-free and model-based kinetic analysis of Poplar fluff (*Populus alba*) pyrolysis process under dynamic conditions. *J. Therm. Anal. Calorim.* **2021**, *143*, 3419–3438. [CrossRef]
61. Khiari, B.; Jeguirim, M. Pyrolysis of grape marc from Tunisian wine industry: Feedstock characterization, thermal degradation and kinetic analysis. *Energies* **2018**, *11*, 730. [CrossRef]

62. Nawaz, A.; Singh, B.; Mishra, R.K.; Kumar, P. Pyrolysis of low-value waste Trapa natans peels: An exploration of thermal decomposition characteristics, kinetic behaviour, and pyrolytic liquid product. *Sustain. Energy Technol. Assess.* **2023**, *56*, 103128. [CrossRef]
63. Roslee, A.N.; Munajat, N.F. Comparative study on pyrolysis behavior and kinetics of two macroalgae biomass (*Ulva* cf. *flexuosa and Hy. edulis) using thermogravimetric analysis. J. Teknol.* **2018**, *80*, 123–130. [CrossRef]
64. Brown, M.E.; Maciejewski, M.; Vyazovkin, S.; Nomen, R.; Sempere, J.; Burnham, A.A.; Opfermann, J.; Strey, R.; Anderson, H.L.; Kemmler, A.; et al. Computational aspects of kinetic analysis: Part A: The ICTAC kinetics project—Data, methods and results. *Thermochim. Acta* **2000**, *355*, 125–143. [CrossRef]
65. Hernowo, P.; Steven, S.; Restiawaty, E.; Bindar, Y. Nature of mathematical model in lignocellulosic biomass pyrolysis process kinetic using volatile state approach. *J. Taiwan Inst. Chem. Eng.* **2022**, *139*, 104520. [CrossRef]
66. Lopez-Velazquez, M.; Santes, V.; Balmaseda, J.; Torres-Garcia, E. Pyrolysis of orange waste: A thermo-kinetic study. *J. Anal. Appl. Pyrolysis* **2013**, *99*, 170–177. [CrossRef]
67. Slopiecka, K.; Bartocci, P.; Fantozzi, F. Thermogravimetric analysis and kinetic study of poplar wood pyrolysis. *Appl. Energy* **2012**, *97*, 491–497. [CrossRef]
68. Di Blasi, C. Modeling intra- and extra-particle processes of wood fast pyrolysis. *AIChE J.* **2002**, *48*, 2386–2397. [CrossRef]
69. Ashwini, K.; Resmi, R.; Reghu, R. Pyrolysis characteristics and kinetic analysis of coconut shell and nutmeg shell for potential source of bioenergy. *Eng. Sci. Technol. Int. J.* **2024**, *50*, 101615. [CrossRef]
70. Suresh Kumar, P.; Edwin, M.; Bensam Raj, J. Pyrolysis characteristics and kinetic analysis of Indian almond fruit biomass by thermogravimetric analysis for the potential source of bioenergy. *Proc. Inst. Mech. Eng. Part E J. Process Mech. Eng.* **2023**, *237*, 290–299. [CrossRef]
71. Colpani, D.; Santos, V.O.; Araujo, R.O.; Lima, V.M.; Tenório, J.A.; Coleti, J.; Chaar, J.S.; de Souza, L.K. Bioenergy potential analysis of Brazil nut biomass residues through pyrolysis: Gas emission, kinetics, and thermodynamic parameters. *Clean. Chem. Eng.* **2022**, *1*, 100002. [CrossRef]
72. Katnić, Đ.; Marinović-Cincović, M.; Porobić, S.J.; Vujčić, I.; Šaponjić, A.; Sikirić, B.; Živojinović, D. Characterization and kinetics of thermal decomposition behavior of plum and fig pomace biomass. *J. Clean. Prod.* **2022**, *352*, 131637. [CrossRef]
73. Espinoza-Montero, P.J.; Méndez-Durazno, C.; Robles Carrillo, N.M.; Ramírez, V.; Chico-Proano, A.; Debut, A. Bioenergy production from Ecuadorian Lignocellulosic Biomass: Comprehensive Characterization, Thermal Behavior, and Pyrolysis Kinetics. *Therm. Behav. Pyrolysis Kinet.* **2024**, *847*, 157521. [CrossRef]
74. Mishra, R.K.; Mohanty, K. Kinetic analysis and pyrolysis behavior of low-value waste lignocellulosic biomass for its bioenergy potential using thermogravimetric analyzer. *Mater. Sci. Energy Technol.* **2021**, *4*, 136–147. [CrossRef]
75. Cao, H.; Xin, Y.; Wang, D.; Yuan, Q. Pyrolysis characteristics of cattle manures using a discrete distributed activation energy model. *Bioresour. Technol.* **2014**, *172*, 219–225. [CrossRef] [PubMed]
76. Gupta, S.; Gupta, G.; Mondal, M. Thermal degradation characteristics, kinetics, thermodynamic, and reaction mechanism analysis of pistachio shell pyrolysis for its bioenergy potential. *Biomass Convers. Biorefinery* **2022**, *12*, 4847–4861. [CrossRef]
77. Chong, C.; Mong, G.; Ng, J.; Chong, W.; Ani, F.; Lam, S.; Ong, H. Pyrolysis characteristics and kinetic studies of horse manure using thermogravimetric analysis. *Energy Convers. Manag.* **2019**, *180*, 1260–1267. [CrossRef]
78. Ye, G.; Luo, H.; Ren, Z.; Ahmad, M.S.; Liu, C.G.; Tawab, A.; Al-Ghafari, A.B.; Omar, U.; Gull, M.; Mehmood, M.A. Evaluating the bioenergy potential of Chinese Liquor-Industry waste through pyrolysis, thermogravimetric, kinetics and evolved gas analyses. *Energy Convers. Manag.* **2018**, *163*, 13–21. [CrossRef]
79. Yuan, X.; He, T.; Cao, H.; Yuan, Q. Cattle manure pyrolysis process: Kinetic and thermodynamic analysis with Isoconversional methods. *Renew. Energy* **2017**, *107*, 489–496. [CrossRef]
80. Xu, Y.; Chen, B. Investigation of thermodynamic parameters in the pyrolysis conversion of biomass and manure to biochars using thermogravimetric analysis. *Bioresour. Technol.* **2013**, *146*, 485–493. [CrossRef]
81. Vlaev, L.T.; Markovska, I.G.; Lyubchev, L.A. Non-Isothermal kinetics of pyrolysis of rice husk. *Thermochim. Acta* **2003**, *406*, 1–7. [CrossRef]
82. Vasudev, V.; Ku, X.; Lin, J. Pyrolysis of Algal Biomass: Determination of the kinetic triplet and thermodynamic analysis. *Bioresour. Technol.* **2020**, *317*, 124007. [CrossRef] [PubMed]
83. Chen, J.; Wang, Y.; Lang, X.; Fan, S. Evaluation of Agricultural Residues Pyrolysis Under Non-Isothermal Conditions: Thermal behaviors, kinetics, and thermodynamics. *Bioresour. Technol.* **2017**, *241*, 340–348. [CrossRef]
84. Malika, A.; Jacques, N.; Jaafar, E.F.; Fatima, B.; Mohammed, A. Pyrolysis investigation of food wastes by TG-MS-DSC technique. *Biomass Convers. Biorefinery* **2016**, *6*, 161–172. [CrossRef]
85. Tahir, M.H.; Mahmood, M.A.; Çakman, G.; Ceylan, S. Pyrolysis of oil extracted safflower seeds: Product evaluation, kinetic and thermodynamic studies. *Bioresour. Technol.* **2020**, *314*, 123699. [CrossRef] [PubMed]
86. Galvan, D.; Orives, J.R.; Coppo, R.L.; Silva, E.T.; Angilelli, K.G.; Borsato, D. Determination of the kinetics and thermodynamics parameters of biodiesel oxidation reaction obtained from an optimized mixture of vegetable oil and animal fat. *Energy Fuels* **2013**, *27*, 6866–6871. [CrossRef]
87. Mallick, D.; Poddar, M.; Mahanta, P.; Moholkar, V. Discernment of synergism in pyrolysis of biomass blends using thermogravimetric analysis. *Bioresour. Technol.* **2018**, *261*, 294–305. [CrossRef]

88. Tian, X.; Dai, L.; Wang, Y.; Zeng, Z.; Zhang, S.; Jiang, L.; Yang, X.; Yue, L.; Liu, Y.; Ruan, R. Influence of torrefaction pretreatment on corncobs: A study on fundamental characteristics, thermal behavior, and kinetic. *Bioresour. Technol.* **2020**, *297*, 122490. [CrossRef]
89. He, Q.; Ding, L.; Gong, Y.; Li, W.; Wei, J.; Yu, G. Effect of torrefaction on pinewood pyrolysis kinetics and thermal behavior using thermogravimetric analysis. *Bioresour. Technol.* **2019**, *280*, 104–111. [CrossRef]
90. Abdullah, N.; Sulaiman, F.; Taib, R.M. Characterization of banana (*Musa* spp.) plantation wastes as a potential renewable energy source. *AIP Conf. Proc.* **2013**, *1528*, 325–330. [CrossRef]
91. Maia, A.A.D.; de Morais, L.C. Kinetic parameters of red pepper waste as biomass to solid biofuel. *Bioresour. Technol.* **2016**, *204*, 157–163. [CrossRef]
92. Turmanova, S.; Genieva, S.; Dimitrova, A.; Vlaev, L. Non-isothermal degradation kinetics of filled with rise husk ash polypropene composites. *Express Polym. Lett.* **2008**, *2*, 133–146. [CrossRef]
93. Kumar, R.; Hynes, N.R.J.; Senthamaraikannan, P.; Saravanakumar, S.; Sanjay, M.R. Physicochemical and thermal properties of *Ceiba pentandra* bark fiber. *J. Nat. Fibers* **2018**, *15*, 822–829. [CrossRef]
94. Pintor-Ibarra, L.F.; Alvarado-Flores, J.J.; Rutiaga-Quiñones, J.G.; Alcaraz-Vera, J.V.; Ávalos-Rodríguez, M.L.; Moreno-Anguiano, O. Caracterización Química y Energética de la Madera de *Prosopis laevigata*: Métodos Químicos y Termogravimétricos. *Molecules* **2024**, *29*, 2587. [CrossRef]
95. Mishra, R.K.; Sahoo, A.; Mohanty, K. Pyrolysis kinetics and synergistic effect in co-pyrolysis of Samanea saman seeds and polyethylene terephthalate using Thermogravimetric Analyser. *Bioresour. Technol.* **2019**, *289*, 121608. [CrossRef] [PubMed]
96. Ghavidel, A.; Bak, M.; Hofmann, T.; Vasilache, V.; Sandu, I. Evaluation of some wood-water relations and chemometric characteristics of recent oak and archaeological oak wood (*Quercus robur*) with archaeometric value. *J. Cult. Herit.* **2021**, *51*, 21–28. [CrossRef]
97. Syed Draman, S.F.; Daik, R.; Abdul Latif, F.; El-Sheikh, S.M. Characterization and thermal decomposition kinetics of kapok (*Ceiba pentandra* L.)-based cellulose. *BioResources* **2014**, *9*, 8–23. [CrossRef]
98. Leal, M.R.; Flores-Sahagun, T.H.S.; Franco, T.S.; Muniz, G.I. *Ceiba speciosa* St. Hill fruit fiber as a potential source for nanocellulose production and reinforcement of polyvinyl acetate composites. *Polym. Compos.* **2021**, *42*, 397–411. [CrossRef]
99. Carranza-Núñez, U.; Vásquez-García, S.R.; Flores-Ramírez, N.; Abdel-Gawwad, H.A.; Rico, J.L.; Arizbe Santiago, A.; Vargas, J.; Cruz-de-León, J. Caracterización fisicoquímica de fibras naturales obtenidas de vainas de semillas de *Ceiba aesculifolia*. *BioResources* **2021**, *16*, 4200–4211. [CrossRef]
100. Sathishkumar, T.P.; Navaneethakrishnan, P.; Shankar, S.; Rajasekar, R. Characterization of new cellulose *Sansevieria ehrenbergii* fibers for polymer composites. *Compos. Interfaces* **2013**, *20*, 575–593. [CrossRef]
101. Moosavinejad, S.M.; Madhoushi, M.; Vakili, M.; Rasouli, D. Evaluation of degradation in chemical compounds of wood in historical buildings using FT-IR and FT-Raman vibrational spectroscopy. *Maderas. Cienc. Y Tecnol.* **2019**, *21*, 381–392. [CrossRef]
102. De Rosa, I.M.; Kenny, J.M.; Maniruzzaman, M.; Moniruzzaman, M.; Monti, M.; Puglia, D.; Santulli, C.; Sarasini, F. Effect of chemical treatments on the mechanical and thermal behaviour of okra (*Abelmoschus esculentus*) fibres. *Compos. Sci. Technol.* **2011**, *71*, 246–254. [CrossRef]
103. Saravanapriya, G.; Shreelavaniya, R. Characterization of *Prosopis juliflora* and its products as a potential feedstock for energy generation. *Pharma Innov.* **2023**, *12*, 1673–1678. [CrossRef]
104. Lehto, J.; Louhelainen, J.; Huttunen, M.; Alén, R. Spectroscopic analysis of hot-water-and dilute-acid-extracted hardwood and softwood chips. *Spectrochim. Acta Part A Mol. Biomol. Spectrosc.* **2017**, *184*, 184–190. [CrossRef]
105. Alvarado Flores, J.J.; Rutiaga Quiñones, J.G.; Ávalos Rodríguez, M.L.; Alcaraz Vera, J.V.; Espino Valencia, J.; Guevara Martínez, S.J.; Márquez Montesino, F.; Alfaro Rosas, A. Thermal Degradation Kinetics and FT-IR Analysis on the Pyrolysis of *Pinus pseudostrobus*, *Pinus leiophylla* and *Pinus montezumae* as Forest Waste in Western Mexico. *Energies* **2020**, *13*, 969. [CrossRef]
106. Xu, F.; Yu, J.; Tesso, T.; Dowell, F.; Wang, D. Qualitative and quantitative analysis of lignocellulosic biomass using infrared techniques: A mini-review. *Appl. Energy* **2013**, *104*, 801–809. [CrossRef]
107. Åkerholm, M.; Hinterstoisser, B.; Salmén, L. Characterization of the crystalline structure of cellulose using static and dynamic FT-IR spectroscopy. *Carbohydr. Res.* **2004**, *339*, 569–578. [CrossRef]
108. Abidi, N.; Cabrales, L.; Haigler, C.H. Changes in the cell wall and cellulose content of developing cotton fibers investigated by FTIR spectroscopy. *Carbohydr. Polym.* **2014**, *100*, 9–16. [CrossRef]
109. Abdullah, M.A.; Afzaal, M.; Ismail, Z.; Ahmad, A.; Nazir, M.S.; Bhat, A.H. Comparative study on structural modification of *Ceiba pentandra* for oil sorption and palm oil mill effluent treatment. *Desalination Water Treat.* **2014**, *54*, 3044–3053. [CrossRef]
110. Herrera-Fernández, A.C.; Carrillo-Parra, A.; Pedraza-Bucio, F.E.; Correa-Méndez, F.; Herrera Bucio, R.; López-Albarrán, P.; Rutiaga-Quiñones, J.G. Densidad, composición química y poder calorífico de la madera de tres especies de encinos (*Quercus candicans*, *Q. laurina* y *Q. rugosa*). *Cienc. Nicolaita* **2018**, 72. [CrossRef]
111. Van Soest, P.V.; Robertson, J.B.; Lewis, B. Methods for dietary fiber, neutral detergent fiber, and nonstarch polysaccharides in relation to animal nutrition. *J. Dairy Sci.* **1991**, *74*, 3583–3597. [CrossRef]
112. Torres Jaramillo, D.; Morales Vélez, S.P.; Quintero Díaz, J.C. Evaluación de pretratamientos químicos sobre materiales lignocelulósicos. *Ingeniare. Rev. Chil. Ing.* **2017**, *25*, 733–743. [CrossRef]
113. *UNE-EN ISO 18134-1*; Biocombustibles Sólidos, Determinación del Contenido de Humedad, Método de Secado en Estufa, Humedad Total, Método de Referencia. Asociación Española de Normalización y Certificación (AENOR): Madrid, Spain, 2016.

114. *UNE-EN 14775;* Biocombustibles Sólidos. Método Para la Determinación del Contenido en Cenizas. Asociación Española de Normalización y Certificación (AENOR), AEN/CTN 164 Biocombustibles Sólidos CONFEMADERA: Madrid, Spain, 2010.
115. *ASTM E872−82;* Standard Test Method for Volatile Matter in the Analysis of Particulate Wood Fuels. ASTM International: West Conshohocken, PA, USA, 2013.
116. García, R.; Pizarro, C.; Lavín, A.G.; Bueno, J.L. Characterization of Spanish biomass wastes for energy use. *Bioresour. Technol.* **2012**, *103*, 249–258. [CrossRef] [PubMed]
117. Arcibar-Orozco, J.A.; Josue, D.-B.; Rios-Hurtado, J.C.; Rangel-Mendez, J.R. Influence of iron content, surface area and charge distribution in the arsenic removal by activated carbons. *Chem. Eng. J.* **2014**, *249*, 201–209. [CrossRef]
118. *UNE-CEN/TS 15104 EX;* Determinación del Contenido de Carbono, Hidrógeno y Nitrógeno. Asociación Española de Normalización y Certificación (AENOR): Madrid, Spain, 2008.
119. *UNE-EN 14918;* Determinación del Poder Calorífico. Asociación Española de Normalización y Certificación (AENOR): Madrid, Spain, 2011.
120. Bhatt, B.P.; Todaria, N.P. Fuelwood characteristics of some mountain trees and shrubs. *Biomass* **1990**, *21*, 233–238. [CrossRef]
121. Kan, T.; Strezov, V.; Evans, T. Effect of the heating rate on the thermochemical behavior and biofuel properties of sewage sludge pyrolysis. *Energy Fuels* **2016**, *30*, 1564–1570. [CrossRef]
122. Fong, M.J.B.; Loy, A.C.M.; Chin, B.L.F.; Lam, M.K.; Yusup, S.; Jawad, Z.A. Catalytic pyrolysis of Chlorella vulgaris: Kinetic and thermodynamic analysis. *Bioresour. Technol.* **2019**, *289*, 121689. [CrossRef]
123. Doyle, C.D. Estimating isothermal life from thermogravimetric data. *J. Appl. Polym. Sci.* **1962**, *6*, 639–642. [CrossRef]
124. Gai, C.; Dong, Y.; Zhang, T. The kinetic analysis of the pyrolysis of agricultural residue under non-isothermal conditions. *Bioresour. Technol.* **2013**, *127*, 298–305. [CrossRef]

molecules

MDPI

Article

Study of the Effect of Cedar Sawdust Content on Physical and Mechanical Properties of Cement Boards

Anas El Hamri *, Yassine Mouhib , Atmane Ourmiche, Mohammed Chigr and Nour-Eddine El Mansouri *

Laboratory of Molecular Chemistry, Materials and Catalysis (LC2MC), Faculty of Science and Technology of Béni-Mellal, University Sultan Moulay Slimane, B.P. 523, Béni-Mellal 23000, Morocco; yassinemouhib.dr@gmail.com (Y.M.); aourmiche@gmail.com (A.O.); chigrm@gmail.com (M.C.)

* Correspondence: anas.elhamri@usms.ac.ma (A.E.H.); n.elmansouri@usms.ma (N.-E.E.M.); Tel.: +212-658-526-504 (A.E.H.); +212-653-457-858 (N.-E.E.M.)

Abstract: The growing demand for sustainable building materials, amid escalating costs, has spurred interest in alternative solutions such as wood cement composites. This study explores the feasibility of producing wood cement boards (WCBs) using locally sourced cedar sawdust as a reinforcing agent. Boards with a thickness of 10 mm and a target density of 1200 kg/m^3 were manufactured under pressures ranging from 2 to 6 MPa for 24 h. Cedar sawdust, used as raw and untreated material, was incorporated into the mixture as a partial substitute for cement in varying proportions, ranging from 10% to 25% (by weight). The WCBs were cured for 28 days under ambient conditions. Physical properties including density, water absorption (WA), and thickness swelling (TS) were assessed, along with mechanical properties through flexural tests. The results showed that increasing cedar sawdust content decreased both density and mechanical performance while increasing WA and TS. Microstructural analysis (SEM and EDS) revealed significant porosity at higher sawdust contents, while lower contents had better matrix–reinforcement cohesion. Additionally, substantial levels of calcium and silicon were detected on the sawdust surface, indicating stabilized cement hydration products. These findings, supported by thermal (TGA and DSC) and FTIR analyses, clearly demonstrate that cement boards with 10% cedar sawdust exhibit favorable properties for non-structural applications, such as wall and partition cladding.

Keywords: cedar sawdust; wood cement boards; bio-based material; mechanical properties; physical properties; cement hydration products

1. Introduction

Recent decades have witnessed a global scientific renaissance aimed at addressing the increasing demands of everyday life. However, this advancement has often compromised natural heritage and ecological stability, especially in developing countries. In these regions, factors such as population growth, overgrazing, logging, and fuelwood collection have led to forest degradation and, ultimately, deforestation. For instance, logging activities have diminished Morocco's cedar forests by 12%, threatening its ecological balance [1]. The cedar, a genus of conifers in the Pinaceae family, is native to the mountains of the western Himalayas and the Mediterranean region [2]. It encompasses several species, including the Atlas cedar (*Cedrus atlantica*), which is endemic to northwest Africa, specifically Morocco and Algeria [3]. The Atlas cedar plays a vital role in Morocco's forestry sector, contributing to 80% of the nation's timber production [4]. Despite its ecological and economic significance, the Atlas cedar population continues to decline due to unsustainable practices. This decline persists even as international organizations like UNESCO recognize Morocco's cedar forests as vital biosphere reserves essential for global biodiversity. These challenges underscore the urgent need for innovative strategies to alleviate pressure on forest resources and enhance sustainable management practices [5]. Such an approach can help preserve forest biodiversity, stabilize ecosystems, mitigate environmental erosion, and combat climate change [6].

The growing interest in sustainable building materials has driven research into the utilization of lignocellulosic waste from agroforestry industries [7]. In Morocco, sawmills generate over 100,000 tons of sawdust annually, yet its utilization remains minimal [8]. Typically, sawdust is used in low-value applications, such as fuel, industrial site cleaning, or bedding for poultry and livestock, which do not fully exploit its potential as a valuable resource. To better harness this potential, recent research has shifted focus toward transforming sawdust into a functional material for construction, particularly through the production of wood cement boards (WCBs) [9].

WCBs are composite materials made by combining wood particles (aggregate), Portland cement (mineral binder), chemical additives, and water under pressure [10,11]. These boards offer an environmentally friendly alternative to traditional wood or asbestos cement boards [12]. Among the various types of mineral-bonded wood composites, cement-bonded particleboard is prominent, consisting of 10–70% wood particles and 30–90% Portland cement binder [13]. Woody material residues are often used as aggregates and include planer shavings; sawmill residues such as slabs, edges, and upholstery; logging residues from furniture match factories; and deforestation residues such as short trunks, broken logs, bent trunks, small crowns and branches, forest thinning, and bark [14]. Over the past few decades, numerous types of composite materials have emerged, including cement-bonded wood wool (CBWW), cement-bonded particleboard (CBPB), and fiber-reinforced cement board [15]. While resin-bonded boards are an alternative in some applications, the high cost of resins and the machinery required for production point in favor of the cement-bonded board [7]. Moreover, replacing organic adhesives like phenols and formaldehyde with cement-based adhesives aligns to manufacturing safe, environmentally-friendly products.

The incorporation of wood particles into the cement matrix inhibits crack propagation, enhancing fracture toughness, ductility, and flexural strength, which allows the board to withstand higher deformation limits, making it more resilient and durable in various applications [16,17]. The quality of these composites is influenced by various factors such as the type of lignocellulosic material used, density, particle size and volume, aspect ratio, mix design, mixing method, and treatment processes [18]. WCBs have a wide range of applications, including exterior wall cladding surfaces, siding panels, roof shakes, fire protection elements, specialized flooring, thermal and acoustic insulation purposes, building renovations, decorative applications, and sound insulation [10]. Their adaptability allows them to be molded into any shape to meet specific needs [6]. Furthermore, WCBs are asbestos-free, are devoid of hazardous and volatile substances, and minimize the spread of dust and toxic gases, ensuring a safe and environmentally friendly manufacturing process [19]. Cement-bonded boards are also highly resistant to various factors including freeze, thaw, fire, water, rot, termites, insects, and fungi attacks [6]. Their accessibility, low density, light weight, rigidity, non-corrosive nature, workability, low cost, and ease of manufacturing make them highly versatile and promising for a wide range of applications [20].

While there is a growing body of literature on the use of sawdust in composite materials, several gaps remain in the research. The careful selection of wood type is crucial for WCB production to ensure good chemical compatibility with cement and minimize chemicals hindering adhesion. Coniferous wood species, such as cedar, are commonly favored due to their superior chemical compatibility with cement [9]. Some wood species may impede cement hydration due to the type and content of extractives they contain [21]. Wood particles can be treated with hot water, sodium hydroxide, or other chemicals to address this. Additionally, mineralizing agents like calcium chloride ($CaCl_2$) can be added to the mix to accelerate cement setting [22]. However, these procedures often increase the cost of production and energy consumption and result in a slower production process [23].

This study introduces a novel application for Atlas cedar residues by utilizing sawdust as a reinforcing material in WCBs. This approach not only enhances the aesthetic value and fragrance of the boards but also addresses the lack of research on the specific use of cedar sawdust. The primary objective of this research is to examine the impact of varying cedar sawdust content on the physical, mechanical, morphological, and thermal properties of WCBs without any additional chemical modifications. This study contributes to the development of sustainable building materials that minimize environmental impact, improve resource efficiency, and promote forest conservation. The insights gained from this research can serve as the basis for optimizing WCBs in diverse non-structural applications, such as wall and partition cladding.

2. Results and Discussion

2.1. Chemical Constituents of Cedar Sawdust

Several challenges still significantly hinder the advancement of wood–cement composites, notably the substantial variability in the compatibility of different wood species with cement constituents and the inhibitory impact that certain wood species have on cement's hydration [24]. Certain wood components, such as holocellulose, lignin, and total extractives, can impede the bond between wood particles and cement, potentially delaying cement hydration or setting time. Conversely, substances like ash can enhance compatibility between these elements [25].

To ensure successful wood–cement composite production, it is crucial to evaluate the chemical composition of wood species and select those with minimal elements hindering the bond between wood and cement molecules. In this study, the suitability of cedar sawdust for cement board manufacturing was evaluated by comparing its chemical composition, based on the literature (Table 1) [26], with that of certain previously used species in these boards. The conifer family, to which cedar belongs, is often preferred for the production of wood–cement boards (WCBs) due to their generally more chemically compatible nature with cement, as they do not contain substances that inhibit its hardening [9].

Table 1. Atlas cedar wood chemical composition found by Quiquandon [26].

Components	Content (%)
Cellulose	50
Hemicelluloses	12.10
Lignin	32
Resin	1.45
Extractives	4.24
Ash	0.31

Wood compositions primarily contain organic chemical compounds such as cellulose, hemicelluloses, and lignin, which collectively constitute over 99% of the total mass [27]. In this context, Pettersen et al. [28], reported that wood generally has a higher proportion of holocellulose than of lignin, extractives, and ash. Similarly, the chemical composition of Atlas cedar wood, as shown in Table 1, indicates a higher quantity of holocellulose (62.10%) compared to lignin (32%), extractives (4.24%), and ash (0.31%). These lignin, holocellulose, and ash contents are comparable to those of other species previously used in wood–cement composite manufacture, such as eucalyptus [11], pine [18], tropical woods [25], and Ochroma pyramidal [9]. However, the extractive content in Atlas cedar wood is lower than in most of these species.

The amount and composition of extractives are specific to each wood species and are the main reason for the prevention of cement hydration, leading to incompatibility between wood and cement [15,29]. Extractives are generally composed of terpenes, fatty acids, tannins, carbohydrates, and inorganic substances [30]. They also contain several organic compounds that form complexes with metal ions present in the cement solution, reducing the concentration of calcium ions (Ca^{2+}), and disrupting the equilibrium of the solution, thereby delaying the onset nucleation of $Ca(OH)_2$ and calcium silicate hydrate (C-S-H) gel [25].

As the content of these elements increases, so does the percentage of anhydrate clinker, resulting in reduced mechanical performance of wood–cement composites [15]. Hence, wood species with low levels of extractives are deemed more suitable for wood–cement composites [29]. For instance, Mandes et al. [18] found that boards produced from eucalyptus wood residues displayed better physical and mechanical properties because of their low extractives content (2.3%) in contrast to other wood sources such as Pinus wood, sugarcane bagasse, and bamboo particles, which had higher extractives contents of 10.8, 12.5, and 9.2%, respectively. Thus, the lower percentage of extractives (4.24%) in Atlas cedar wood is in line with the expected results of this study.

In previous studies, attempts were made to improve board properties by removing certain extractives to expedite the cement setting and hardening process. However, this treatment did not always yield beneficial results. For instance, Setter et al. [9] utilized Ochroma pyramidal wood with a comparable extractives content (5.78%) and applied treatments involving cold water, hot water, and NaOH solution. However, the outcomes did not demonstrate any enhancement in the final quality of the produced boards, whether in terms of physical or mechanical properties. Consequently, these findings have favored the use of cedar sawdust in its untreated form, reducing energy consumption and expediting the production process.

XRF analysis of the cement unveiled its chemical composition, comprising various metal oxides, as detailed in Table 2. Chowdhury et al. [31] highlighted that these oxides are also present in wood ash in significant quantities. This suggests that the cementitious properties of wood ash enhance the role of cement in the production of wood–cement composites, as pozzolanic reactions occur between wood ash (pozzolan), water, and calcium hydroxide released by cement hydration, increasing the amount of the cement bonding phase (C-S-H) and improving cement bond strength [25]. However, considering the composition outlined in Table 1, Atlas cedar contains a limited amount of ash (0.31%), thereby reducing its potential to improve the properties of the produced boards.

Table 2. Cement composition and its physical properties.

Chemical Composition		**Physical Properties**	
SiO_2 (%)	20.43	Specific surface Blaine (cm^2/g)	4500
Al_2O_3 (%)	4.21	Initial setting (min)	100
Fe_2O_3 (%)	0.43	Final setting (min)	130
CaO (%)	66.03	Refusal at 45 μm sieve (%)	1.69
MgO (%)	1.40	Refusal at 90 μm sieve (%)	0.18
SO_3 (%)	3.35	Apparent density (g/L)	3047
K_2O (%)	0.54	Loss on ignition (%)	3.45
Na_2O (%)	0.14		
		Bogue's Compounds	
C_3S	84.59	C_2S	5.22
C_3A	10.43	C_4AF	1.30

2.2. *Physical Properties*

2.2.1. Density

The results of the physical and mechanical properties of the manufactured boards are shown in Table 3, while Table 4 reports their variance analysis. The change in density concerning cedar sawdust content, illustrated in Figure 1, indicates an average value ranging from 1093.81 to 1432.51 kg/m^3. The lowest density was obtained for boards with a 10% sawdust content (WCB 1), while the highest density was observed in boards with a 25% sawdust content (WCB 6). The variance analysis for density revealed significant differences among all levels, indicating a noticeable effect of cedar sawdust content on board density.

As depicted in Figure 1, as the amount of cedar sawdust in the boards increased, there was a discernible drop in density. This trend aligns with results from some prior research, where an increase in the percentage of wood particles or fibers in wood–cement composites resulted in decreased density [32–34]. This decrease can be attributed to the higher proportion of cedar wood in the mixture, which has a much lower density (218 kg/m^3) compared to cement (3047 kg/m^3). In other words, increasing the amount of sawdust increases the total volume of the mixture required to produce a particleboard of a given thickness. It is essential to note that the mechanical and physical properties of particleboard can be influenced by board density and particle size [35].

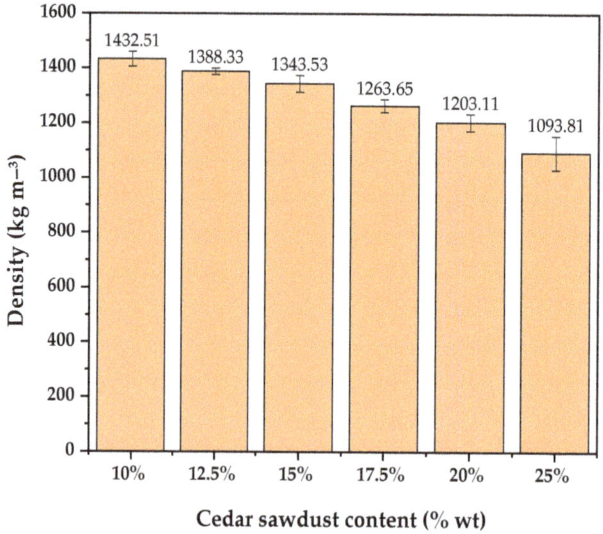

Figure 1. Variation in density with sawdust content.

Similarly, in the current study, it was observed that the board with the highest density value at 10% cedar sawdust content exhibited the highest modulus of rupture (6.94 MPa) and modulus of elasticity (3181.33 MPa), while the lower mechanical properties of the board prepared with 25% cedar sawdust may be attributed to its lower density. Except WCB 6, all of the produced boards meet the requirements for HZ-type BISON boards, which specify values not less than 1200 kg/m^3 [36].

Table 3. Mean values obtained for physical and mechanical properties of cedar–cement boards.

Board	Density (kg/m^3)	WA 2 h (%)	WA 24 h (%)	TS 2 h (%)	TS 24 h (%)	MOR (MPa)	MOE (MPa)
WCB 1	1432.51 (27) ^{a}	11.11 (0.59) ^{a}	12.35 (2.58) ^{a}	0.27 (0.04) ^{a}	0.36 (0.09) ^{a}	6.94 (0.67) ^{a}	3181.33 (286) ^{a}
WCB 2	1388.33 (11) ^{ab}	13.01 (1.79) ^{ab}	13.70 (1.67) ^{ab}	0.39 (0.04) ^{b}	0.49 (0.11) ^{a}	6.47 (0.75) ^{ab}	2634.16 (188) ^{b}
WCB 3	1343.53 (31)^b	13.82 (1.24) ^{b}	16.49 (2.31) ^{b}	0.43 (0.04) ^{bc}	0.65 (0.06) ^{b}	5.21 (0.96) ^{bc}	2197.83 (271) ^{b}
WCB 4	1263.65 (23) ^{c}	14.82 (0.55) ^{b}	16.07 (1.32) ^{b}	0.48 (0.02) ^{c}	0.66 (0.06) ^{b}	4.19 (0.82) ^{cd}	1655.33 (350) ^{c}
WCB 5	1203.11 (31) ^{d}	17.77 (1.44) ^{c}	21.07 (2.12) ^{c}	0.67 (0.04) ^{d}	1.16 (0.12) ^{c}	3.65 (1.06) ^{d}	1608.80 (414) ^{c}
WCB 6	1093.81 (62) ^{e}	23.06 (1.45) ^{d}	27.75 (2.69) ^{d}	0.71 (0.05) ^{d}	1.22 (0.11) ^{c}	1.27 (0.74) ^{e}	476.83 (259) ^{d}
BISON board type HZ [36]	1200	--	--	0.8	1.2–1.8	9	3000

Means followed by the same letter in the same column are statistically equal according to the Tukey test with a probability of 95%. Values in parenthesis are standard deviations.

Table 4. Variance analysis of physical and mechanical properties of cedar–cement boards.

Source	Density	WA 2 h	WA 24 h	TS 2 h	TS 24 h	MOR	MOE
Sawdust content	***	***	***	***	***	***	***

*** (significant on $\alpha < 0.01$).

2.2.2. Water Absorption and Thickness Swelling

During the lifespan of wood and its derivatives, exposure to weathering and various decomposition factors, both biotic and abiotic, significantly impacts their performance [37]. These factors, along with others, contribute to increased wood permeability and decreased resistance to water [38], underscoring the importance of assessing the durability of produced boards against such conditions. Water absorption (WA) and thickness swelling (TS) are crucial physical properties utilized to evaluate the dimensional stability of boards, particularly those intended for outdoor use under extreme humidity conditions [34,39].

This study investigates how sawdust content affects the WA and TS characteristics of the boards. Figure 2 provides the average %WA and %TS values after immersing the samples in water for both 2 h and 24 h. Notably, boards with 10% sawdust (WCB 1) display the lowest WA and TS values across both immersion durations, while those containing 25% sawdust (WCB 6) exhibit the highest values. Specifically, Figure 2a shows WA values ranging from 11.11% to 23.06% after 2 h of immersion and from 12.35% to 27.75% after 24 h. However, Figure 2b reveals that average TS values vary from 0.27% to 0.71% after 2 h of immersion, and from 0.36% to 1.22% after 24 h. The increase in both WA and TS with immersion time is evident, as prolonged immersion increases the likelihood of water permeating the board's interiors, heightening their instability. This phenomenon can be attributed to the inherent hygroscopic properties of wood fibers, which possess a considerable affinity for water [40]. This aligns with findings from Setter et al. [9], where increasing the soaking time of WCBs from 2 h to 72 h resulted in similar trends.

Additionally, statistical analysis of the results in Table 3 reveals that these properties intensify with higher sawdust content in the boards, particularly noticeable for WCB 5 and WCB 6. Other researchers have reported similar trends. For instance, Ali et al. [41] observed an increase in both WA and TS as the sawdust content in cement boards ranged from 20% to 40%.

Likewise, Bilcati et al. [42] conducted an evaluation of cement composites with sawdust content between 10% and 20%, supporting these findings. In addition to the hygroscopic nature of lignocellulosic materials and the composition of its hemicellulose-rich fibers, which are inherently more hydrophilic [43], this type of behavior can also be explained by the wood particles' porosity [41]. Sotannede et al. [34] pointed out that the difficulty in compressing and the presence of void spaces in the boards enhance water absorption. Hence the high proportion of wood particles within the board increases the rate of pores accessible to water, accounting for the low densities recorded at high sawdust content.

Conversely, increasing the percentage of cement and decreasing the sawdust content reduces the water absorption by the boards [34]. This effect occurs because the binder fills the pore holes that would otherwise result from the wood's porosity [19]. For sawdust content ranging from 10% to 20%, the WA values observed in this study are lower than those reported by other researchers. For example, Olorunnisola et al. [44] used eucalyptus wood particles as reinforcement, while Bilcati et al. [42] assessed lightweight cement-based composites reinforced with Pinus sawdust. Conversely, for a 25% sawdust content (cement–wood ratio of 3:1), several studies found WA values lower than those observed in this study (27.75%). For instance, Ogunjobi et al. [19], using Anogeissus leiocarpus wood particles, reported an average value of 24.95%. Similarly, Adelusi et al. [45] and Amiandamhen and Izekor [46] found values of 24.45% and 21.59%, respectively, using Gmelina arborea sawdust in cement-bonded boards.

The variation in results observed for the 25% sawdust content compared to previous studies could be attributed to particle size effects. Smaller dimensions lead to increased surface area of the particles, making it challenging for the cement to adequately cover them all. To ensure the production of high-quality cement-bonded boards with good dimensional stability, it is recommended to use mixed-size or coarse particles [13]. Mendes et al. further substantiated this point by confirming that particle size and geometry significantly impact the physical properties of the cement-bonded boards, with larger particles yielding better performance [18]. Similarly, Sotannde et al. demonstrated that composites made from a mixture of sawdust and flakes outperformed those made from each component separately [34]. In this study, particle size diversity was limited (ranging from 2 mm to 4 mm). Additionally, there may have been an incompatibility between wood and cement, resulting in internal free space at the particle–cement interface, which potentially led to significant moisture absorption. Frybort et al. highlighted that the presence of extractives is one of the primary causes of this incompatibility [15].

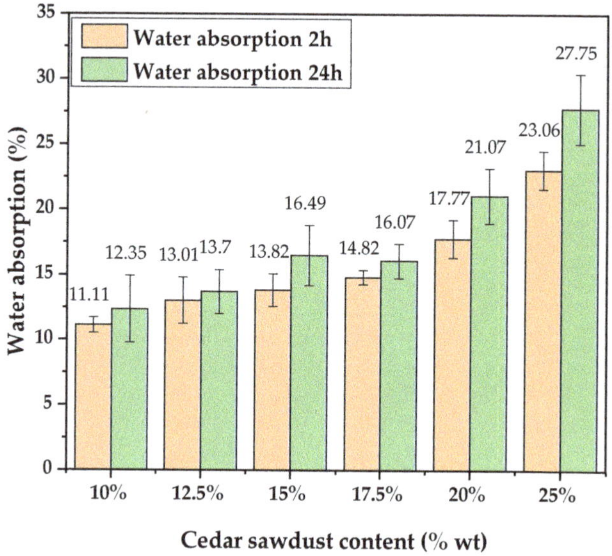

(a)

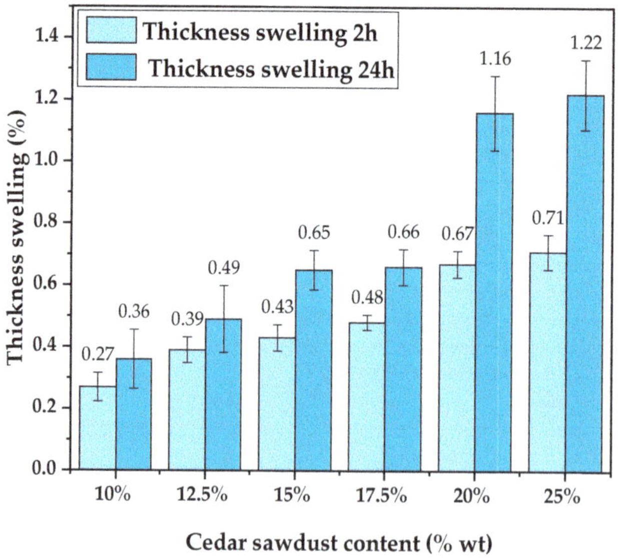

(b)

Figure 2. (**a**) Variation in WA with cedar sawdust content; (**b**) Variation in TS with cedar sawdust content.

In prior research endeavors, chemical and/or physical pretreatment involving additives has often been employed during the manufacturing process of cement-bonded boards to enhance their dimensional stability. For example, Sotannde et al. explored the impact of incorporating three different chemical additives ($CaCl_2$, $MgCl_2$, $AlCl_3$) on the WA and TS of CBPB derived from Afzelia Africana wood residues. Their findings revealed that the introduction of chemical additives led to a reduction in water absorption of the boards by up to 16% [34]. Similar conclusions have been reached by other studies as well [41,47]. The average TS values obtained in this study closely align with the findings of most of the previously mentioned works [41,44–46]. They also conform to the normative standards outlined by both the CBPB manufacturer Cetris® [48], and the requirement set by BISON for an HZ-type board [36] which stipulates a maximum of 1.8% TS. However, only the WA values of WCB 1, WCB 2, and WCB 3 are close to those specified by the manufacturer Cetris® (16%).

2.3. Mechanical Properties

The flexural mechanical properties of the produced boards were evaluated through measurements of the modulus of rupture (MOR) and the modulus of elasticity (MOE). As illustrated in Figure 3, MOR values ranged from 1.27 to 6.94 MPa, while the MOE values varied from 476.83 to 3181.33 MPa. A thorough statistical analysis of the results in Table 3 indicates that cedar sawdust content significantly affects mechanical properties. Notably, there is a discernible negative correlation between sawdust content and board strength: as cedar sawdust content increases, mechanical properties decrease. Boards containing 10% cedar sawdust (WCB1) exhibited the highest values of MOR (6.94 MPa) and MOE (3181.33 MPa), aligning with previous research on cementitious composites using various lignocellulosic reinforcements, such as eucalyptus [44], oil palm veins [49], and rattan particles [50], where optimal mechanical properties were similarly achieved at 10% content. However, other studies have reported maximum mechanical properties with reinforcement contents exceeding 10% [51,52]. These results can be attributed to several factors, including improved compatibility between particles and the cementitious matrix, effective particle dispersion within the cement matrix, and the pressing cycle during board production, which minimized voids as much as possible [40].

Beyond 10% cedar sawdust content, mechanical properties declined, with a more pronounced decrease observed when content exceeded 20%. This trend mirrors findings from some previous studies. For instance, Ali et al. noted a 17% to 48% decrease in strength properties when sawdust content increased from 20% to 30% and 40% in cement particleboards [41]. Similarly, Ayrilmis et al. noted a decline in the mechanical properties of cement-bonded composites with increased palm particle main vein content beyond 10%, attributing this to increased composite porosity and reduced particle–matrix interfacial area [49]. The significant decrease in mechanical properties observed beyond 20% cedar sawdust content for WCB 6 can be attributed to inadequate compaction and the lower density of the wood–cement composites produced with high contents [33]. This phenomenon is linked to the attainment of high porosity due to a low cement-to-wood ratio, resulting in diminished MOE and MOR values [15]. At this juncture, the cement becomes insufficient to bind the wood particles, leading to maximum porosity, as evidenced by the low density (1093.81 kg/m^3) recorded for boards with 25% cedar sawdust content.

Conversely, applying higher pressures during manufacturing leads to the densification of the wood, elimination of gaps, and stronger matrix–reinforcement connections, thereby enhancing strength and stiffness [53]. The findings of this study are comparable to those reported by Setter et al., who observed MOR values ranging from 3.08 to 3.58 MPa and MOE values spanning from 2197 to 2739 MPa when using Ochroma pyramidal wood particles as reinforcement in WCBs [9]. It is worth noting that the maximum mechanical property values of cement-bonded boards vary across different studies, primarily influenced by factors such as the aspect ratio, size, orientation, and interfacial adhesion of the particles used as reinforcement. Ghofrani et al. observed that for the same 25% reinforcement content, the MOR value obtained when incorporating hardwood fibers (8.1 MPa) was lower than that obtained with rice stalk fibers as reinforcement (8.6 MPa). They attributed this difference to the aspect ratio of the fibers, which is significantly higher in rice stalk fibers, enabling better stress transfer between the matrix and the fibers [52]. In another study, Nasser et al. achieved MOR and MOE values of 11.78 MPa and 5667 MPa, respectively, for cement boards produced using the prunings of certain Saudi Arabian tree species, highlighting the importance of employing mixed or coarse particle sizes to achieve high mechanical properties [13]. Similarly, Olorunnisola observed that boards prepared with 10% rattan particles of varying sizes exhibited a higher MOR value (6.8 MPa) compared to those with uniform particle size (4.3 MPa) [54].

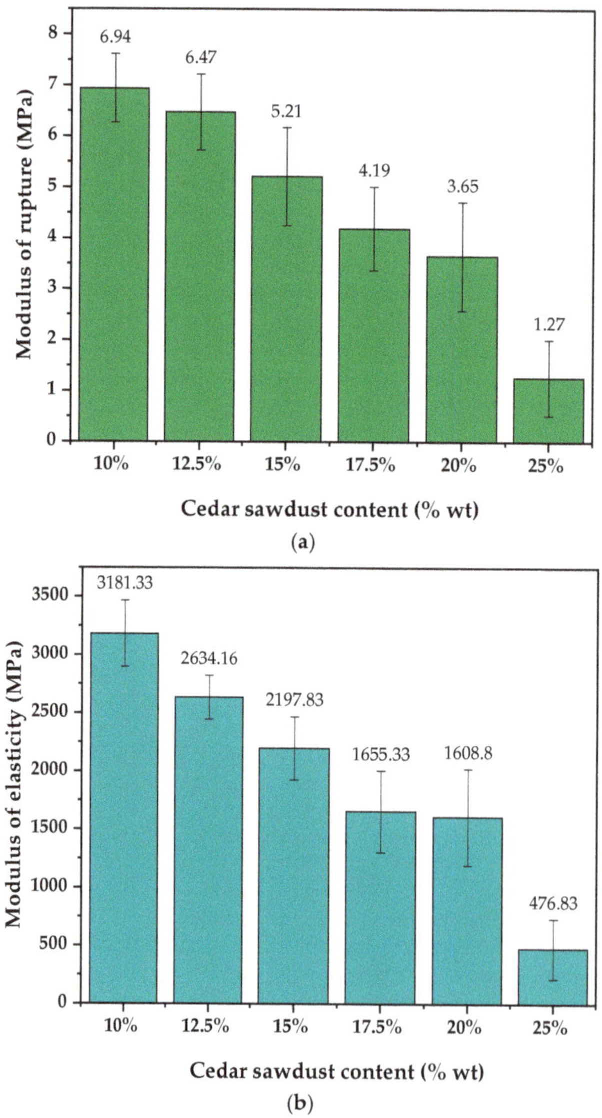

Figure 3. (**a**) Variation in MOR with cedar sawdust content; (**b**) Variation in MOE with cedar sawdust content.

Alternatively, in a study exploring the effect of strand orientation on the physical and mechanical properties of cement boards, Cabral et al. found that the best performance was observed with boards containing 35% parallel-oriented eucalyptus wood strands, with MOR and MOE values reaching approximately 15 MPa and 7072 MPa, respectively [55]. While cedar sawdust demonstrated effectiveness as reinforcement in WCBs, the mechanical properties achieved in this study were not on par with those reported in most previous research. This shortfall can be attributed to certain characteristics of the reinforcement employed, such as its uniform particle size, non-fibrous nature, low aspect ratio, and lack of orientation. These factors contribute to inadequate physical contact between particles.

To address these limitations and enhance the findings, it is essential to improve the properties of the reinforcement by adjusting particle size and orientation during the manufacturing process. Based on the findings of this study, none of the boards met the minimum bending strength (MOR) requirement of 9 MPa set by BISON for HZ-type boards. However, boards containing 10% cedar sawdust content surpassed the MOE requirements of 3000 MPa for HZ-type boards [36]. Despite not meeting structural standards, these boards remain suitable for applications where direct structural support is unnecessary, such as interior or exterior cladding for thermal and acoustic insulation purposes.

2.4. Microstructural Analysis

2.4.1. Scanning Electron Microscopy (SEM)

Scanning electron microscopy (SEM) analysis stands as the predominant technique for scrutinizing fiber–matrix interactions at fracture surfaces and matrix distributions in plant fiber composites [56]. In this study, SEM was employed to unveil the internal microstructure and alterations on the fracture surface of the examined boards after the post-bending test, offering insights into some physical and mechanical properties obtained. Specifically, the analysis was conducted on two samples from each of the boards containing 10% and 25% cedar sawdust, with images captured at various magnifications, as represented in Figure 4.

The fracture surface unveils failure occurring either at the particle–matrix interface or within the cementitious matrix, with samples displaying fractured ends of sawdust particles and instances of pull-out from the matrix (Figure 4A), confirming that the interface is indeed the weakest part of the board, potentially due to the diminutive size of the particles used (2–4 mm). This mode of failure mirrors observations made by Amiandamhen et al. in cementitious composites containing wood residues, attributing it to the small size of the wood particles adversely affecting interfacial adhesion [56,57].

Samples containing 10% cedar sawdust content exhibit notable adhesion between the sawdust particles and the cement matrix, as evidenced by matrix encrustation on the surfaces of the sawdust particles (Figure 4B,C), indicating well-distributed particles devoid of agglomeration. The presence of matrix encrustation on fiber surfaces and the appearance of fractured ends of pulled-out fibers are clear signs of successful fiber reinforcement [58]. Furthermore, at this content, the matrix appears more compact and denser with reduced amounts of air pores at the interface (Figure 4D), fostering better adhesion between cedar sawdust and cement and resulting in superior physical and mechanical performance. Prior research attributes this cohesion to the formation of cement hydration products, which serve as anchors between reinforcement and the cement matrix, enhancing inter-particle conditioning in the composite [57,59,60]. Additionally, the quality of the cement significantly impacts the formation of these compounds, with studies demonstrating that the use of 52.5R Super White Cement generates more hydration products, thereby bolstering physical–mechanical properties due to its higher CaO and SiO_2 content compared to 32.5R Portland cement [61].

Conversely, samples containing 25% cedar sawdust exhibit an escalation in porosity with the increase in cedar sawdust content (Figure 4E,F). The structure reveals the presence of air pores surrounding the wood particles at the interface (Figure 4E), while the cementitious matrix appears less dense and more hollow (Figure 4F). This suggests a weaker bond strength, with the cementitious matrix seemingly inadequate to fully bind all the wood particles together. This aligns with earlier research showing that higher aggregate content at low cement content results in inadequate hydration products, attributed to unreacted CaO and SiO_2 [57]. Furthermore, Nasser et al. stated that voids within the particleboard may originate from air within the wood particles [13]. This assertion finds support in the microstructure of cedar sawdust illustrated in Figure 4B, validating their conclusions.

In summary, the porous microstructure of the used reinforcement affects board stability, with an increase in its content in the mixture explaining lower density, MOR, and MOE values, as well as higher water absorption observed.

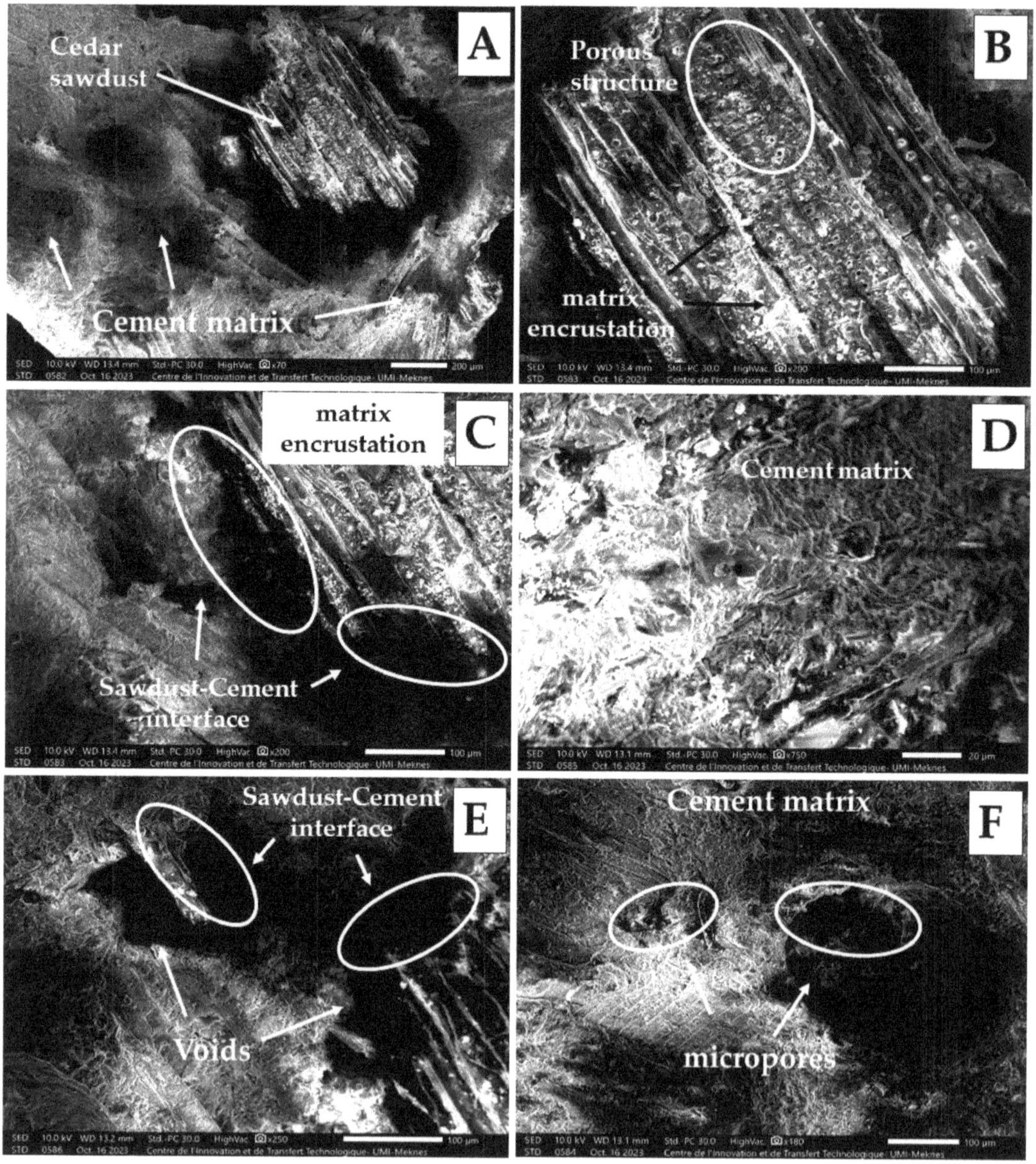

Figure 4. SEM micrographs of WCB samples: (**A**,**B**) the fractured surface; (**C**,**D**) board samples containing 10% cedar sawdust; (**E**,**F**) board samples containing 25% cedar sawdust.

2.4.2. Energy-Dispersive X-ray Spectroscopy (EDS)

Energy-dispersive X-ray spectroscopy (EDS) analysis was conducted on the cedar sawdust particles, as well as at two locations of the fracture surface of the board containing 10% cedar sawdust: on the surface of the cement matrix and the surface of the sawdust reinforcement. Figure 5A–C illustrate the EDS spectra obtained, respectively, while Table 5 provides a summary of the corresponding chemical weight composition.

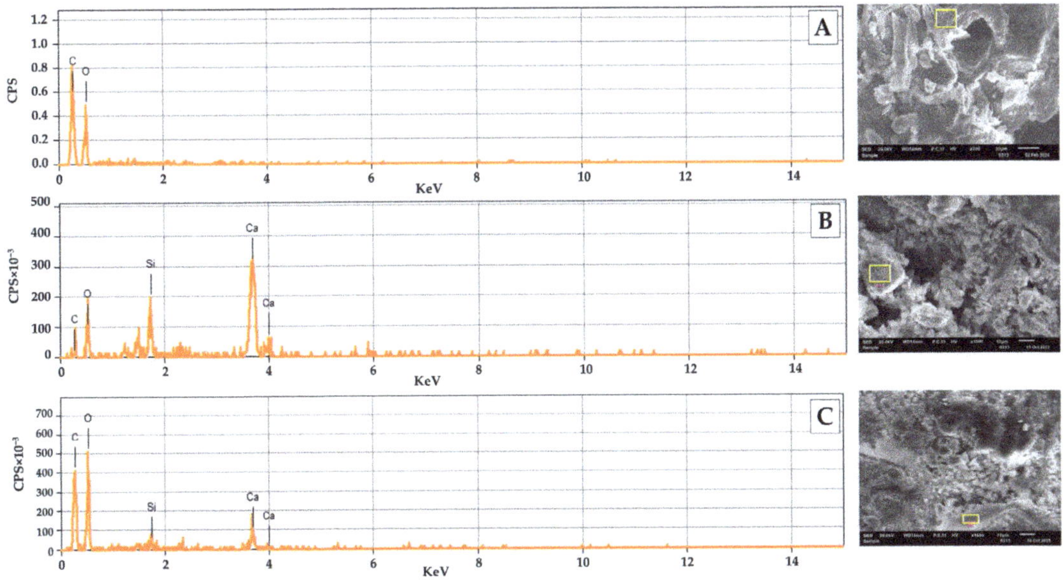

Figure 5. EDS spectra and analyzed points of (**A**) Cedar sawdust; (**B**) Cement matrix; and (**C**) Reinforcement.

Table 5. Estimated elemental chemical composition for the composites.

	Detection of Chemical Elements (% by Weight)				Ca/Si
	C	O	Si	Ca	
Cedar sawdust	52.38	47.62	-	-	-
Surface of the cement matrix	8.28	51.12	8.97	31.63	3.52
Surface of reinforcement	41.18	53.57	1.04	4.21	4.05

Figure 5A reveals that cedar sawdust particles consist primarily of oxygen (O) and carbon (C), which are key components of lignocellulosic materials, while Figure 5B indicates that the cement matrix surrounding the reinforcement contains inorganic elements, such as calcium (Ca) and silicon (Si), as its main constituents. These findings align with the X-ray fluorescence (XRF) analysis presented earlier. The predominant element (Ca) originates from lime (CaO), while the silicon is derived from silica (SiO_2). Some minerals, such as Al, Fe, and Mg, were not detected prominently due to the low proportions of alumina (Al_2O_3), hematite (Fe_2O_3), and magnesium oxide (MgO) in this type of cement.

Figure 5C reveals that the chemical composition of the reinforcement displays elevated levels of oxygen (O) and carbon (C) in comparison to the cement matrix, attributed mainly to the organic nature of the cedar sawdust particles utilized in board production. Additionally, the surface of the reinforcement exhibits a moderate presence of calcium and a relatively low quantity of silicon. This outcome is consistent with previous studies. For instance, Miranda et al. observed low amounts of Ca, Si, Al, and Mg on the surface of *Pinus* spp.

Particles attribute this to the precipitation of materials from cement hydration on the cell walls of these particles [12].

Similarly, Brasileiro et al. noted a significant rise in Ca concentration on the surface of coir pith particles used as reinforcement in cementitious composites, attributing it to the ability of lignocellulosic materials to stabilize this element on their surfaces, along with the high mobility of Ca ions in the cement hydration solution. Consequently, there is an accumulation of cement hydration products, represented mainly by calcium hydroxide ($Ca(OH)_2$) and hydrated calcium silicate (C-S-H), at the reinforcement–cement interface [59]. Aras et al. reported that the formation of these products increases as the amount of CaO and SiO_2 components in the cement rises, strengthening the bonds between wood and matrix and improving the mechanical and physical properties of cement boards [61].

In addition to hydrogen bonds, cement hydration products contribute to the creation of a physical anchor between vegetal particles and the paste, due to the porosity of these materials. Furthermore, the analysis indicates a notable variation in the Ca/Si ratio at the fracture surface. Specifically, it increases from 3.52 at the fracture surface of the cement matrix to 4.05 at the surface of the reinforcement. This observation suggests that the localized augmentation in Ca/Si ratios at interfaces means that these regions facilitated the accumulation of more hydration products, such as CH crystals, thereby fostering the development of interlocking structures between wood and cement. Consequently, this phenomenon contributes to enhanced bond strength and elevated mechanical strength [60].

2.5. Fourier Transform Infrared Spectroscopy (FTIR)

Figure 6 displays the FTIR spectra for cedar sawdust, Super White Cement, and boards with different cedar sawdust contents (10%, 15%, 20%, and 25%). These results reveal several significant absorption bands that explain the findings of the MEB, EDS, and thermal analyses.

The FTIR spectra of the cement show intense peaks characteristic of the C-O bond in calcium carbonate ($CO_3{}^{2-}$), represented by an asymmetric stretching vibration mode at approximately 1420 cm^{-1} and a bending vibration mode at 875 cm^{-1} [62]. The band at around 925 cm^{-1} is related to the asymmetric stretching vibration of the silicate group (Si-O) in SiO_4^{4-}, associated with the presence of the C_3S and C_2S mineral phases, and the band at 1130 cm^{-1} is attributed to the vibrations of the sulfate group (S-O) in SiO_4^{2-}. The presence of these bands indicates the existence of calcium silicates and calcium sulfate such as gypsum and anhydrite [63]. The small peaks in the range of 712 to 780 cm^{-1} correspond to the symmetric stretching vibration of Si-O-Si, and those in the range of 600 to 660 cm^{-1} are attributed to the bending vibration of Si-O-Si [62].

The FTIR spectra of the cedar sawdust reveal the presence of a broad band located at 3330 cm^{-1}, corresponding to the stretching vibrations of bonded O-H groups from the three main polymers present in wood (cellulose, hemicellulose, and lignin). The band at 2910 cm^{-1} is attributed to the asymmetric stretching vibration of aliphatic C-H bonds in methyl (CH_3) and methylene (CH_2) groups of holocellulose and lignin. The small bands at 1590 cm^{-1} and 1505 cm^{-1} correspond to the symmetric elongations of C=C bonds in aromatic rings, attributed to lignin, while the broad band located at 1019 cm^{-1} is attributed to the stretching vibration of the C-O bond of the primary alcohol in carbohydrates and the C-O-bond of ether in cellulose, or the CH_3-O bond in lignin and the CH_3-O- bond in ester [3,64].

The FTIR spectra of all the boards show a strong presence of the characteristic peaks of calcium carbonate at 1420 cm^{-1} and 875 cm^{-1}. The intensity of these peaks increases with the cedar sawdust content in the boards, reaching its highest level at a cedar sawdust content of 20%. Govin observed similar results, noting that the intensity of the broad band associated with the C-O bond vibrations in calcium carbonate increased as the poplar content in the boards increased [65]. This increased formation of carbonates can be explained by the reaction of organic compounds present in the wood with the calcium in the cement, thus promoting the formation of calcium carbonates. This result corroborates the increase in carbonation observed during the thermal analysis.

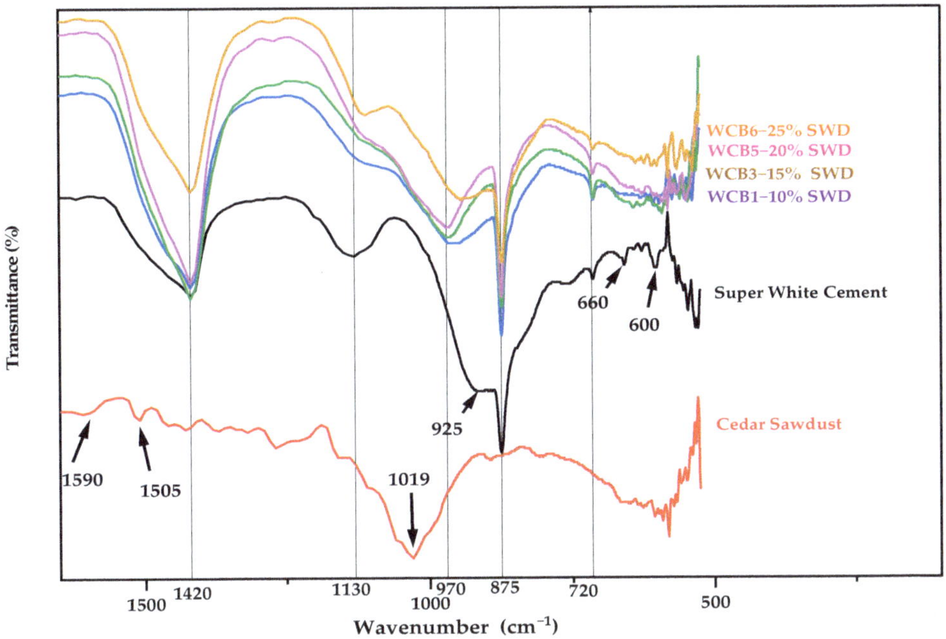

(**a**)

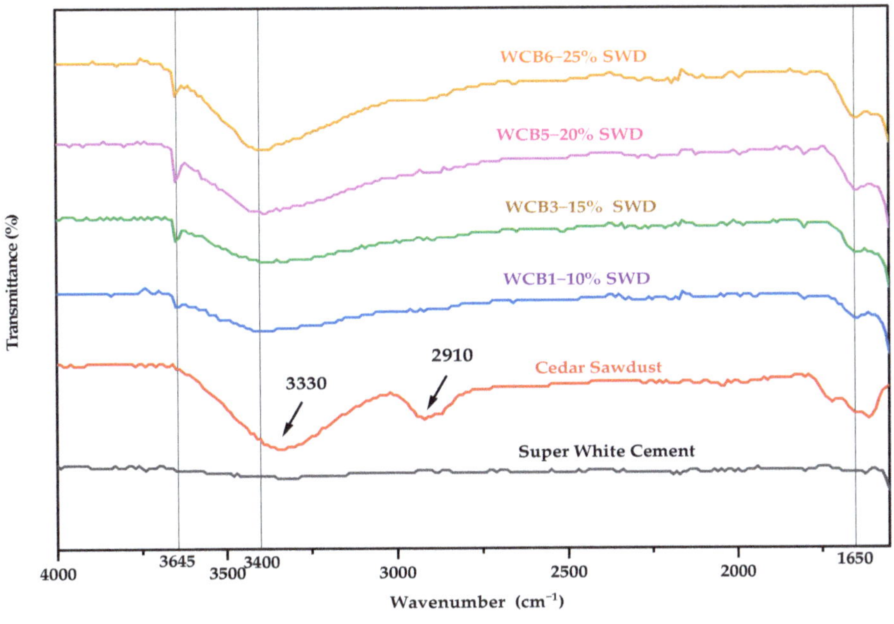

(**b**)

Figure 6. FTIR spectra of samples: (**a**) 1600–500 cm^{-1} (**b**) 4000–1600 cm^{-1}.

Furthermore, the spectra reveal the increasing appearance of broad bands at 3400 cm^{-1} and 1650 cm^{-1}, corresponding to the stretching and bending vibrations of bonded -OH groups and H_2O, respectively [62]. The intensity of these bands rises with higher cedar sawdust content in the boards. This enhancement is likely due to the hydrophilic nature of organic compounds in the cedar sawdust, which promote the formation of calcium hydroxides (portlandite) during cement hydration, thereby increasing the presence of -OH groups.

The spectra reveal prominent bands at approximately 975 cm^{-1}, associated with Si–O asymmetric stretching vibration. These bands confirm the formation of calcium silicate hydrate (C–S–H) gel, which is essential for the strength of the concrete [66]. Previous studies suggest that the peak associated with C-S-H gel in Portland cement systems represents a shift of the previously observed peak at 923 cm^{-1} in cement spectra to higher wave numbers [67–69]. Mollah links this shift to the polymerization of SiO_4^{4-} silicate units during the formation of C-S-H gel [69]. Additionally, isolated hydroxide ion vibrations were found to exhibit a distinctive fine band feature at 3645 cm. This band facilitates the monitoring of portlandite ($Ca(OH)_2$) formation [63]. The intensity of this band slightly increases with sawdust content, suggesting that cedar sawdust promotes the precipitation of portlandite. This observation supports the noted increase in carbonation up to 20% cedar sawdust content.

2.6. Thermal Analysis

Figure 7 presents the TG/DTG and DSC analysis results for samples of Super White Cement (Figure 7a), cedar sawdust (Figure 7b), and the WCB 1 board containing 10% cedar sawdust (Figure 7c). The thermal behavior of Super White Cement begins at 60 °C, as shown in Figure 7a, with a total mass loss of 5%. Previous studies [63,70,71] attribute this mass loss to (i) dehydration of crystal-bonded water molecules, primarily due to cement's hygroscopic nature and the presence of pre-hydrated phases, occurring up to 275 °C; (ii) dihydroxylation of calcium hydroxide ($Ca(OH)_2$), as detailed in Equation (1), marked by a moderate-intensity endothermic peak at 433 °C; and (iii) the decarbonation of calcium carbonate ($CaCO_3$), as described in Equation (2), indicated by a sharp endothermic peak at 762 °C.

$$Ca(OH)_2 \rightarrow CaO + H_2O, \quad (1)$$

$$CaCO_3 \rightarrow CaO + CO_2, \quad (2)$$

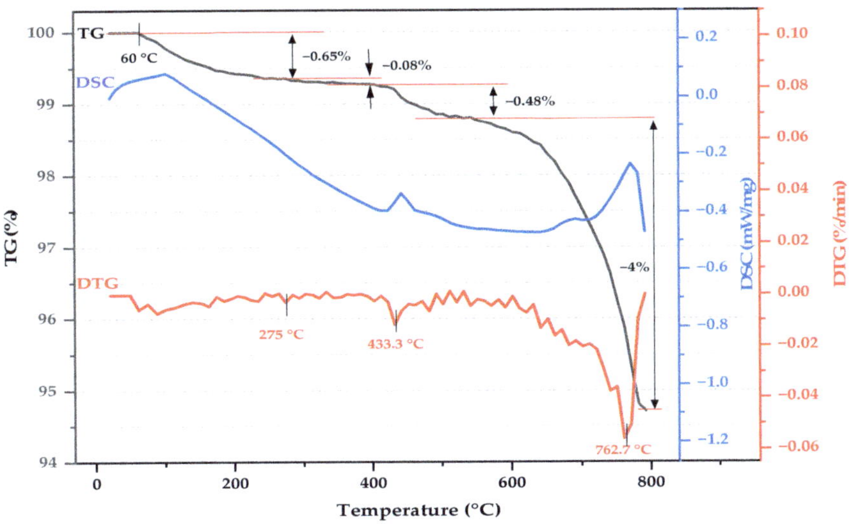

(a)

Figure 7. *Cont.*

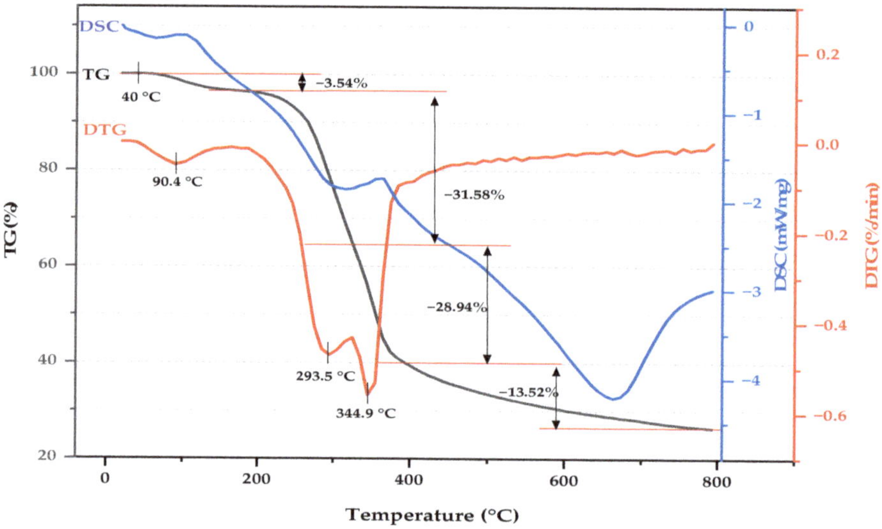

(**b**)

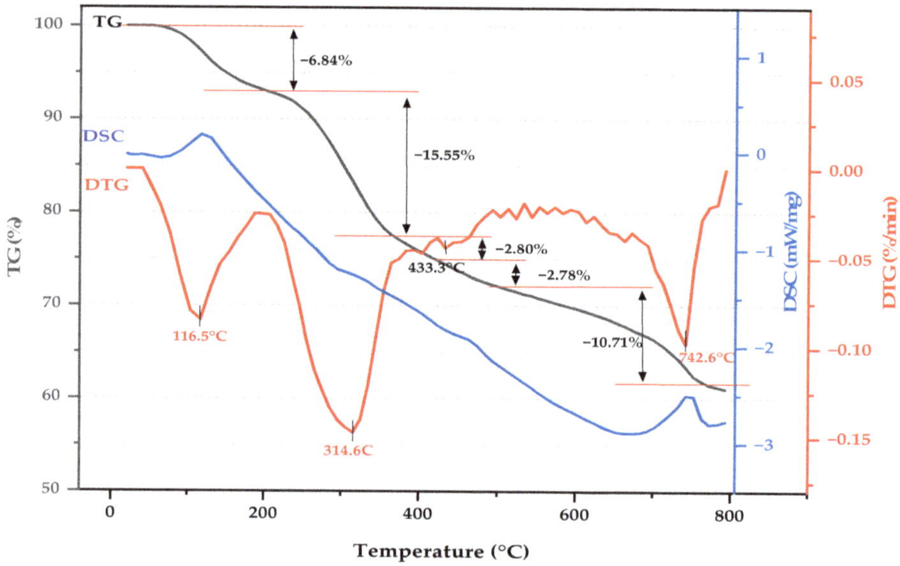

(**c**)

Figure 7. Thermal analysis (TG, DTG, and DSC) of (**a**) Super White Cement, (**b**) Cedar sawdust, and (**c**) WCB (10% SWD).

The TG/DTG and DSC results in Figure 7b indicate that cedar sawdust degradation starts at 40 °C, with a total mass loss exceeding 77%, occurring in three distinct stages. In the first weight loss stage, a 3.54% mass loss is observed, accompanied by a low-intensity endothermic peak at 90.4 °C. This peak is associated with the evaporation of water and light volatile substances from cedar wood. The second stage features two closely connected intense peaks: the first peak, observed at 293 °C, corresponds to a 31% mass loss, primarily due to the decomposition of hemicellulose and minor cellulose dehydration. The second peak, located at 345 °C, shows a 29% mass loss attributed to the thermal decomposition of cellulose [72]. According to Kim et al., wood components decompose at different temperatures: hemicellulose between 150 and 350 °C, cellulose between 275 and 350 °C, and lignin between 250 and 500 °C [73]. The final phase, starting at 390 °C, reveals a 13% mass loss, likely due to carbonaceous residues' combustion and lignin's thermal degradation, which is relatively stable and degrades over a wider temperature range [72].

The TG/DTG and DSC curves of the WCB 1 shown in Figure 7c reveal a total mass loss of 38%, occurring in several stages marked by significant endothermic effects. The initial endothermic effect observed between 40 and 190 °C is attributed to the dehydration of interstitial water and the moisture present in cement and sawdust, as well as to the thermal decomposition of calcium silicate hydrate (C–S–H) and ettringite [74,75]. According to the findings of Yel et al. [71], the largest mass loss (15%) observed between 190 and 390 °C, is attributed to the decomposition of wood components, during which the most intense endothermic peak occurs, consistent with findings from cedar sawdust analysis in Figure 7b. Between 420 and 490 °C, a low-intensity thermal effect is associated with the dehydration of calcium hydroxide ($Ca(OH)_2$), accompanied by a measured mass loss (approximately 2.78%) that is slightly lower than reported in other studies [71,74]. Petkova et al. suggest that this discrepancy may indicate a lower presence of portlandite crystals, potentially leading to the increased amount of C-S-H phases in the samples and enhanced strength properties [63].

The decarbonization of calcium carbonate ($CaCO_3$) occurs between 620 and 800 °C, marked by an intense endothermic peak and a mass loss exceeding 10%. Similar observations have been noted in previous studies on cement-bonded particleboards made from various species [71,74,75]. The higher mass loss (10.7%) compared to raw cement (4%) in this temperature range suggests differences in the amount and crystallization of carbonates formed. Studies on accelerated carbonation in particleboards, including Cabral et al.'s research [74], indicate a higher mass loss in carbonated samples due to the decomposition of well-crystallized calcium carbonate compared to poorly crystallized forms in control samples. Additionally, Hassan et al. highlight the conversion of portlandite to $CaCO_3$ in the presence of organic fibers such as plant fibers [75], suggesting that cedar sawdust promotes increased formation of well-crystallized calcium carbonate within the boards. Similar results were observed by Govin for cementitious pastes containing poplar wood [65].

3. Materials and Methods

3.1. Materials

The materials used in the production of the boards included cedar sawdust as a reinforcing agent, Super White Cement (SWC) as a binder, and an appropriate amount of water. Figure 8 illustrates the morphologies and images of the cedar sawdust and cement employed in this study.

3.1.1. Cedar Sawdust

The raw material used in this study comprised wood sawdust sourced from woodworking workshops in Khénifra, Morocco, where Atlas cedar wood is commonly processed. Following the cutting of cedar logs, the resulting residues underwent collection under industrial conditions, requiring the removal of flakes and the sorting of the fraction suitable for the study. The sorted sawdust particles then underwent classification using 2 mm and 4 mm diameter sieves. Particles that passed through the 4 mm mesh but were retained by the 2 mm mesh were chosen for the study. Any particles larger than 4 mm were trimmed before reuse, while those smaller than 2 mm were deemed insignificant and excluded. The acquired sawdust particles underwent air-drying for two weeks until reaching a moisture content of 12%. Subsequently, their density was determined, and they were stored in plastic bags for future utilization.

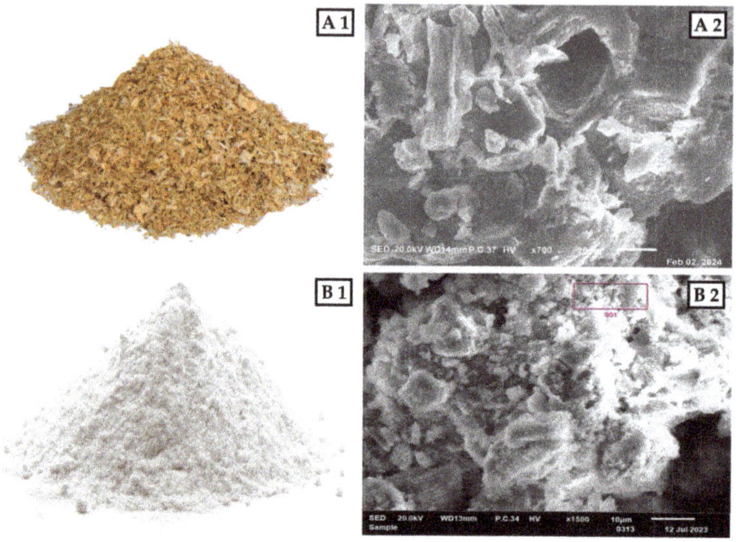

Figure 8. (**A1**) Image of cedar sawdust; (**A2**) SEM image of cedar sawdust; (**B1**) Image of SWC; (**B2**) SEM image of SWC.

3.1.2. Super White Cement (SWC)

The Super White Cement (SWC, CEM 52.5R) utilized as a binder was a commercial product from the manufacturer Çimsa (Ataşehir, Turkey). The cement was characterized following the procedures outlined in NM 10.1.005 (2008) and EN 196-2-3-6 standards [76–79]. The characterization process involved analyzing the chemical composition of cement samples using X-ray fluorescence spectroscopy. Each sample, weighing 10 g, was compacted inside a metal ring using an APM device, alongside specific admixture tablets. The resulting pellet was then inserted into an XRF spectrometer (ARL 9900, Thermo Fisher Scientific, Waltham, MA, USA) for analysis. Based on the chemical composition of the SWC, Bogue's equations were employed to calculate the percentage of anhydrous cement compounds [80]. Additionally, cement characterization encompassed the determination of physical properties, mainly specific surface area, setting time, fineness, density, and loss on ignition. The obtained cement composition and its physical properties are displayed in Table 2.

3.2. Boards Production

The production of the boards was based on a single variable, namely sawdust content by weight. Six different levels of cedar sawdust content were employed: 10%, 12.5%, 15%, 17.5%, 20%, and 25%, with four repetitions for each cedar sawdust content level, resulting in a total of 24 panels. The quantities of components and the designations of the prepared panels are indicated in Table 6. Methods from prior studies were employed to calculate the required amounts of cement and cedar sawdust [22,44], aiming to produce boards with a density of 1200 kg/m^3 and nominal dimensions of 28 cm × 28 cm × 1 cm. The specifics of the components and designations of the prepared boards are provided in Table 6, along with the necessary water amounts calculated using Equation (3), developed by Simatupang [81], as follows:

$$\text{Water requirement (mL)} = 0.35\,Z + (0.30 - r) \times H, \quad (3)$$

where Z (g) represents the cement weight, r (%) indicates the moisture content of the sawdust, and H (g) denotes the sawdust weight.

Table 6. The production plan of wood cement boards.

Board Types	Raw Material		
	Sawdust (wt%)	Cement (wt%)	Water (g)
WCB 1	10	90	313
WCB 2	12.5	87.5	309
WCB 3	15	85	305
WCB 4	17.5	82.5	301
WCB 5	20	80	297
WCB 6	25	75	289

The board production process is illustrated in Figure 9. Following weighing, the cement and cedar sawdust quantities were meticulously dry-mixed in a plastic bowl before adding the required water amount. The mixture was uniformly stirred for 10 min until achieving a homogeneous paste, free of lumps. The resulting paste was poured into a detachable iron mold designed for creating boards with dimensions of 28 cm × 28 cm × 1 cm. Using a wooden ruler, the paste was evenly distributed to create a flat, homogeneous mat. To aid in demolding and prevent adherence of the board to the mold's interior, the mat was covered with a polyethylene sheet before placing the metal cap. The paste then underwent 24 h of cold pressing in a manual hydraulic press, with pressure varying between 2 and 6 MPa, to reach a final board thickness of 10 mm. Once pressed, the boards were detached from the mold and stored in polyethylene bags for 28 days to prevent moisture loss, complete the cement hydration, and allow it to fully harden. Subsequently, the boards underwent finishing operations using sandpaper, including sanding the surfaces and edges (Figure 10a). Then, they were cut using circular saws based on the principles specified in the European Standards to obtain specimens suitable for mechanical and physical tests, with the remaining parts utilized in other tests.

Figure 9. The production steps of WCB.

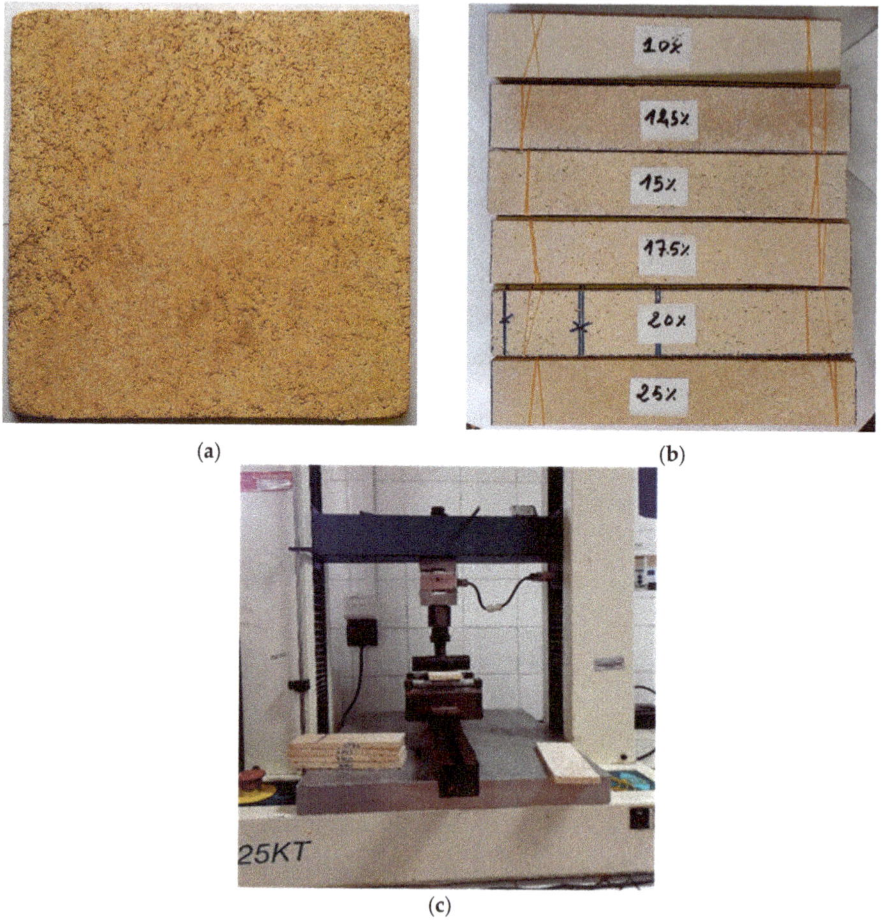

Figure 10. (**a**) WCB5 after finishing operations; (**b**) MOR-MOE test samples; (**c**) The universal testing machine used for mechanical characterization.

3.3. *Physical and Mechanical Characterization of the Boards*

The physical tests encompassed the measurement of density, water absorption (WA), and thickness swelling (TS). A precision balance and a digital caliper were employed to measure the mass and dimensions of the samples following the PNM EN 325 (2012) standard [82]. For density calculations, eight specimens of each content with dimensions of 50×50 mm^2 were used. Average values were obtained by calculating the ratio between the weight (g) and volume (mm^3) of each test specimen according to the EN 323 (1993) standard [83]. The WA and TS were evaluated successively after both 2 h and 24 h immersion in water. Tests were manually conducted on square specimens, similar to those used for density measurements, following the procedures described in the EN 317 (1993) standard [84]. Eight samples were tested for each cedar sawdust content level by immersing them in deionized water under room conditions. The initial (W_1) and final (W_2) weights before and after immersing were measured, and the WA was calculated using the following formula:

$$\text{Water absorption} = (W_2 - W_1)/W_1 \times 100. \quad (4)$$

Similarly, for the initial (T1) and final (T2) thicknesses, the TS was calculated using the formula:

$$\text{Thickness swelling} = (T_2 - T_1)/T_1 \times 100. \quad (5)$$

The assessment of the mechanical properties involved three-point bending tests conducted on rectangular specimens measuring $25 \times 5 \times 1$ cm^3 (Figure 10b). For each cedar sawdust content level, six samples were individually mounted on a universal testing machine (H25KT, Tinius Olsen, Horsham, PA, USA) equipped with a 25 KT compression/tension sensor (Figure 10c).

A perpendicular bending load was applied to the center of each sample at a speed of 10 mm/min using an electromechanical motor until failure occurred. The modulus of rupture (MOR) and the modulus of elasticity (MOE) were determined following the EN 310 (1993) standard [85].

3.4. Microstructural Analysis of the Boards

Microstructural analysis of the produced boards was conducted via scanning electron microscopy (SEM) using a JSM-IT500HR/LA microscope (JEOL, Tokyo, Japan). This analysis was conducted to investigate structural variations in the boards. It also examined the differences in the fracture surfaces between boards containing 10% and 25% cedar sawdust after bending tests. The SEM analysis was performed at an accelerating voltage of 10 kV and a working distance of 13.4 mm in a low-vacuum mode (LV-SEM). The SEM instrument is coupled with Energy-dispersive X-ray spectroscopy (EDS), which was utilized to examine variations in the elemental composition of the samples.

3.5. Fourier Transform Infrared Spectroscopy Analysis

Fourier transform infrared spectroscopy (FTIR) was employed to observe chemical reactions and identify functional groups of cedar sawdust, Super White Cement, and various WCB samples. Spectral data were acquired using a Nicolet iS10 FTIR spectrophotometer (Thermo Fisher Scientific, Waltham, MA, USA), utilizing the Attenuated Total Reflectance (ATR) technique. Pellets were prepared from 300 mg of finely ground sample powders, ensuring an accurate representation of each material. Spectra were recorded over a broad wavelength range spanning from 500 to 4000 cm^{-1}, with a high spectral resolution of 4 cm^{-1} and an average of 32 co-addition scans.

3.6. Thermal Analysis of the Boards

Thermogravimetric analysis (TGA), derivative thermogravimetry (DTG), and differential scanning calorimetry (DSC) were performed to investigate the thermal behavior and stability profiles of cedar sawdust, Super White Cement, and related WCBs. The measurements were conducted using a simultaneous thermal analyzer, specifically the Netzsch STA 449 F5 Jupiter (NETZSCH Gerätebau GmbH, Selb, Germany). Operational parameters for TG/DTG-DSC measurements included a sample mass of 15.0 ± 5.0 mg and an Al_2O_3 sample crucible with a volume of 85 µL. Heating was carried out under static conditions in an argon atmosphere, with a temperature range spanning from 25 °C to 800 °C and a heating rate of 10 °C/min.

3.7. Statistical Analysis

Statistical analysis was carried out on the data obtained during the experiment, including a one-way analysis of variance (ANOVA), to assess the relative importance of various sources of variation in the physical and mechanical property results. The ANOVA revealed statistically significant differences between the levels of all properties. Subsequently, mean values were compared using Tukey's test at a 95% confidence level.

4. Conclusions

This study highlights the feasibility of using cedar sawdust as a reinforcing agent in wood–cement boards, providing an effective approach for recycling bio-waste and advancing eco-friendly construction materials. Incorporating cedar sawdust improves production efficiency and environmental benefits, with optimal physical and mechanical performance achieved at a sawdust content of 10%. The key findings from this research include the following:

- Atlas cedar sawdust, with lower extractives compared to other species, enhances board properties without significantly affecting the cement setting.
- Physical properties indicate that board density decreases with increasing cedar sawdust content, reaching its lowest at 25% sawdust. Concurrently, water absorption and thickness swelling increase with higher sawdust content due to its hygroscopic nature and resultant porosity.
- Mechanical properties reveal that boards with 10% cedar sawdust content exhibit the highest modulus of rupture (MOR) and modulus of elasticity (MOE). In contrast, increasing the sawdust content beyond this level results in lower mechanical performance due to higher porosity and insufficient cement bonding, making the boards less suitable for structural applications.
- Morphological analysis shows that lower sawdust content (10%) provides better adhesion between the sawdust and the cement matrix, resulting in higher density and improved mechanical properties. In contrast, boards with higher sawdust content (25%) display increased porosity and weaker particle–matrix bonds, contributing to reduced strength and increased water absorption and thickness swelling.
- EDS analysis indicates an increased calcium-to-silicon ratio at the particle–matrix interface, suggesting enhanced bonding for boards with 10% cedar sawdust due to the accumulation of cement hydration products, mainly calcium hydroxide ($Ca(OH)_2$) and calcium silicate hydrate (C-S-H).
- FTIR spectroscopy results show an increase in calcium carbonate and hydroxide formation, with higher sawdust content, indicating improved carbonation and C-S-H gel formation.
- Thermal analysis confirms the consistent formation of well-crystallized calcium carbonate and a lower presence of portlandite crystals, contributing to increased C-S-H phases in the boards.

This research underscores the potential of using locally sourced bio-waste and highlights the need for further investigations to optimize board performances. Future studies will focus on strategies to decrease porosity at higher sawdust contents and the use of pretreated cedar sawdust, and explore chemical additives to reduce setting time and improve the interface bonding quality. Addressing these aspects will enhance the sustainability and practical application of wood–cement boards, promoting more environmentally friendly construction practices.

Author Contributions: Conceptualization, A.E.H. and N.-E.E.M.; methodology, A.E.H. and N.-E.E.M.; software, A.E.H.; validation, A.E.H., Y.M. and N.-E.E.M.; formal analysis, A.E.H., Y.M. and A.O.; investigation, A.E.H.; resources, A.E.H., Y.M., A.O. and N.-E.E.M.; data curation, A.E.H. and Y.M.; writing—original draft preparation, A.E.H. and N.-E.E.M.; writing—review and editing, A.E.H., Y.M. and N.-E.E.M.; visualization, A.E.H. and N.-E.E.M.; supervision, N.-E.E.M.; project administration, M.C. and N.-E.E.M.; funding acquisition, A.E.H. and N.-E.E.M. All authors have read and agreed to the published version of the manuscript.

Funding: This research received no external funding.

Institutional Review Board Statement: Not applicable.

Informed Consent Statement: Not applicable.

Data Availability Statement: Data are contained within the article.

Acknowledgments: The authors would like to thank the company CIMAT-Béni-Mellal and the CTIBA center for their continuous support.

Conflicts of Interest: The authors declare no conflicts of interest.

Abbreviations

WCB	Wood-Cement Board
WA	Water Absorption
TS	Thickness Swelling
SEM	Scanning Electron Microscopy
EDS	Energy-Dispersive X-ray Spectroscopy
XRF	X-ray Fluorescence
TG/DTG—DSC	Thermal analysis
FTIR	Fourier Transformed Infrared
CBPB	Cement-Bonded Particleboard
SWC	Super White Cement
WCB 1	Boards with 10% Cedar sawdust
WCB 2	Boards with 12.5% Cedar sawdust
WCB 3	Boards with 15% Cedar sawdust
WCB 4	Boards with 17.5% Cedar sawdust
WCB 5	Boards with 20% Cedar sawdust
WCB 6	Boards with 25% Cedar sawdust
C–S–H	Calcium silicate hydrate
SWD	Cedar sawdust

References

1. Haboudane, D. El Mustapha Bahri Deforestation Detection and Monitoring in Cedar Forests of the Moroccan Middle-Atlas Mountains. In Proceedings of the 2007 IEEE International Geoscience and Remote Sensing Symposium, Barcelona, Spain, 23–28 July 2007; IEEE: Barcelona, Spain, 2007; pp. 4327–4330.
2. Farjon, A. *Pinaceae: Drawings and Descriptions of the Genera Abies, Cedrus, Pseudolarix, Keteleeria, Nothotsuga, Tsuga, Cathaya, Pseudotsuga, Larix and Picea*; Koeltz: Königstein, Germany, 1990; ISBN 978-3-87429-298-6.
3. Bouramdane, Y.; Fellak, S.; El Mansouri, F.; Boukir, A. Impact of Natural Degradation on the Aged Lignocellulose Fibers of Moroccan Cedar Softwood: Structural Elucidation by Infrared Spectroscopy (ATR-FTIR) and X-ray Diffraction (XRD). *Fermentation* **2022**, *8*, 698. [CrossRef]
4. Azzouzi, K.E.; Keller, R. Propriétés technologiques du bois de cèdre de l'Atlas (Cedrus atlantica Manetti). *Mediterr. For.* **1998**, *19*, 11–33.
5. Ellatifi, M. *L'Economie de la Forêt et des Produits Forestiers au Maroc Bilan et Perspectives*; 1. Aufl.; Presses Académiques Francophones: Saarbrücken, Germany, 2014; ISBN 978-3-8381-7547-8.
6. Adelusi, E.A.; Olaoye, K.O.; Adebawo, F.G. Strength and Dimensional Stability of Cement-Bonded Boards Manufactured from Mixture of Ceiba Pentandra and Gmelina Arborea Sawdust. *JERR* **2019**, *8*, 1–10. [CrossRef]
7. Ajayi, B.; Badejo, S. Effects of Board Density on Bending Strength and Internal Bond of Cement-Bonded Flakeboards. *J. Trop. For. Sci.* **2005**, *17*, 228–234.
8. Rihani, M.; Rihani, M. *Design and Application of an Innovative Composting Unit for the Effective Treatment of Sludge and Other Biodegradable Organic Waste in Morocco, Annex 4 2006*; University of Chouaib Doukkali: El Jadida, Morocco, 2006; pp. 1–6.
9. Setter, C.; de Melo, R.R.; do Carmo, J.F.; Stangerlin, D.M.; Pimenta, A.S. Cement Boards Reinforced with Wood Sawdust: An Option for Sustainable Construction. *SN Appl. Sci.* **2020**, *2*, 1631. [CrossRef]
10. Donmez Cavdar, A.; Yel, H.; Boran, S.; Pesman, E. Cement Type Composite Panels Manufactured Using Paper Mill Sludge as Filler. *Constr. Build. Mater.* **2017**, *142*, 410–416. [CrossRef]
11. De Souza, M.J.C.; de Melo, R.R.; Guimarães, J.B.; Carnaval, T.K.B.D.A.; Pimenta, A.S.; Mascarenhas, A.R.P. Wood–Cement Boards with Addition of Coconut Husk. *Wood Mater. Sci. Eng.* **2022**, *17*, 617–626. [CrossRef]
12. Miranda de Lima, A.J.; Iwakiri, S.; Satyanarayana, K.G.; Lomelí-Ramírez, M.G. Studies on the Durability of Wood-Cement Particleboards Produced with Residues of *Pinus* spp., Silica Fume, and Rice Husk Ash. *BioRes* **2020**, *15*, 3064–3086. [CrossRef]
13. Nasser, R.A.; Al-Mefarrej, H.A.; Abdel-Aal, M.A.; Alshahrani, T.S. Effects of Tree Species and Wood Particle Size on the Properties of Cement-Bonded Particleboard Manufacturing from Tree Prunings. *J. Environ. Biol.* **2014**, *35*, 961–971.
14. Islam, M.R.; Rahman, F.; Islam, M.N.; Rana, M.N.; Nath, S.K.; Ashaduzzaman, M.; Shams, M.I. Cement-Bonded Lignocellulosic Panel (CLP): A Promising Environmental Friendly Construction Material for Conservation of Forest Resources. In *Handbook of Ecomaterials*; Martínez, L.M.T., Kharissova, O.V., Kharisov, B.I., Eds.; Springer International Publishing: Cham, Switzerland, 2018; pp. 1–13. ISBN 978-3-319-48281-1.
15. Frybort, S.; Mauritz, R.; Teischinger, A.; Müller, U. Cement Bonded Composites—A Mechanical Review. *BioRes* **2008**, *3*, 602–626. [CrossRef]
16. Wolfe, G. Cement-Bonded Wood Composites as an Engineering Material. In *The Use of Recycled Wood and Paper in Building Applications*; Forest Products Society: LaGrange, GA, USA, 1996; pp. 84–91.

17. Moslemi, A.A.; Souza, M.; Geimer, R. Accelerated Ageing of Cement-Bonded Particleboard. In *Inorganic Bonded Wood and Fiber Composite Materials*; Forest Products Research Society: Madison, WI, USA, 1995; pp. 83–88.
18. Mendes, R.F.; Vilela, A.P.; Farrapo, C.L.; Mendes, J.F.; Denzin Tonoli, G.H.; Mendes, L.M. Lignocellulosic Residues in Cement-Bonded Panels. In *Sustainable and Nonconventional Construction Materials Using Inorganic Bonded Fiber Composites*; Elsevier: Amsterdam, The Netherlands, 2017; pp. 3–16. ISBN 978-0-08-102001-2.
19. Ogunjobi, K.M.; Ajibade, M.A.; Gakenou, O.F.; Gbande, S. Physical and mechanical properties of cement-bonded particle board produced from Anogeissus leiocarpus (DC.) Guill and Perr wood species. *Afr. J. Agric. Technol. Environ.* **2019**, *8*, 192–199.
20. Adedeji, Y. Sustainable Housing in Developing Nations: The Use of Agrowaste Composite Panels for Walls. *Built Human Environ. Rev.* **2011**, *4*, 36–47.
21. Jorge, F.C.; Pereira, C.; Ferreira, J.M.F. Wood-Cement Composites: A Review. *Holz Roh Werkst.* **2004**, *62*, 370–377. [CrossRef]
22. Dotun, A.O.; Olalekan, O.S.; Olugbenga, A.L.; Emmanuel, M.A. Physical and Mechanical Properties Evaluation of Corncob and Sawdust Cement Bonded Ceiling Boards. *JERA* **2019**, *42*, 65–75. [CrossRef]
23. Pirayesh, H.; Saadatnia, M.A. Effects of Accelerating Agent and SiO_2 Nano-Powder on the Mechanical and Physical Characteristics of Nut Shell–Cement Based Composites. *Mater. Struct.* **2016**, *49*, 3435–3443. [CrossRef]
24. Dong, Z.; Yu, Y.; Song, P.; Ma, L.; Lu, F. Effects of Two Types of Waste Wood Species on the Hydration Characteristic of Portland Cement. *ACT* **2016**, *14*, 13–20. [CrossRef]
25. Antwi-Boasiako, C.; Ofosuhene, L.; Boadu, K.B. Suitability of Sawdust from Three Tropical Timbers for Wood-Cement Composites. *J. Sustain. For.* **2018**, *37*, 414–428. [CrossRef]
26. Quiquandon, B. *Le Bois de Cèdre (Cedrus atlantica) Provenant des Reboisements Français*; CTB: Paris, France, 1976; p. 31. Available online: https://infodoc.agroparistech.fr/index.php?lvl=notice_display&id=119406 (accessed on 6 September 2024).
27. Sanaev, V.G.; Zaprudnov, V.I.; Gorbacheva, G.A.; Oblivin, A.N. Factors Affecting the Quality of Wood-Cement Composites. *Bull. Transilv. Univ. Brasov. Ser. II For. Wood Ind. Agric. Food Eng.* **2016**, *9*, 63–70.
28. Rowell, R. (Ed.) *The Chemistry of Solid Wood*; Advances in Chemistry; American Chemical Society: Washington, DC, USA, 1984; Volume 207, ISBN 978-0-8412-0796-7.
29. Na, B.; Wang, Z.; Wang, H.; Lu, X. Wood-Cement Compatibility Review. *Wood Res.* **2014**, *59*, 813–826.
30. Okino, E.Y.; Souza, M.R.; Santana, M.A.; Alves, M.V.S.; de Sousa, M.E.; E Teixeira, D. Cement-Bonded Wood Particleboard with a Mixture of Eucalypt and Rubberwood. *Cem. Concr. Compos.* **2004**, *26*, 729–734. [CrossRef]
31. Chowdhury, S.; Maniar, A.; Suganya, O.M. Strength Development in Concrete with Wood Ash Blended Cement and Use of Soft Computing Models to Predict Strength Parameters. *J. Adv. Res.* **2015**, *6*, 907–913. [CrossRef] [PubMed]
32. Atoyebi, O.D.; Awolusi, T.F.; Davies, I.E.E. Artificial Neural Network Evaluation of Cement-Bonded Particle Board Produced from Red Iron Wood (*Lophira alata*) Sawdust and Palm Kernel Shell Residues. *Case Stud. Constr. Mater.* **2018**, *9*, e00185. [CrossRef]
33. He, P.; Hossain, U.; Poon, C.S.; Tsang, D.C. Mechanical, Durability and Environmental Aspects of Magnesium Oxychloride Cement Boards Incorporating Waste Wood. *J. Clean. Prod.* **2019**, *207*, 391–399. [CrossRef]
34. Sotannde, O.A.; Oluwadare, A.O.; Ogedoh, O.; Adeogun, P.F. Evaluation of Cement-Bonded Particle Board Produced Fromafzelia Africanawood Residues. *J. Eng. Sci. Technol.* **2012**, *7*, 732–743.
35. Farag, E.; Alshebani, M.; Elhrari, W.; Klash, A.; Shebani, A. Production of Particleboard Using Olive Stone Waste for Interior Design. *J. Build. Eng.* **2020**, *29*, 101119. [CrossRef]
36. BISON. *Wood-Cement Board Report*; British Research Establishemen (Bre): Hertfordshire, UK, 1978; p. 10.
37. Keržič, E.; Humar, M. Studies on the Material Resistance and Moisture Dynamics of Wood after Artificial and Natural Weathering. *Wood Mater. Sci. Eng.* **2022**, *17*, 551–557. [CrossRef]
38. Žlahtič-Zupanc, M.; Lesar, B.; Humar, M. Changes in Moisture Performance of Wood after Weathering. *Constr. Build. Mater.* **2018**, *193*, 529–538. [CrossRef]
39. Mancera, C.; El Mansouri, N.-E.; Pelach, M.A.; Francesc, F.; Salvadó, J. Feasibility of Incorporating Treated Lignins in Fiberboards Made from Agricultural Waste. *Waste Manag.* **2012**, *32*, 1962–1967. [CrossRef]
40. Mahzabin, M.S.; Hamid, R.; Badaruzzaman, W.H.W. Evaluation of Chemicals Incorporated Wood Fibre Cement Matrix Properties. *J. Eng. Sci. Technol.* **2013**, *8*, 385–398. [CrossRef]
41. Ali, I.M.; Nasr, M.S.; Naje, A.S. Enhancement of Cured Cement Using Environmental Waste: Particleboards Incorporating Nano Slag. *Open Eng.* **2020**, *10*, 273–281. [CrossRef]
42. Bilcati, G.K.; Lengowski, E.C.; Jumes, J.J.; Andrade, F.H.; Nervis, I.R.; De Souza, I.D.S.C.; Langaro, E. Light Weight Cementitious Composites Reinforced with *Pinus* spp. Sawdust. *CCT* **2021**, *38*, 26684. [CrossRef]
43. Bilcati, G.K.; Matoski, A.; Trianoski, R.; Lengowski, E. Uso Potencial de La Fibra de Curauá (Ananas acutifolius) Para La Fabricación de Compuestos Cementicios. *Rev. Ing. Constr.* **2018**, *33*, 155–160. [CrossRef]
44. Olorunnisola, A.O.; Agrawal, S.P. Influences of Particle Size and Cement Content on Properties of Cement-Bonded Composites from Eucalyptus Veneer Waste. In Proceedings of the International Inorganic-Bonded Fiber Composites Conference (IIBCC), Aalborg, Denmark, 21–24 September 2010.
45. Adelusi, E.A.; Olaoye, K.O.; Adelusi, F.T.; Adedokun, S.A. Physico-Mechanical Properties of Cement-Bonded Boards Produced from Mixture of Corn Cob Particles and Gmelina Arborea Sawdust. *J. For. Environ. Sci.* **2021**, *37*, 79–89. [CrossRef]
46. Amiandamhen, S.O.; Izekor, D.N. Effect of Wood Particle Geometry and Pre-Treatments on the Strength and Sorption Properties of Cement-Bonded Particle Boards. *JANS* **2013**, *5*, 318–322. [CrossRef]

47. Olorunnisola, A.O. Effects of Husk Particle Size and Calcium Chloride on Strength and Sorption Properties of Coconut Husk–Cement Composites. *Ind. Crops Prod.* **2009**, *29*, 495–501. [CrossRef]
48. Cetris. Available online: https://www.cetris.cz/en/downcnt/?file=pagedata_en/download/158_2.pdf?1700133830 (accessed on 6 September 2024).
49. Ayrilmis, N.; Hosseinihashemi, S.K.; Karimi, M.; Kargarfard, A.; Ashtiani, H.S. Technological Properties of Cement-Bonded Composite Board Produced with the Main Veins of Oil Palm (*Elaeis guineensis*) Particles. *BioResources* **2017**, *12*, 3583–3600. [CrossRef]
50. Olorunnisola; Pitman; William, M. Strength Properties and Potential Uses of Rattan–Cement Composites. *J. Bamboo Ratt.* **2005**, *4*, 343–352. [CrossRef]
51. Ashori, A.; Tabarsa, T.; Sepahvand, S. Cement-Bonded Composite Boards Made from Poplar Strands. *Constr. Build. Mater.* **2012**, *26*, 131–134. [CrossRef]
52. Ghofrani, M.; Mokaram, K.N.; Ashori, A.; Torkaman, J. Fiber-Cement Composite Using Rice Stalk Fiber and Rice Husk Ash: Mechanical and Physical Properties. *J. Compos. Mater.* **2015**, *49*, 3317–3322. [CrossRef]
53. Bejó, L.; Takáts, P.; Vass, N. Development of Cement Bonded Composite Beams. *Acta Silv. Lign. Hung* **2005**, *1*, 111–119. [CrossRef]
54. Olorunnisola, A.O. Effects of Particle Geometry and Chemical Accelerator on Strength Properties of Rattan–Cement Composites. *Afr. J. Sci. Technol.* **2007**, *8*, 22–27.
55. Cabral, M.R.; Nakanishi, E.Y.; Santos, S.F.; Christoforo, A.L.; Fiorelli, J. Orientation Effect on the Physical and Mechanical Properties of Strand Cement Boards. *Constr. Build. Mater.* **2021**, *275*, 122121. [CrossRef]
56. Amiandamhen, S.O.; Meincken, M.; Tyhoda, L. Natural Fibre Modification and Its Influence on Fibre-Matrix Interfacial Properties in Biocomposite Materials. *Fibers Polym.* **2020**, *21*, 677–689. [CrossRef]
57. Amiandamhen, S.O.; Adamopoulos, S.; Adl-Zarrabi, B.; Yin, H.; Norén, J. Recycling Sawmilling Wood Chips, Biomass Combustion Residues, and Tyre Fibres into Cement-Bonded Composites: Properties of Composites and Life Cycle Analysis. *Constr. Build. Mater.* **2021**, *297*, 123781. [CrossRef]
58. Savastano, H.; Warden, P.G.; Coutts, R.S.P. Microstructure and Mechanical Properties of Waste Fibre–Cement Composites. *Cem. Concr. Compos.* **2005**, *27*, 583–592. [CrossRef]
59. Brasileiro, G.A.M.; Vieira, J.A.R.; Barreto, L.S. Use of Coir Pith Particles in Composites with Portland Cement. *J. Environ. Manag.* **2013**, *131*, 228–238. [CrossRef]
60. Wei, Y.M.; Fujii, T.; Hiramatsu, Y.; Miyatake, A.; Yoshinaga, S.; Fujii, T.; Tomita, B. A Preliminary Investigation on Microstructural Characteristics of Interfacial Zone between Cement and Exploded Wood Fiber Strand by Using SEM-EDS. *J. Wood Sci.* **2004**, *50*, 327–336. [CrossRef]
61. Aras, U.; Kalaycıoğlu, H.; Yel, H.; Kuştaş, S. Utilization of Olive Mill Solid Waste in the Manufacturing of Cement-Bonded Particleboard. *J. Build. Eng.* **2022**, *49*, 104055. [CrossRef]
62. Tawfik, A.; Mohammed, M.S.; Abd-El-Raoof, F.; Aggor, F.S.; Ahmed, E.M. Heat-Treated Portland Cement Pastes Incorporating Superabsorbent Hydrogels for Precast Applications. *Interceram—Int. Ceram. Rev.* **2018**, *67*, 30–37. [CrossRef]
63. Petkova, V.; Stoyanov, V.; Kostova, B.; Kostov-Kytin, V.; Kalinkin, A.; Zvereva, I.; Tzvetanova, Y. Crystal-Chemical and Thermal Properties of Decorative Cement Composites. *Materials* **2021**, *14*, 4793. [CrossRef]
64. Boukir, A.; Mehyaoui, I.; Fellak, S.; Asia, L.; Doumenq, P. The Effect of the Natural Degradation Process on the Cellulose Structure of Moroccan Hardwood Fiber: A Survey on Spectroscopy and Structural Properties. *Mediterr.J.Chem.* **2019**, *8*, 179–190. [CrossRef]
65. Govin, A. *Aspects Physico-Chimiques de L'interaction Bois–Ciment. Modification de L'hydratation du Ciment Par le Bois*; Ecole Nationale Supérieure des Mines de Saint-Etienne: Saint-Étienne, France, 2004.
66. Jain, B.; Sancheti, G.; Jain, V. FTIR Analysis of Silica Fume and Iron Dust Added Concrete. *Mater. Today Proc.* **2022**, *60*, 777–781. [CrossRef]
67. Peschard, A.; Govin, A.; Grosseau, P.; Guilhot, B.; Guyonnet, R. Effect of Polysaccharides on the Hydration of Cement Paste at Early Ages. *Cem. Concr. Res.* **2004**, *34*, 2153–2158. [CrossRef]
68. Bensted, J.; Varma, S.P. Some Applications of Infrared and Raman Spectroscopy in Cement Chemistry. Part 3-Hydration of Portland Cement and Its Constituents. *Cem. Technol.* **1974**, *5*, 440–445.
69. Mollah, M.Y.A.; Yu, W.; Schennach, R.; Cocke, D.L. A Fourier Transform Infrared Spectroscopic Investigation of the Early Hydration of Portland Cement and the Influence of Sodium Lignosulfonate. *Cem. Concr. Res.* **2000**, *30*, 267–273. [CrossRef]
70. Lima, P.R.L.; Santos, D.O.J.; Fontes, C.M.A.; Barros, J.A.O.; Toledo Filho, R.D. Deflection Hardening of Sustainable Fiber–Cement Composites. *Green Mater.* **2016**, *4*, 18–30. [CrossRef]
71. Yel, H.; Cavdar, A.D.; Torun, S.B. Effect of Press Temperature on Some Properties of Cement Bonded Particleboard. *Maderas Cienc. Tecnol.* **2020**, *22*, 83–92. [CrossRef]
72. Eseyin, A.E.; Steele, P.H.; Pittman, C.U.; Ekpenyong, K.I.; Soni, B. TGA Torrefaction Kinetics of Cedar Wood. *J. Biofuels* **2016**, *7*, 20. [CrossRef]
73. Kim, H.-S.; Kim, S.; Kim, H.-J.; Yang, H.-S. Thermal Properties of Bio-Flour-Filled Polyolefin Composites with Different Compatibilizing Agent Type and Content. *Thermochim. Acta* **2006**, *451*, 181–188. [CrossRef]
74. Cabral, M.R.; Nakanishi, E.Y.; Dos Santos, V.; Gauss, C.; Dos Santos, S.F.; Fiorelli, J. Evaluation of Accelerated Carbonation Curing in Cement-Bonded Balsa Particleboard. *Mater. Struct.* **2018**, *51*, 52. [CrossRef]

75. Hasan, K.M.F.; Horváth, P.G.; Alpár, T. Development of Lignocellulosic Fiber Reinforced Cement Composite Panels Using Semi-Dry Technology. *Cellulose* **2021**, *28*, 3631–3645. [CrossRef]
76. *NM 10.1.005;* Liants Hydrauliques—Techniques Des Essais. SNIMA: Rabat, Morocco, 2008.
77. *NF EN 196-2;* Méthodes d'essais Des Ciments—Partie 2: Analyse Chimique Des Ciments. AFNOR Editions: Saint-Denis, France, 2006.
78. *NF EN 196-3;* Méthodes d'essais Des Ciments—Partie 3: Détermination Du Temps de Prise et de La Stabilité. AFNOR Editions: Saint-Denis, France, 2006.
79. *NF EN 196-6;* Méthodes d'essai Des Ciments—Détermination de La Finesse. AFNOR Editions: Saint-Denis, France, 2006.
80. Neville, A.M.; Brooks, J.J. *Concrete Technology*, 2nd ed.; Prentice Hall: Harlow, UK, 2010; ISBN 978-0-273-73219-8.
81. Simatupang, M.H. Der Wasserbedarf bei der Herstellung zementgebundener Holzspanplatten. *Holz Als Roh-Und Werkst.* **1979**, *37*, 379–382. [CrossRef]
82. *PNM EN 325;* Panneaux à Base de Bois-Détermination Des Dimensions Des Éprouvettes. IMANOR Editions: Rabat, Morocco, 2012.
83. *EN 323;* Wood-Based Panels. Determination of Density, European Committee for Standardization: Brussels, Belgium, 1993.
84. *EN 317;* Particleboards and Fiberboards. Determination of Swelling in Thickness after Immersion in Water, European Committee for Standardization: Brussels, Belgium, 1993.
85. *EN 310;* Wood-Based Panels. Determination of Modulus of Elasticity in Bending and of Bending Strength, European Committee for Standardization: Brussels, Belgium, 1993.

molecules

Article

Production of Long-Fiber Pulp from Enset Plant Residues by Soda Pulping

Hanna Berhanu Lemma [1,*], Friedrich Steffen [1,*], Abubeker Yimam Ali [2] and Bodo Saake [1]

[1] Institute of Wood Science, University of Hamburg, Haidkrugsweg 1, 22885 Barsbuettel-Willinghusen, Germany; bodo.saake@uni-hamburg.de
[2] Addis Adaba Institute of Technology, Addis Adaba University, 2QR7+584, King George VI St, Addis Ababa 1000, Ethiopia; abubeker.yimam@aau.edu.et
* Correspondence: berhanu.hanna@gmail.com (H.B.L.); friedrich.steffen@uni-hamburg.de (F.S.); Tel.: +49-40-822459-211 (F.S.)

Abstract: This paper evaluates the modification of fiber morphology and the strength property development of paper from Enset fiber as a function of soda pulping conditions and refining energy. Soda pulping was conducted at pulping temperatures between 160 and 180 °C. The NaOH charge was 16, 20, and 24% based on the initial raw material. The beating of pulp was conducted using a Jokro mill. The refining of pulp was conducted in a laboratory refiner at different refining intensities. The mild Jokro mill beating was not effective on Enset fiber pulp. On the other hand, the laboratory refiner effectively refined the pulp. The fiber morphology was altered in the way of improving the paper formation and strength. The beating degree of the pulp was increased to about 49 °SR. The tensile index was enhanced to around 80 Nm/g using a refining energy input of 250 kwh/t. From the results, it can be concluded that Enset fiber pulp is suitable for packaging papers due to its high strength level. On the other hand, Enset fiber can be a potential raw material for specialty papers like filter paper and tea bags because of its high porosity.

Keywords: non-wood; alternative fiber sources; Enset fiber; soda pulping; refining; papermaking; paper strength; fiber morphology; fiber length

1. Introduction

Even if digitalization has led to a drastic decrease in graphic paper demand, the paper and pulp industry is still growing as packaging and tissue papers are filling the shrinking graphic paper market share gap. Various non-wood plants and agricultural residues are already being used to satisfy the pulp and paper fiber industry demand.

Enset plant fiber is one of the readily available agricultural residues in southern and southwestern parts of Ethiopia, with increasing interest being generated in natural fiber research. It is a compact yellowish-white cellular bunch of predominantly long cells (fibers), vessels, and parenchyma. The length of the cellular fiber obtained after dissociation of the bundle varied from 2.6 to 11.4 mm, and it had a width varying from 15.3 to 18.1 µm, a lumen width varying from 7.0 to 11.2 µm, and a cell wall thickness varying from 2.3 to 4.6 µm. The fiber comprises a high proportion of holocellulose (87.5%), of which 60% is cellulose. The amount of lignin is relatively low (10.5%). However, a strong delignification condition may be needed to delignify the fiber effectively due to the lower S/G ratio in the lignin monomeric composition (H:G:S, 1:0.7:0.8). The mineral components of the fiber are dominated by potassium, whereas, in the micronutrient, iron is present in significant amounts. Due to its promising characteristics, Enset fiber can be considered a good candidate as a long-fiber source in the pulp and paper industry. Berhanu et al. [1] reported a pulp yield of around 70%, with a low kappa number of around five and excellent sheet paper strength properties during the ethanol alkaline pulping of Enset fiber without further bleaching and a pulp refining stage. The major influencing factor in the pulping experiment was the NaOH charge, followed by temperature, and reaction time had little effect on the resulting pulp [1,2].

Soda pulping is the preferred process for the pulping of annual plants [3]. It uses sodium hydroxide (NaOH) as an active chemical under high pressure and temperatures to allow fiber swelling, and it increases the better penetration of the cooking chemicals [4]. Besides delignification, the peeling reaction of glucosidic bonds could be of significant concern during soda pulping since there are considerable yield losses due to carbohydrate degradation by the peeling reaction [3]. Previous studies have shown that the amounts of dissolved carbohydrates in the cooking liquor increase rapidly during the initial delignification of the cooking. The dissolution of carbohydrates then decreases during the bulk delignification phase, growing again in the final delignification [3]. Thus, to reduce the degradation of carbohydrates and promote delignification, a proper alkali charge and reaction temperature should be selected.

The other vital phase in the paper industries is pulp refining or pulp beating. Pulp refining is a mechanical treatment step to improve the paper formation and paper mechanical properties. Compressive, shear, and bending forces are randomly and repeatedly undertaken on the fibers during refining. This mechanical action reduces fiber stiffness and increases fiber flexibility. Moreover, the surfaces of the fibers are modified so it improves the fiber-to-fiber bonds in the paper [5–9]. Refining causes significant changes in the fiber structure, such as cutting, the swelling of fiber walls, external and internal fibrillation, increased flexibility, and curling [10]. Those refining effects lead to changes in density, porosity, formation, and other essential paper properties [5,11].

Refining also lowers the pulp drainage capability. That means it reduces the paper production rate and increases the energy consumption. One of the essential features of pulp is the beating degree. It is the susceptibility of the pulp to refining that is key as refining energy is required to reach a specific water retention value or tensile index. The mechanical action of refining causes cavity formations and delamination of the fiber wall, which allows the water to penetrate through the fiber wall, resulting in a more significant swelling potential [11]. The water makes the fiber much more flexible and relevant to breaking hydrogen bonds. Whenever the fibers are thoroughly wetted, refining disrupts the crystallinity of fibers, and more water can be added to the cellulose fibers. This phenomenon is known as internal fibrillation, and it is among the main reasons for improving the strength properties of paper.

External fibrillation is, on the other hand, when fibrils are peeled off from the surface of the fiber surface while still attached. Fibrillation exposes the cellulose fibers and increases the fiber surface area to improve the fiber-to-fiber bonding of the final sheet. The fibrils enhance the bonding between fibers through mechanisms such as mechanical interlocking, leading to paper with higher tensile strength. Furthermore, the surface area of the fiber increases and becomes more flexible to fit in around each other. These increase the bonding surface area and lead to a denser sheet [7]. However, due to the reduction in the strength of individual fibers, external fibrillation can negatively influence the tear strength [11].

The fibers also change their geometric shape and fibril alignment along their length, and the fibers flatten or collapse. Fiber curl and kinks are induced or straightened. Moreover, little dislocations and compressions are generated or diminished in the fiber [7].

Fiber cutting and breakage also occur during refining. This contributes to the changes in fiber length distribution and decreases in the mean fiber length. In addition, the proportion of fines might be increased. Fines have pros and cons regarding the strength of the final paper, and they are conditional to their shape and type [11]. The primary fines are present in the pulp before refining. They have an adverse effect on the strength of the paper. Secondary fines are created during refining, and they are mainly composed of fibrils. They improve the strength of the paper by increasing the effective contact area of the bonded fibers, which can help establish contact between close fibers [6]. Generally, some fiber morphological properties are modified in a way to improve paper properties, but others deteriorate with refining. All these changes that co-occur during refining are irreversible. The morphology of the fiber and the refining conditions governs the extent of the changes in the pulp. The refining conditions are the design of the refining equipment and its operating variables, such as the consistency, intensity, and refining time [7,12].

The refining equipment can be conical or disk-type refiners with a parallel flow to the bar crossings. Refining power is a measure of the power input to the motors of the refiner based on the amount of pulp processed. It is an indirect measure of the energy used in cutting and fibrillating the pulp fibers, although these processes consume only a small percentage of the power. These values are essential in the design and economic calculations. The specific refining energy is commonly expressed in kilowatt hours per ton of pulp processed (kWh/t) [7,13].

The different pulp originating from various fiber sources have varying fiber lengths, cell wall thicknesses, and widths to their central lumen. They respond accordingly to a given type of refining and set of refining conditions, and not all fibers must receive the same treatment. Some fibers are fibrillated more by one kind of beating than another. Thus, the right refining type and intensity should be selected based on the type of fiber and desired paper properties.

In the present paper, the effect of soda pulping parameters and the type and severity of refining on the morphology of the fiber and the physical paper properties were studied.

2. Results and Discussion

2.1. Laboratory-Scale Soda Pulping and Jokro Mill Beating

In the first series of experiments, the soda pulping of Enset fiber was investigated using a laboratory-scale rotating digester. The pulping temperature varied between 160 and 180 °C. The NaOH charge was 16%, 20%, and 24% based on oven-dry raw material. All other pulping parameters were kept constant. The pulp yield, the kappa number, and the carbohydrate composition of the soda pulps from the pulping in the laboratory-scale digester are presented in Figure 1.

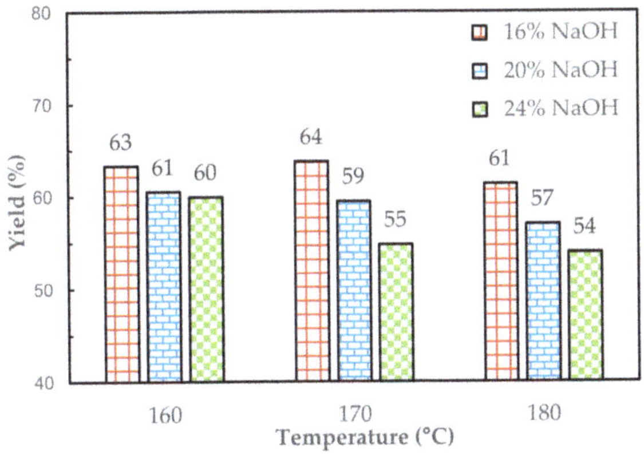

(a)

Figure 1. *Cont.*

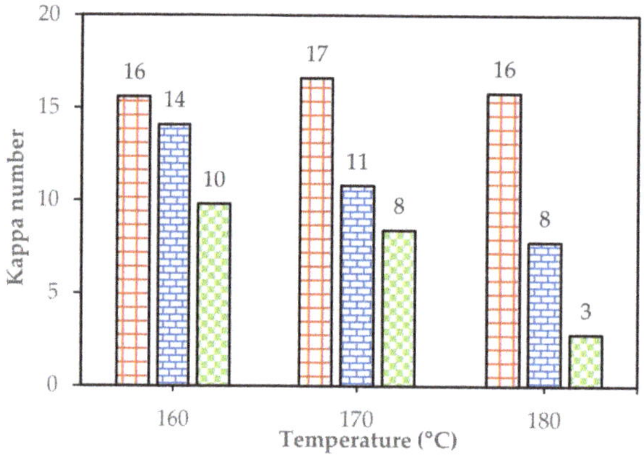

(**b**)

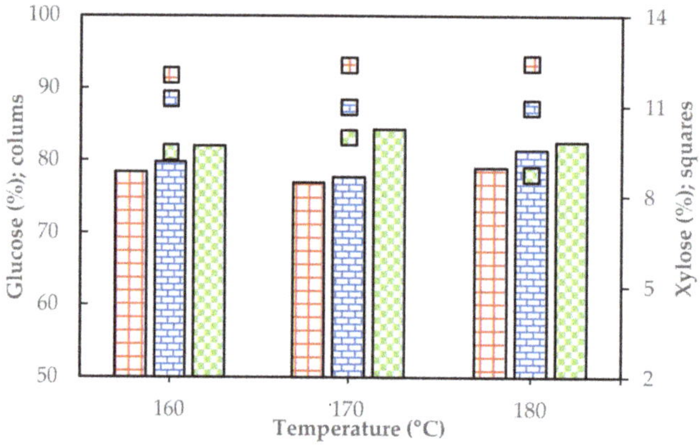

(**c**)

Figure 1. The pulp yield (**a**), kappa number (**b**), and carbohydrate composition (**c**) of the soda pulps from the pulping of Enset fiber at different temperatures and different NaOH charges.

The pulp yield and the kappa number dropped with the pulping condition severity, as shown in Figure 1a,b. This was due to the degradation and dissolution of mainly lignin, as expressed in the decrease in kappa number. However, under such strong alkaline conditions, the degradation and dissolution of cellulose and hemicellulose also occurred. The latter can be clearly seen from the graph in Figure 1c, where the xylose content of the pulp was at the lowest level when the highest NaOH charge of 24% was applied.

The NaOH charge was the dominant factor in the soda pulping of Enset fiber compared to the pulping temperature [1]. The effect of temperature was found to be statistically insignificant (p-value < 0.08), whereas the NaOH charge was found to be statistically significant (p-value < 0.01). As can be clearly seen in Figure 1a, the effect of temperature on the pulp yield was not significant for pulping at a lower NaOH charge and was significant for pulping at a higher NaOH charge. The same scenario was obtained in the statistically significant result in the kappa number and carbohydrate composition (Table S1).

Berhanu et al. [1] reported a similar situation in the ethanol alkaline pulping of Enset fiber, where the alkali charge was the dominant factor compared to other pulping parameters. This might be because the peeling reaction can occur whenever the alkaline liquor is in contact with the lignocellulosic material, even at low temperatures, as stated by Ek et al. [3].

The fiber length distribution and fiber morphology of the selected pulp was measured before and after beating in a Jokro mill for 4 min, 20 min, and 30 min. The fiber length distribution data for pulps made at 160 °C, 170 °C, and 180 °C at a 16% NaOH charge are presented in Figure 2.

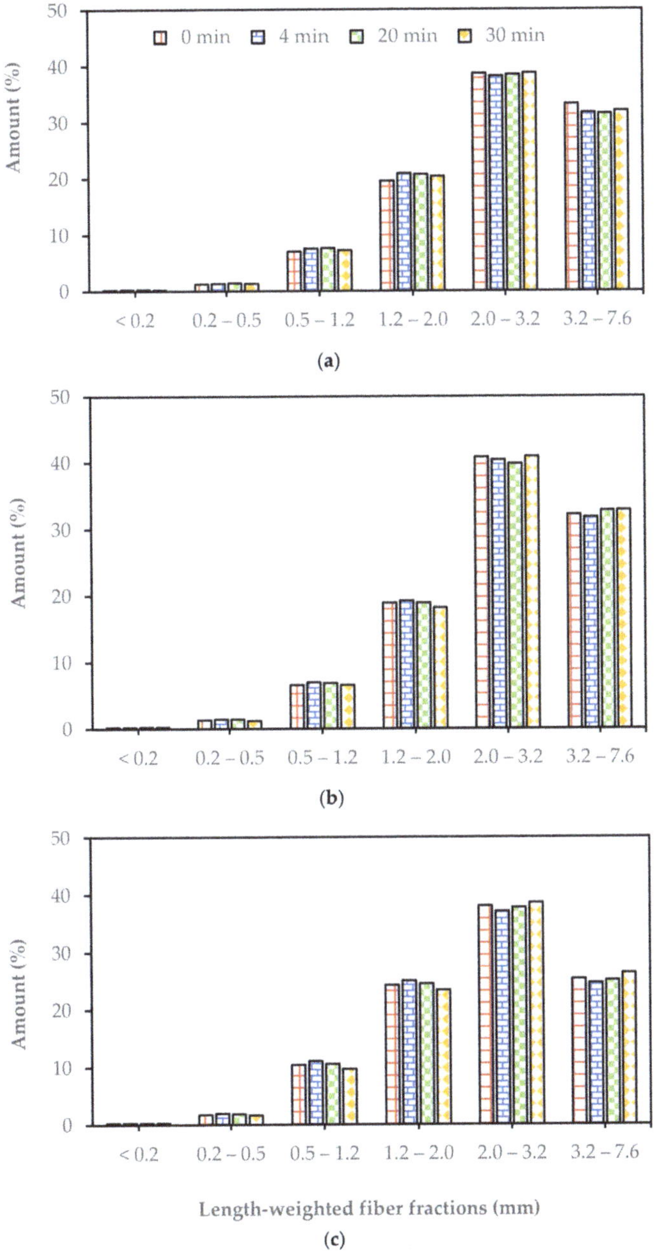

Figure 2. Fiber length distribution data for the pulps made at 160 °C (**a**), 170 °C (**b**), and 180 °C (**c**) at a 16% NaOH charge at beating times of 0–30 min.

The pulp comprised fine particles and short and long fibers. The average fiber length of the Enset fiber soda pulp was 2.8 mm, which was not altered due to Jokro mill beating, as can be seen in Table 1. The soda pulp from Enset fiber had a longer average fiber length and lower average fiber width than most agricultural residues, such as wheat straw and rye straw [14]. These morphological properties resulted in a higher slenderness ratio of Enset soda pulp, making the fiber a more suitable raw material for paper industries with higher strength properties [14,15]. The content of primary fiber fines in the Enset fiber was also lower than that in the wheat straw and rye straw. The fines present in the pulp before refining adversely affect the strength properties of the final paper [11].

Table 1. Fiber morphology of the Enset soda pulp prepared at 160 °C and 16% NaOH using the rotating digester and Jokro mill beating.

Beating Time (min)	Fiber Length (mm)	Fiber Width (µm)	CWT	Fines Fraction (%)	Fibrillation (%)
0	2.8	17.5	7.0	5.0	9.7
4	2.8	17.7	6.4	5.0	9.6
20	2.8	17.7	6.1	5.7	9.7
30	2.8	17.7	6.6	5.4	9.6

The long fibers (>2 mm) were 63% of the pulp, and the fine particles (<0.2 mm) were less than 2% of the pulp. Figure 2 shows that was no significant change in the fiber length distribution due to Jokro mill beating. Statistical analysis also demonstrated the ineffectiveness of Jokro mill for Enset fiber pulp (Table S2). In addition to the fiber length, the other fiber morphological properties were not altered meaningfully and significantly due to Jokro mill beating, as shown in Table 1.

The strength property of the paper sheets was measured for all nine samples with a 15 L rotary digester to see the possible effect of Jokro milling at different beating times. The strength properties of the pulps are presented in Figure S1 (Supplementary Materials).

The Schopper-Riegler value ranged from 17 to 36 °SR without any defined correlation with the beating time or pulping condition. The tensile and burst indexes of the unbeaten pulp produced by 16% NaOH were much higher than the other two unbeaten pulp produced by a 20% and 24% NaOH charge. In contrast, the tear index was much lower for the unbeaten soda pulp produced by milder pulping conditions, as shown in Figure S1. The paper from Enset fiber has higher strength properties than Bagasse, wheat straw, and rye straw. As explained earlier, this can be due to the superiority of the morphological property of Enset fibers [16–18]. Unlike wheat straw, the beating time, beating degree, and strength properties of Enset fiber do not correlate during Jokro mill beating [17]. The porosity (gurley) and specific volume of unbeaten pulp produced by milder conditions were much lower than those produced by severe conditions (Figure S1). The porosity of the hand sheet from unbeaten Enset fiber pulp was much higher than that of the hand sheet from wheat straw and bagasse. This can be explained by its lower amount of primary fine particles in the pulp, lower fibrillation, and longer fiber length [17,18].

Moreover, ununiform fiber distribution in the formed paper sheet was visible to the naked eye, and some of the fibers were even observable. This indicates that the mild Jokro mill treatment is unsuitable for Enset fiber. Therefore, a Voith LR 40 refiner had to be used to see the full development of the strength potential of Enset fiber. For this experiment, larger quantities of pulp were required. Upscaled pulping was performed in a large paddle digester simulating the horizontal tube digester usually applied for the pulping of the annual plants in the industries.

2.2. Pilot-Scale Soda Pulping

Figure 3 shows the yield, kappa number, and carbohydrate composition of the pulps from a paddle digester compared to a small electrical digester at the same cooking conditions. The pulp yield from the paddle digester was 60% for mild and 54% for severe pulping, and the corresponding kappa numbers were 9.9 and 3, respectively (Figure 3a,b). On the other hand, the yield was 65% and 55%, and the kappa number was 16.6 and 8.4 for the small electrical digester at the same pulping conditions. The type of digester did not significantly affect the pulp yield (p value < 0.26). On the contrary, the kappa number and the total lignin content were much lower for the paddle digester (Figure 3b). The total lignin content for the pulp was between 1.6% and 1% for the paddle cooker and 7% to 1.6% for the electrical digester at the same cooking conditions. The large-scale pulping in the paddle cooker was effective in selective delignification (p value < 0.05) due to the high mixing efficiency in the paddle reactor.

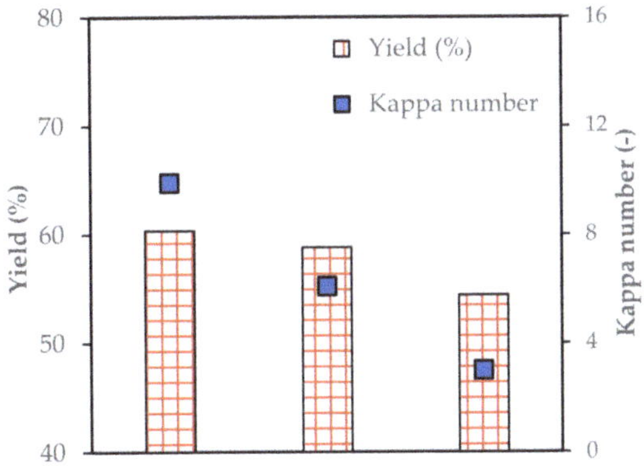

(a)

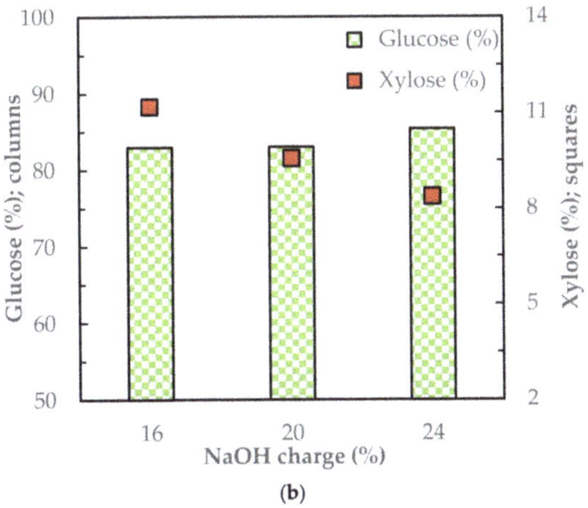

(b)

Figure 3. The pulp yield, kappa number (**a**), and carbohydrate composition (**b**) of the soda pulps from pulping at different NaOH charges using a paddle digester.

The carbohydrate composition of the pulp was evaluated because it affects the final characteristics of the paper. Paper with higher hemicellulose and cellulose content show higher tensile strength because they are responsible for the fiber-to-fiber bonding in the paper. In addition, the bonding is improved by hemicellulose through its ability to hold water during processing [19]. On the other hand, lignin prevents the formation of fiber-to-fiber bonds in paper and reduces paper strength [12]. The result shows that the paddle reactor removed xylose more effectively than the small electrical digester. The amount of xylose was from 11% to 8.4% for pulps from the paddle reactor, whereas it was 12% to 10% in the electrical digester at the same cooking conditions. The reactor type significantly affected the removal of xylose in the pulp (p-value < 0.05). On the other hand, the glucose proportion was comparable to the pulp from a small electrical reactor.

2.3. Refining and Fiber Morphology

The pulp from the paddle digester was refined in a laboratory refiner (Voith LR 40) at different refining energy levels up to 250 kwh/t. The average fiber length of the pulp decreased with the pulping condition severity and refining energy, as shown in Figure 4a. The difference between the average fiber length of the unrefined pulp in mild and severe conditions was 0.2 mm. The variation became 0.8 mm at the refining energy level of 250 kwh/t, as shown in Figure 4a. The result clearly indicates that the pulping severity affects the average fiber length and makes the fiber susceptible to refining.

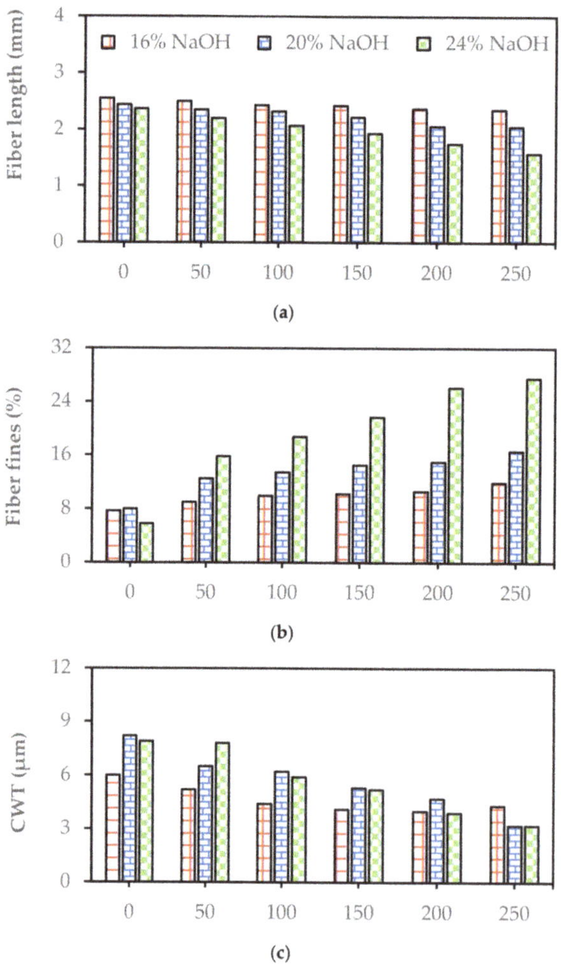

Figure 4. *Cont.*

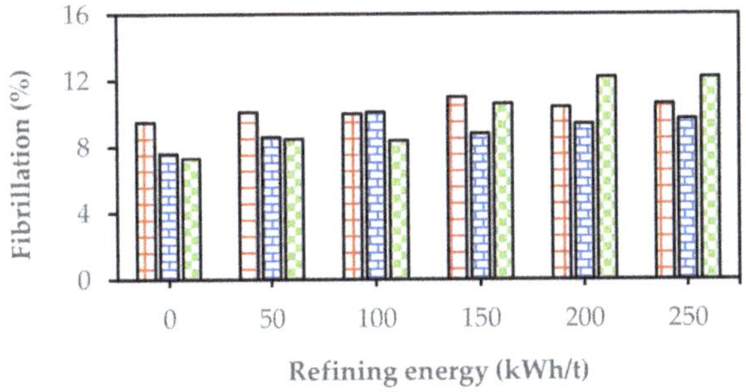

(d)

Figure 4. The average fiber length (**a**), fiber fines (**b**), CWT (**c**), and fibrillation (**d**) of the soda pulps from pulping at different NaOH charges depending on the refining energy.

On the other hand, the fine fraction in the pulp increased with increasing pulping severity and refining energy (Figure 4b). It is clear from the slope of the graphs in Figure 4b that the increasing rate of the fine fraction in severe conditions was higher than the other two pulps due to the weakness of the fiber in resisting the refining effect.

Figure 5 shows the impact of the pulping severity and refining intensity on the fiber length distribution. The proportion of long fiber (>2 mm) decreased while the proportion of short fiber (<1.2 mm) increased with increasing soda pulping conditions and pulp refining energy. The longest fiber fractions (>3.2 mm) of the unrefined pulp decreased from 25.7% to 20.8% with pulping severity. The proportion of long fiber was also remarkably reduced with refining energy. It declined at a higher rate for the pulp produced with severe pulping. It decreased by 5.3%, 6.5%, and 13.5% at 16%, 20%, and 24% NaOH charges, respectively. On the other hand, the fine fraction (<0.5 mm) increased by 0.43%, 1.2%, and 3.6%, respectively (Figure 6). The weakness of the fiber due to severe pulping conditions facilitated the formation of fine particles. Statistical analysis of the fiber length distribution proved that the fiber length distributions depended on both pulping and refining conditions (Table S2). Moreover, the pulping severity and refining condition interaction significantly affected the fiber length distribution. The Enset fiber had a wide range of fiber length distribution compared to other agricultural residues for instant EFB fiber, which only ranged from 0.2 mm to 1.6 mm, and the wheat straw ranged from 0.1 to 1.7 mm [20].

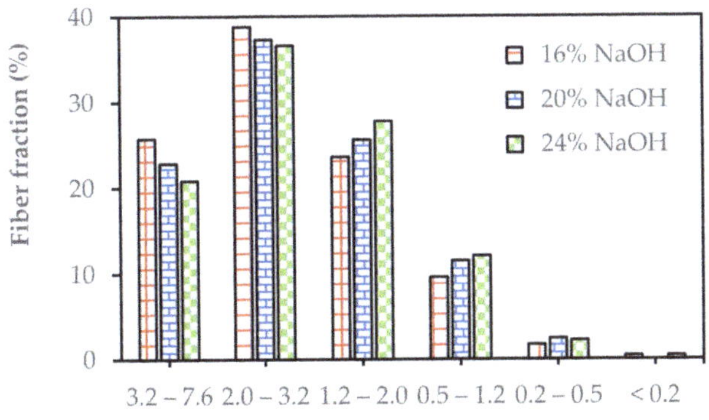

(a)

Figure 5. *Cont.*

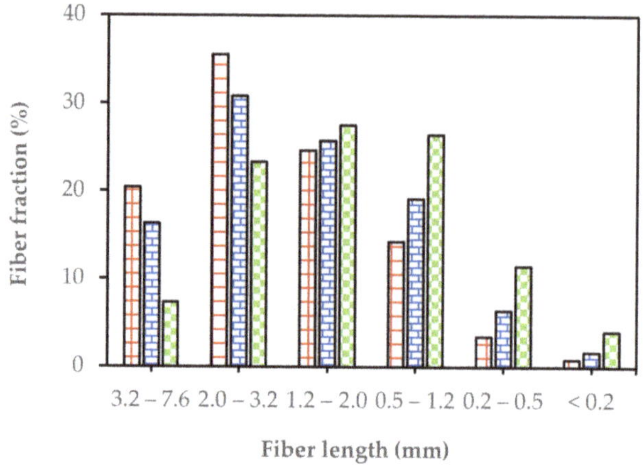

(**b**)

Figure 5. Fiber length distribution of the soda pulps from pulping at different NaOH charges depending on the refining energy: (**a**) 0 kWh/t and (**b**) 250 kWh/t.

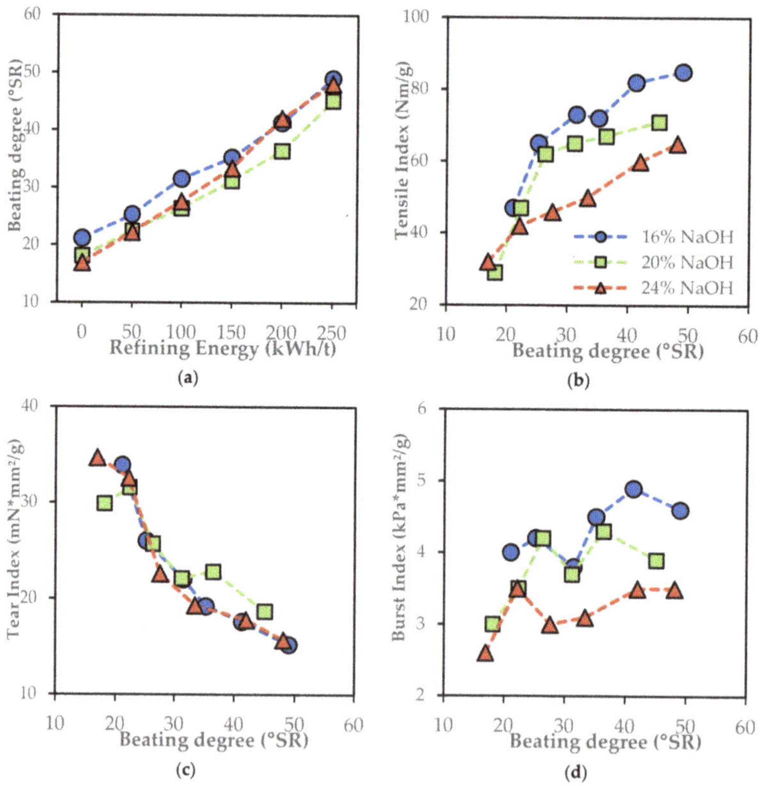

Figure 6. The beating degree (**a**) and strength properties of the soda pulps from pulping at different NaOH charges depending on the refining energy: (**b**) tensile index; (**c**) tear index; and (**d**) burst index.

The refining intensity indirectly affects the fiber width, fiber surface area (CSA), and cell wall thickness (CWT). As reported by Berhanu et al. [2], the cell wall thickness of the raw Enset fiber was in the range of 2.3 to 4 μm. The cell wall thickness increased to 6 μm for pulping when using a low NaOH charge of 16% and to 8 μm when using a high NaOH charge of 24% (Figure 4c). The increase in CWT was due to the swelling effect of the increasing NaOH charge on the fiber cell. The fiber width and surface area decreased with refining due to the deformation of the cell wall and the lumen. They decreased with refining due to the pressing of two opposite walls and a collapsing of the fibers, or due to the lumen being eliminated partly or entirely. The other crucial parameter was fibrillation, which makes the fiber flexible and ready to form fiber-to-fiber bonds in the paper. This improves paper formation, where the fibrillation of the pulp increases with refining (Figure 4d). The curl index decreased with increasing refining energy. Fiber curls are directly related to the tensile index and influence the stretchability of a paper, so they are more critical on sack paper [12].

2.4. Refining and Paper Strength Property

The beating degree and paper strength properties strongly depend on the original structure of the fiber and the morphological modification of the fibers during refining [12]. In addition to various interrelated fiber properties, the strength of the hand sheet paper is also affected by pulping techniques. Therefore, it is not easy to find correlations that justify every scenario [21]. The primary targets of refining are to improve the strength and formation of paper. The extent of refining is monitored by measuring the beating degree (Shopper Riegler SR° pulp freeness) and the strength properties of hand sheets [7].

The refining of Enset fiber pulp using a Voith refiner showed apparent enhancement in SR° and in the strength properties of the paper, as shown in Figure 6a. The °SR value of unrefined Enset fiber pulp at different NaOH charges was 17 to 21. The °SR linearly increased with increasing refining energy. At the refining energy of 250 kWh/t, the °SR reached 45 to 49. The increase in °SR value was related to the surface conditions and swelling of the fibers.

Moreover, the internal and external fibrillation and fine formation was the main reason for the increment in the Shopper Riegler value [22]. Olejnik et al. [5] also reported an increment in the water retention value of fiber with increasing refining energy. Compared to bagasse and wheat straw, Enset fibers are resistant to refining. The SR° value of wheat straw has already reached 55 and bagasse soda pulp has reached 75 at 120 kwh/t and 200 kwh/t refining, respectively [17,18].

Paper sheets were then formed using unrefined and refined pulps to analyze the improvement of the strength properties of the paper. Figure 6b–d shows the strength property of the paper plotted against the beating degree.

The tensile index and burst index of the paper sheet increased with increasing SR°, while the tear index decreased (Figure 6b–d). The tensile index was improved through increasing bonding potential due to fibrillation and fiber flexibility [23]. The fibrillation of the individual fiber sourced a better fiber-to-fiber contact area. The reduction in average fiber length negatively affected the tensile strength of the paper. However, the reduction in fiber width and cell wall thickness compensates for the strength drop [12]. The secondary fine particles are also suitable for the tensile and burst index [10]. Salehi et al. [18] reported the maximum specific energy that the bagasse pulp tolerates before the strength start deterioration is 163 kwh/t. Unlike bagasse, the strength development in the Enset fiber pulp showed a potential increment beyond a refining energy of 250 kwh/t.

The higher tensile strength in the pulp produced by 16% NaOH can also be correlated with its high xylose content. Omer et al. [23] reported the good bonding ability of the fiber with high hemicellulose content. High hemicellulose content in the pulp makes the fiber more flexible and fibrillated during refining [12,19]. As reported by Olejnik et al. [5], the tensile strength of paper from eucalyptus tends to be constant after the refining energy of

200 kwh/t. However, in the case of Enset fiber pulp, the tensile index still has the potential to increase if the refining energy increases beyond 250 kwh/t.

The tear index dropped with refining because it depends on the individual fiber length, strength, degree of inter-fiber bonding, and fiber orientation in the paper (Figure 6d; [7]). The tear index is lower when the fibers are short, weak, and poorly bonded. It is more dependent on fiber breakage than bond breakage. Longer and stronger fibers provide higher tear strength because of the work needed to pull the fibers from the sheet [24].

The porosity and specific volume of the sheet decrease with SR° because when the fibers are more flexible, denser papers are produced [7,18]. In addition, the CWT and fiber width reduction make the paper more compact each other and result in lower specific volume. The increasing ratio of small non-fiber fragments and fine particles fills the paper and reduces porosity [10].

Uncorrelated changes were clearly observed because the paper-making pulp was a highly heterogeneous lignocellulosic material, and there was a wide range of fiber length distribution in the pulp compared to pulp from other annual plants, as shown in Figure 6 [20,25]. As a result, no satisfactory mathematical equations unambiguously predict pulp properties based on pulping and refining parameters [5].

3. Materials and Methods

3.1. Raw Material Characterization

Enset fiber bundles were collected from southwestern Ethiopia, Wolkite, and Wolayita. The samples were air-dried and cut into 2–5 cm-long pieces. Extensive chemical composition analysis was conducted in our previous study regarding extractive content, ash content, carbohydrate composition, and lignin content [26]. The chemical compositions of the Enset fiber are summarized in Table 2.

Table 2. Summary of the chemical compositions of the Enset fiber [27].

Extractives (%)	Petrol ether	0.1
	Acetone	0.9
	Water	3.4
	Σ	**4.4**
Lignin (%)	Klason lignin	10.8
	Acid-soluble	2.5
	Σ	**13.3**
Monosaccharides (%)	Rhamnose, mannose, arabinose, galactose	2.4
	Xylose	10.7
	Glucose	58.9
	Σ	**72.1**
Ash (%)	Silica	2.6
	Σ	**5.4**

3.2. Soda Pulping of the Enset Fiber

A total of 300 g (oven-dry) of Enset fiber was used for soda pulping in a 15-Liter laboratory rotating digester at a maximum temperature of 160 °C, 170 °C, and 180 °C. Sodium hydroxide charges were 16%, 20%, and 24% based on the oven-dry weight of the raw material. The solid-to-liquid ratio was kept at 1:4 for all experiments. The maximum temperature was attained within 60 min and kept there for another 60 min.

Three pulping conditions were selected for large-scale pulping in an 80-Liter, steam-heated horizontal reactor with a mixer rotated at 0.12–0.13 rpm. The pulping was performed at 170 °C and at NaOH charges of 16%, 20%, and 24%.

The resulting pulps from both reactors were washed and dewatered by centrifugation in a spinner (Thomas Centri 776 SEK, Thomas, Neunkirchen, Germany) for 10 min at 2800 rpm. The pulp yield was analyzed after homogenizing of the pulp for 10 min in a rotary stirrer (Hobart, Offenburg, Germany) and reported as a percentage of the initial weight of fiber.

3.3. Chemical Composition of the Enset Fiber Soda Pulp

A 200 mg soda pulp sample was air-dried and milled with a disc mill (SM 200, Retsch, Haan, Germany) to determine the monomeric carbohydrate composition from the two-stage acid hydrolysis filtrates by Borate-AEC using a Dionex Ultimate 3000 (Dionex, Sunnyvale, CA, USA) as described in Lorenz et al. [27]. Specifically, 200 mg of oven-dry material were hydrolyzed with 2 mL of 72% sulfuric acid at 30 °C for exactly one hour to break up the crystalline structure of the cellulose. Thereafter, the sulfuric acid was diluted by adding 56 mL of water, and then the samples were put in an autoclave at 120 °C and 1.2 bar for 40 min to degrade the remaining oligosaccharide chains into monomeric sugars. After cooling, the samples were filtered on a N°4-sintered glass crucible. The hydrolysis residue (HR) was oven-dried and gravimetrically determined, and it represented the acid-insoluble lignin content, which is analogous to Klason lignin. The quantitative and qualitative carbohydrate composition in the hydrolysates was analyzed by borate-HPAEC analysis. Dionex Ultimate 3000 and an anion exchange resin MCL Gel CA08F (Mitsubishi Chemical, Tokyo, Japan), in a column with the dimensions of 5 × 120 mm (Omnifit) (packed at 65 °C), were used for chromatographic separation. The mobile phase consisted of two potassium tetraborate/boric acid buffers in water: 0.3 M (A; pH 8.6) and 0.9 M (B; pH 9.5). Separation was performed at 65 °C with a flow rate of 0.7 mL/min. The elution program started with 90% A and 10% B and changed to 10% A and 90% B within 35 min. This rate was kept constant for 8 min and changed back to 90% A and 10% B within 7 min. Prior to detection, a post-column derivatization by Cubicinchoninate (0.35 mL/min) was applied at 105 °C in a Teflon® coil of 30 m length and 0.3 mm diameter. The analytes were detected at 560 nm via UV/VIS-detection. The contents of the carbohydrates and lignin were determined as triplicates. The standard deviation (SD) was between 0.2% and 1.7% for the carbohydrate content and less than 0.5% for the lignin content (HR).

The UV-Spectrophotometer LAMBDA 650 (PerkinElmer, Waltham, MA, USA) at a wavelength of 205 nm was used to determine the acid-soluble lignin content according to Maekawa et al. [28]. The Klason lignin content was calculated from the solid residue [26]. Kappa numbers were determined using the TAPPI standard T236 om-99 (1999) [29].

3.4. Pulp Beating and Refining

The never dried pulp from the 15-Liter rotating digester was beaten using a Jokro mill (FRANK-PTI, Birkenau, Germany) according to the German standard Zellcheming V/5/60 (1960) [30]. The mechanical action was conducted between the inner and outer walls of the mill at a principle of centrifugal force of 150 rpm for 4 min, 20 min, and 30 min. The fiber morphology and sheet paper strength were measured before and after beating.

The pulps from the large-scale reactor were subjected to refining using a fully automated laboratory refiner (Votih LR 40, FRANK-PTI) with disc refiner headsets (standard plate for long fibers). The full refiner was operated under a refining intensity of 1 J/m and a rotation speed of 2000 rpm. The refining energy range was from 0 kwh/t to 250 kwh/t. A pulp consistency of 3% was kept for each batch of samples.

After beating or refining, the pulp was disintegrated using a standard pulp disintegrator based on ISO 5263-2:2004 (2004) [31] to disperse the pulp fiber into the solution. The beating degree was determined with a Schopper-Riegler (°SR) freeness tester type SR1 (Karl Schröder KG, Weinheim, Germany). The resulting pulp with different degrees of refining was analyzed using a Kajani Fiber lab (Metso, Helsinki, Finland) to determine the fiber length distribution and fiber morphology characterization, such as the average fiber length, the amount of fine material, fiber width, and the fiber curl and kink index.

3.5. Analysis of the Paper Strength

Laboratory handsheets of about 80 g/m^2 basis weight were prepared using a Rapid-Koethen sheetformer (FRANK-PTI, Birkenau, Germany). The resulting paper was acclimatized at 23 ± 1 °C and 55 ± 2% relative humidity overnight before testing its strength properties.

The compression, burst, tensile, and tear strength were measured according to DIN 54518:2004 (2004) [32], ISO 2759:2014 (2014) [33], ISO 1924-2:2009 (2009) [34], and ISO 1974:2012-09 (2012) [35], respectively, using instruments manufactured by FRANK-PTI GmbH (Birkenau, Germany). The results were normalized by dividing with the grammage of each paper sheet.

4. Conclusions

The Enset fiber soda pulp shows superior quality for the paper industry compared to other annual plants and agricultural residues. It is a better suited raw material compared to wheat straw and bagasse. The soda pulp from Enset fiber was effectively refined by a Voith LR 40 refiner, and the strength of the paper was improved.

One main new achievement was the observation that the beating of Enset fiber pulp in a Jokro mill had almost no effect on the fiber morphology and the development of paper strength. This put emphasis on the fact that the mechanisms in Jokro mill and laboratory refiner beating are completely different and therefore have a different impact on the fiber. It was found that fibers are more squeezed and maybe crushed in Jokro mill beating, while the fibers are actually cut in laboratory refining. This can be seen in Figure 5a,b, where the fraction of "long" fibers (3.2–7.6 mm) decreased from >20% to less than 10% after refining. Only the cutting of fibers in laboratory refining led to a change in fiber and paper properties, while the squeezing and crushing in Jokro mill beating had almost no effect on fiber and paper properties.

Based on this result, the fiber can be used for packaging or sack paper due to its high tensile strength and curl index. Due to its porosity, Enset fiber pulp can also be used for specialty paper, such as tea bags or as filter paper. Sanitary purposes are also another possibility because of its high-water permeability due to higher hemicellulose content. Depending on the intended product being made, there should be always a trade-off between various properties during pulping and refining stages.

Supplementary Materials: The following supporting information can be downloaded at: https://www.mdpi.com/article/10.3390/molecules29204874/s1. Figure S1: Selected physical paper properties of pulps from laboratory-scale pulping with a 15 L rotary digester at different temperatures and NaOH charges; Figure S2: Process scheme—from the Enset plant to Enset fibers to Enset paper grade pulp; Figure S3: Correlation between Xylose content and beating degree of unrefined pulp at different NaOH charges (large-scale pulping at 170 °C); Figure S4: Example Chromatogram from carbohydrate analysis via borate-HPAEC analysis; Table S1: Analysis of variance (Anova) of the pulping data (pulp yield and kappa number) at laboratory scale with a rotary digester; Table S2: Analysis of variance (Anova) of fiber morphology data, and the proportion of different fiber fractions after pulping at laboratory scale with a rotary digester.

Author Contributions: Conceptualization, H.B.L., A.Y.A. and B.S.; methodology, H.B.L. and F.S.; writing—original draft preparation, H.B.L. and B.S.; writing—review and editing, F.S. and B.S.; visualization, H.B.L. and F.S.; supervision, B.S.; project administration, B.S. and A.Y.A.; funding acquisition, H.B.L., A.Y.A. and B.S. All authors have read and agreed to the published version of the manuscript.

Funding: This research was funded by the Schlumberger Foundation.

Informed Consent Statement: Not applicable.

Data Availability Statement: The data presented in this study are available from the corresponding authors upon request.

Acknowledgments: The authors gratefully acknowledge Andreas Schreiber for his valuable assistance during laboratory work.

Conflicts of Interest: Author Hanna Berhanu Lemma was employed by the company Sappi Paper. As part of her scholarship from the Schlumberger Foundation, the author Hanna Berhanu Lemma was employed at the University of Hamburg during her two-year post-doctoral research stay in Hamburg. The remaining authors—Friedrich Steffen, Abubeker Yimam Ali, and Bodo Saake—declare that their research was conducted in the absence of any commercial or financial relationship that could be construed as a potential conflict of interest.

References

1. Berhanu, H.; Kiflie, Z.; Neiva, D.; Gominho, J.; Feleke, S.; Yimam, A.; Pereira, H. Optimization of ethanol-alkali delignification of false banana (*Ensete ventricosum*) fibers for pulp production using response surface methodology. *Ind. Crops Prod.* **2018**, *126*, 426–433. [CrossRef]
2. Berhanu, H.; Kiflie, Z.; Miranda, I.; Lourenço, A.; Ferreira, J.; Feleke, S.; Yimam, A.; Pereira, H. Characterization of crop residues from false banana/Ensete ventricosum/in Ethiopia in view of a full-resource valorization. *PLoS ONE* **2018**, *13*, e0199422. [CrossRef] [PubMed]
3. Ek, M.; Gellerstedt, G.; Henriksson, G. *Pulping Chemistry and Technology*; De Gruyter: Berlin, Germany, 2009; Volume 2. [CrossRef]
4. Heitner, C.; Dimmel, D.; Schmidt, J. *Lignin and Lignans: Advances in Chemistry*; Taylor and Francis: Boca Raton, FL, USA, 2010. [CrossRef]
5. Olejnik, K. Impact of pulp consistency on refining process conducted under constant intensity determined by SEL and SEC factors. *BioResources* **2013**, *8*, 3212–3230. [CrossRef]
6. Bajpai, P. Chapter 14—Production of Dissolving Grade Pulp. In *Biermann's Handbook of Pulp and Paper*, 3rd ed.; Bajpai, P., Ed.; Elsevier: Amsterdam, The Netherlands, 2018. [CrossRef]
7. Bajpai, P. Chapter 16—Pulp Cleaning, Screening, and Fractionation. In *Biermann's Handbook of Pulp and Paper, Volume 1, Raw Material and Pulp Making*; Bajpai, P., Ed.; Elsevier: Amsterdam, The Netherlands, 2019. [CrossRef]
8. Danielewicz, D.; Surma-Ślusarska, B. Miscanthus × giganteus stalks as a potential non-wood raw material for the pulp and paper industry. Influence of pulping and beating conditions on the fiber and paper properties. *Ind. Crops Prod.* **2019**, *141*, 111744. [CrossRef]
9. Danielewicz, D.; Kmiotek, M.; Surma-Ślusarska, B. Comparison of Some Properties of Selected Non-Wood Plants and Wood Species and Unbleached Kraft Pulps from These Materials. *J. Nat. Fibers* **2021**, *18*, 1296–1306. [CrossRef]
10. Chen, T.; Xie, Y.; Wei, Q.; Wang, X.; Hagman, O.; Karlsson, O.; Liu, J. Effect of refining on physical properties and paper strength of Pinus massoniana and China fir cellulose fibers. *BioResources* **2016**, *11*, 7839–7848. [CrossRef]
11. Motamedian, H.R.; Halilovic, A.E.; Kulachenko, A. Mechanisms of strength and stiffness improvement of paper after PFI refining with a focus on the effect of fines. *Cellulose* **2019**, *26*, 4099–4124. [CrossRef]
12. Karlsson, H. *Fiber Guide—Fiber Analysis and Process Applications in the Pulp and Paper Industry*; AB Lorentzen & Wettre: Kista, Sweden, 2006.
13. Gharehkhani, S.; Sadeghinezhad, E.; Kazi, S.N.; Yarmand, H.; Badarudin, A.; Safaei, M.R.; Mohd Zubir, M.N. Basic effects of pulp refining on fiber properties—A review. *Carb. Pol.* **2015**, *115*, 785–803. [CrossRef] [PubMed]
14. Hintz, H.L. Paper: Pulping and Bleaching. In *Encyclopedia of Materials: Science and Technology*; Buschow, K.H.J., Flemings, M.C., Kramer, E.J., Veyssière, P., Eds.; Elsevier: Amsterdam, The Netherlands, 2001.
15. Salehi, K.; Kordsachia, O.; Patt, R. Comparison of MEA/AQ, soda and soda/AQ pulping of wheat and rye straw. *Ind. Crops Prod.* **2014**, *52*, 603–610. [CrossRef]
16. Lemma, H.B.; Kiflie, Z.; Feleke, S.; Yimam, A. Chemical and Morphological Analysis of Enset (*Ensete ventricosum*) Fiber, Leaf, and Pseudo stem. *Lignocellulose* **2016**, *5*, 139–151.
17. Salehi, K.; Latibari, A.J.; Kordsachia, O.; Saake, B. The potential of bagasse soda pulp as a strength enhancer for old corrugated pulp. *Appita J.* **2017**, *70*, 371–377.
18. Salehi, K.; Kordsachia, O.; Saake, B. The potential of wheat straw high yield MEA pulp for enhancing strength properties of recycled paper. *BioResources* **2017**, *12*, 8255–8271. [CrossRef]
19. Iwamoto, S.; Abe, K.; Yano, H. The Effect of Hemicelluloses on Wood Pulp Nanofibrillation and Nanofiber Network Characteristics. *Biomacromolecules* **2008**, *9*, 1022–1026. [CrossRef] [PubMed]
20. McKean, W.T.; Jacobs, R.S. Wheat Straw as a Paper Fiber Source. Available online: https://www.paperonweb.com/Documents/WheatStrawAsPaperFiber.pdf (accessed on 26 February 2024).
21. Seth, R.S. *Fiber Quality Factors in Paper Making—I The Importance of Fiber Length and Strength*; Materials Research Society Symposium Proceedings; Cambridge University Press: Cambridge, UK, 1990; Volume 197, pp. 125–141.
22. Gonzalo, A.; Bimbela, F.; Sánchez, J.L.; Labidi, J.; Marín, F.; Arauzo, J. Evaluation of different agricultural residues as raw materials for pulp and paper production using a semichemical process. *J. Clean. Prod.* **2017**, *156*, 184–193. [CrossRef]
23. Omer, S.H.; Khider, T.O.; Elzaki, O.T.; Mohieldin, S.D.; Shomeina, S.K. Application of soda-AQ pulping to agricultural waste (okra stalks) from Sudan. *BMC Chem. Eng.* **2019**, *1*, 6. [CrossRef]
24. Fadavi, F. Effect of fiber fractionation on refineability and strength properties of wheat straw soda-AQ pulp. *Lignocellul. J.* **2012**, *1*, 153–163.

25. Chibudike, H.O.; Mgbachiuzo, E.; Ndukwe, N.A.; Obi, N.I.; Adeyoju, O.A.; Chibudike, E.C. A study on mea-pulping of ten (10) nigerian cultivated agro-based fiber for the production of pulp stock paper. *IEEE-SEM* **2020**, *8*, 112–121.
26. Lemma, H.B.; Freund, C.; Yimam, A.; Steffen, F.; Saake, B. Prehydrolysis soda pulping of Enset fiber for production of dissolving grade pulp and biogas. *RSC Adv.* **2023**, *13*, 4314–4323. [CrossRef] [PubMed]
27. Lorenz, D.; Erasmy, N.; Akil, Y.; Saake, B. A new method for the quantification of monosaccharides, uronic acids and oligosaccharides in partially hydrolyzed xylans by HPAEC-UV/VIS. *Carbohydr. Polym.* **2016**, *140*, 181–187. [CrossRef] [PubMed]
28. Maekawa, E.; Ichizawa, T.; Koshijima, T. An evaluation of the acid-soluble lignin determination in analyses of lignin by the sulfuric acid method. *J. Wood Chem. Technol.* **2006**, *9*, 549–567. [CrossRef]
29. *TAPPI/ANSI T 236 om-13;* Kappa Number of Pulp TAPPI/ANSI T 236 om-13. TAPPI: Peachtree Corners, GA, USA, 1999.
30. Zellcheming. *Zellcheming Standard V/5/60;* Zellcheming: Frankfurt, Germany, 1960.
31. *ISO 5263-2:2004;* Pulps—Laboratory Wet Disintegration Part 2: Disintegration of Mechanical Pulps at 20 Degrees. ISO: Geneva, Switzerland, 2004.
32. *DIN 54518:2004;* Testing of Paper and Board—Compression Strength, Short Span Test. DIN: Berlin, Germany, 2004.
33. *ISO 2759:2014;* Board—Determination of Bursting Strength. ISO: Geneva, Switzerland, 2014.
34. *NEN-EN-ISO 1924-2:2009 en;* Paper and Board—Determination of Tensile Properties—Part 2: Constant Rate of Elongation Method (20 mm/min). ISO: Geneva, Switzerland, 2009.
35. *DIN EN ISO 1974:2012-09;* Paper—Determination of Tearing Resistance—Elmendorf Method. DIN: Berlin, Germany, 2012.

molecules

MDPI

Article

The Fungi–Bacteria Interaction Mechanism of Microbial Consortium During Efficient Lignin Degradation Based on Metabolomics Analysis

Wen Zhang [1,*,†], Yilei Wen [1,†], Zhequan Wang [1], Chenyang Diao [1] and Zhiwei Liu [2]

1 Key Laboratory of Pollution Exposure and Health Intervention of Zhejiang Province, College of Biology and Environmental Engineering, Zhejiang Shuren University, Hangzhou 310015, China; yilei0430@126.com (Y.W.); 13757575939@163.com (Z.W.); x_yang24@163.com (C.D.)

2 College of Environment and Chemical Engineering, Yanshan University, Qinhuangdao 066004, China; zwliu@ysu.edu.cn

* Correspondence: gillianjay@126.com; Tel./Fax: +86-0571-88297098

† These authors contributed equally to this work.

Abstract: Microbial consortium degradation technology can improve the degradation efficiency and adaptability through fungi–bacteria synergism, but the mechanism of the fungi–bacteria interaction is still unclear, making it difficult to optimize the degradation process. The microbial consortium J-6, with high lignin degradation efficiency and strong environmental adaptability, was obtained in our previous research. In this study, the fungi–bacteria interacting mechanism of the microbial consortium J-6 was inferred based on metabolomics technology. The results showed that the positive interaction between fungi and bacteria could improve the efficiency of lignin degradation. The metabolites released by fungi, especially betanidin and ergosterol, had an impact on bacterial metabolism, promoted the degradation of macromolecules, and significantly increased the lignin degradation efficiency. Metabolites released by bacteria, especially L-phenylalanine and taurine, played a key role in fungal metabolism, leading to more complete degradation. The interaction mechanism of chemical currencies exchange between fungi and bacteria during lignin degradation obtained in this study can provide theoretical guidance for microbial consortium degradation technology.

Keywords: lignin degradation; microbial consortium; metabolomics analysis; interaction mechanism

1. Introduction

The lignocellulosic biomass has wide sources on Earth. The annual production of lignocellulose can reach 170 billion tons; however, only 3% of lignocellulose can be effectively utilized in the circular bioeconomy [1]. The effective utilization of lignocellulosic biomass can not only reduce environmental pollution and carbon emissions but also produce bioethanol and other fuels to relieve fossil energy shortages [2]. Lignin, as a major component of plant cell walls, not only provides strength and resistance to lignocellulosic biomass but also protects carbohydrates from enzyme digestion. The degradation of lignin is a key step in the utilization of lignocellulosic resources [3]. However, lignin, as a macromolecular aromatic polymer, is difficult to degrade due to its complex structure [4]. At present, the degradation process of lignin has been a research hotspot in the field of global carbon cycle and biomass resource utilization.

Lignin can be degraded by many methods, such as biodegradation, catalytic degradation [5,6], solvent degradation [7], and so on. By comparison, microbial degradation has the advantages of low energy consumption, low cost, environmental friendliness, and safe operation. However, single fungi always grow and degrade slowly, while bacterial degradation has the disadvantages of incomplete degradation and low degradation efficiency [8]. As a result, the research on the lignin degradation system of fungi–bacteria mixed microorganisms has attracted a lot of attention. Iimura et al. [9] found five species of bacteria of the genus *Methylobacterium* that coexisted with white-rot fungi from decayed wood, and their research showed that the lignin degradation rate was increased by the fungi–bacteria collaboration. Zhong et al. [10] also revealed the effectiveness of fungi–bacteria synergistic pretreatment for the lignin degradation process. In our preliminary research, four efficient lignin-degrading microbial consortia were screened, among which the lignin degradation efficiency by microbial consortium J-6 could reach 54% in 2 d. Moreover, microbial consortium J-6 could effectively and selectively degrade lignin in the biomass degradation system (selectivity value = 3.59), which proved that fungi and bacteria could effectively increase the degradation efficiency of lignin in lignin degradation systems through positive interactions [11].

The degradation mechanism of lignin by the single strain has been extensively investigated [12]. However, research on the degradation mechanism of the microbial consortium to degrade lignin is insufficient. Although there were reports on the possible ways of fungi–bacteria interactions in the microbial consortium during lignin degradation, the mechanisms of fungi–bacteria interaction during lignin degradation still remain unclear. For example, Wilhelm et al. [13] reported that the fungi carried out the primary degradation of lignin, and the bacteria continued to metabolize and synthesize their cells subsequently. In the study of lignin decomposition processes in soil, Lacerda-Junior et al. [14] found that key bacterial (Bordetella and Streptomyces) and fungal (Aspergillus) members may play synergistic eco-physiological roles in the decomposition of lignin-derived compounds through the carriage of most of the genes encoding auxiliary activity (AA) family of lignin-modifying enzymes. It is worth noting that these studies lacked exploration of interacting substances in microbial consortiums, resulting in unclear mechanisms of interaction.

Currently, the mechanism of interactions among microorganisms in the microbial consortium has attracted extensive attention. For example, substances released by microorganisms may become signal transducers that activate or inhibit gene expression or biological activity, thereby altering the behavior and growth of other microorganisms. Amin et al. [15] found that the indoleacetic acid secreted by *sulfitobacter* species could stimulate the division of diatom cells and promote the growth of algae. Song et al. [16] used metabolomics analysis to study the interacting substance between chlorella vulgaris and *microcystis*. Bruger et al. [17]'s research suggests that bacteria produce and release small hormone-like signaling molecules called autoinducers (AI) into the environment and that adjacent microbial cells detect and respond to these molecules to change gene expression and coordinate the behavior of the entire community. Shibl et al. [18] identified a key chemical currency in the regulation of microbial interactions in different ecosystems.

Meanwhile, the use of metabolomics technology has made positive progress in terms of the mechanism of microbial lignin degradation in recent years [19,20]. Metabolomics refers to the identification and quantitative analysis of all metabolite components in a particular tissue or cell of an organism [21]. At present, metabolite composition analysis is mainly achieved by liquid chromatography–mass spectrometry (LC-MS), gas chromatography–mass spectrometry (GC-MS), nuclear magnetic resonance (NMR), Fourier transform infrared spectroscopy (FTIR), capillary electrophoresis–mass spectrometry (CE-MS), LC-MS/MS (liquid chromatography–tandem mass spectrometry) and so on [22,23]. The rapidly developing and widely-used LC-MS/MS technology can meet the requirements of general quantitative analysis accuracy, with good reproducibility and linear range [24]. LC-MS/MS can help find the changes of different metabolites and metabolic characteristics in the fungi–bacteria co-culture system to infer the fungi–bacteria interaction and its influence on each other's growth and lignin degradation process.

In this research, the metabolic profiles of fungi and bacteria were analyzed using metabolomics technology based on LC-MS/MS. The fungi–bacteria interaction in the co-culture system and the effects of interaction on the microbial growth and lignin degradation process were analyzed by variation in metabolic properties and screening of differential variables. The key interaction substances were verified by the complementation experiments. This research can elucidate the mutualism mechanism of lignin degradation by microbial consortium and clarify the exchange of substances between fungi and bacteria.

2. Results and Discussion

2.1. Metabolomics Analysis of Different Lignin Degradation Systems

Metabolomics studies are expected to obtain data on the abundance of large amounts of metabolites in biological degradation systems, and the abundance contains information about the biological phenomenon under research, which is manifested in (co-) variations in the abundance of metabolites [25]. In this research, three groups containing fungal samples (labeled F), bacterial samples (labeled B), and original J-6 samples (labeled 6) were set up for comparison, and metabolomics analysis was carried out. To ensure the output of high-quality data, quality control of the assay process was performed. The TIC plots of the experimentally detected samples (Figure S1) showed good sample quality, good experimental methodology, and good system stability. Through the analysis of the above data, the key interaction substances produced in the process of microbial degradation of lignin were speculated, and the fungi–bacteria interaction mechanisms were analyzed.

2.1.1. Metabolite Classification Statistics

The kinds and proportions of metabolites found in the samples were determined by statistical analysis, with the result displayed in Table 1. The metabolites are mainly composed of organoheterocyclic compounds (31.469%), organic acids and derivatives (24.126%), lipids and lipid-like molecules (14.336%) and benzenoids (8.042%). Similarly, Zhu et al. [26] analyzed the mechanism of lignin degradation by Aspergillus fumigatus G-13 and obtained that the types of differential metabolites were mainly lipids and lipid-like molecules, organic acids and their derivatives, and organoheterocyclic compounds. The results of this study were consistent with their research. The result also indicated that lignin might be depolymerized by extracellular oxidative enzymes into simple monomeric forms such as benzene ring-type compounds such as coumarin alcohol, coniferyl alcohol, and sinapyl alcohol [27], which were subsequently metabolized by various intracellular enzymes into the intermediates, such as syringic acid, vanillic acid, and p-hydroxybenzoic acid [28].

2.1.2. Principal Component Analysis (PCA)

The first and second principal component scores were shown by the horizontal coordinate PC1 and the vertical coordinate PC2, respectively, in the scatter plots of PCA scores (Figure 1A,B). The more similar the types and contents of metabolites in the samples were, the closer the distribution of the sample points [29]. The result showed that the degree of dispersion of the sample points was higher in the fungi–bacteria consortium than that in the single fungal consortium or single bacterial consortium sample points. This clearly indicated that the degree of difference in metabolic levels was overall greater in the fungi–bacteria mixed consortium compared to the single fungal consortium or single bacterial consortium.

Table 1. Metabolite classification and proportion for all samples (including the J-6 consortium and J-6 bacterial/fungal consortium).

Super Class	Percentage
Organoheterocyclic compounds	31.47
Organic acids and derivatives	24.13
Lipids and lipid-like molecules	14.34
Benzenoids	8.04
Organic oxygen compounds	7.69
Nucleosides, nucleotides, and analogues	4.55
Organic nitrogen compounds	3.85
Phenylpropanoids and polyketides	3.15
Alkaloids and derivatives	2.10
Organic 1,3-dipolar compounds	0.35

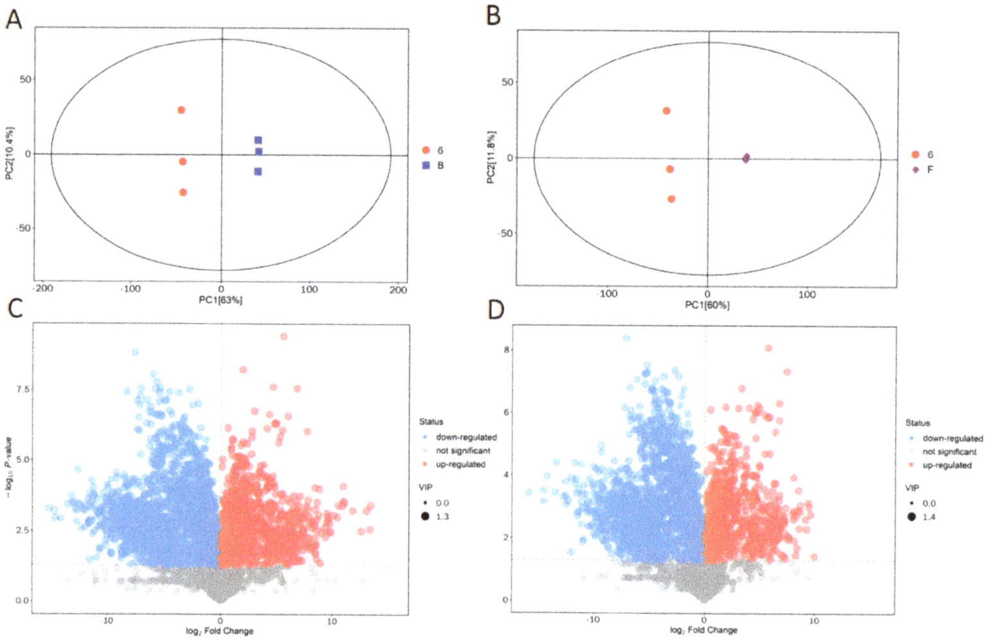

Figure 1. Basic data analysis of metabolomics: (**A**) Score scatter plot of PCA model for J-6 consortium vs. J-6 bacterial consortium; (**B**) score scatter plot of PCA model for J-6 consortium vs. J-6 fungal consortium; (**C**) volcano plot for J-6 consortium vs. J-6 bacterial consortium; (**D**) volcano plot for J-6 consortium vs. J-6 fungal consortium.

2.1.3. Screening of Differential Metabolites

The volcano plots were created by combining the findings of the univariate and multivariate statistical analyses to test for different metabolites (Figure 1C,D). Every point on the volcano plot represented a peak. The vertical coordinate indicated the p-value of the Student's t-test (derived from the negative of the logarithm of base 10) and the horizontal coordinate indicated the fold change of the group comparing each substance. The scatter size corresponded to the value of the variable importance projection (VIP) in the OPLS-DA model, where a larger scatter indicated a higher VIP value. Red points represented significantly up-regulated metabolites, blue points represented significantly down-regulated metabolites, and gray points represented metabolites that were not significantly different [30]. The volcano plot results showed that there were many types of differential metabolites in different metabolic systems, and further analysis of the key substances was needed.

In addition, hierarchical clustering analysis (HCA) was performed in order to discover the characteristics of metabolite variation among experimental groups (Figure 2). The highly expressed substances in each group illustrated in the clustered heatmaps include the substances in Table 2, thus suggesting that these substances may be biologically similar/complementary in outcome and function and may be involved together in the same metabolic processes.

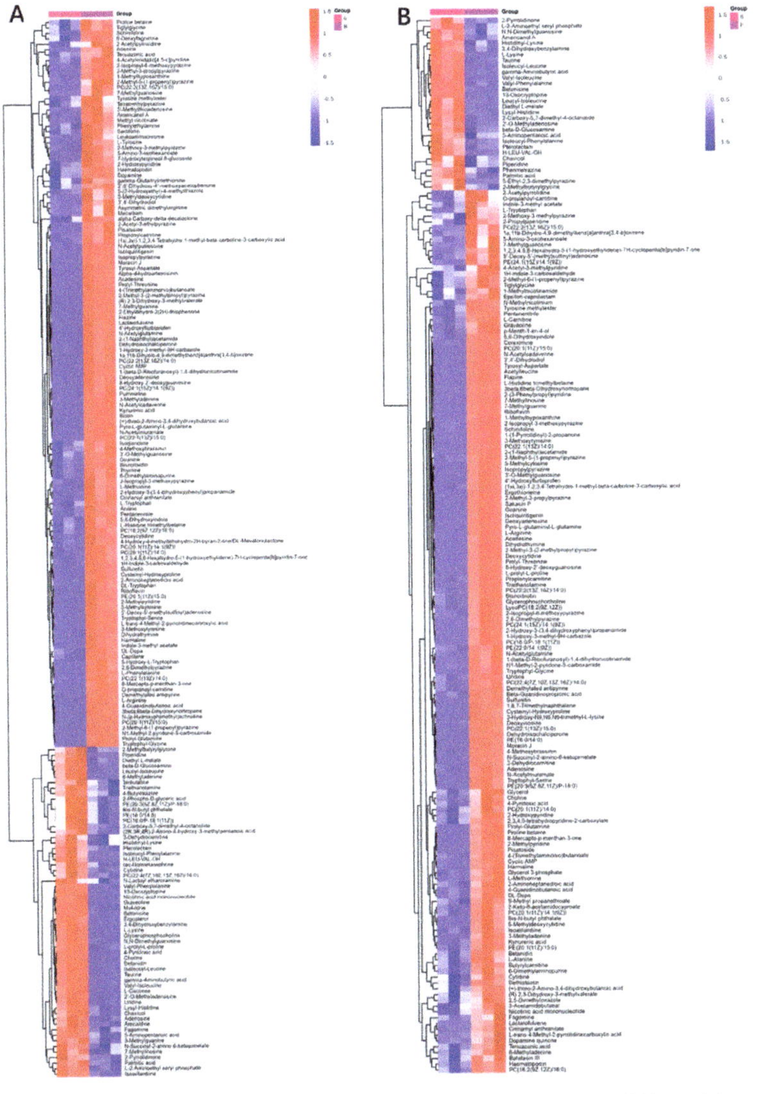

Figure 2. Heatmap of hierarchical clustering analysis: (**A**) J-6 consortium vs. J-6 bacterial consortium; (**B**) J-6 consortium vs. J-6 fungal consortium.

Table 2. Major differential metabolites in lignin degradation systems.

J-6 and J-6 Bacterial Consortium			J-6 and J-6 Fungal Consortium		
Name	**Fold-Change**	***p*-Value**	**Name**	**Fold-Change**	***p*-Value**
Betanidin	2113.76	0.00028	Taurine	4.83	0.000036
4-Pyridoxic acid	1536.90	0.00039	L-Phenylalanine	2.11	0.031
Ergosterol	243.13	0.0016			
Graveoline	229.67	0.00051			
Nicotinic acid mononucleotide	135.94	0.0019			
Glycerophosphocholine	71.14	0.0012			

2.1.4. Matchstick Analysis of Differential Metabolites

The matchstick plots (Figure 3) demonstrated differential metabolites with a large degree of change. The expression of corresponding enzyme genes might be strongly activated or strongly inhibited (* represents significance). The horizontal axis of the figure represented the change multiple after logarithmic conversion, while the depth of the dot color indicated the VIP value [31]. This analysis showed that the substances that were significantly up-regulated in the mixed system compared to the bacterial system were betanidin, 4-pyridoxic acid, 13-oxocryptopine, graveoline, choline, and so on. The substances that were significantly up-regulated in the mixed system compared to the fungal system were 13-oxocryptopine, taurine, gamma-aminobutyric acid, betonicine, palmitic acid, beta-D-glucosamine, and so on.

2.1.5. KEGG (Kyoto Encyclopedia of Genes and Genomes) Annotation and Enrichment Analysis of Differential Metabolites

KEGG pathway map contains reaction/interaction networks for species metabolism, genetic information processing, and so on. It not only provides all possible metabolic pathways for the interconversion of biochemicals but also contains comprehensive annotations of the enzymes that catalyze each step [32]. All pathways mapped by the corresponding species differential metabolites were collated, differential metabolites were labeled on KEGG pathway maps, and information such as the flow of up-/down-regulated significant differential metabolites and their metabolic responses was also collated.

The enrichment results of differential metabolites in the KEGG metabolic pathway were presented in differential abundance score graphs (Figure 4A,B). The DA Score (differential abundance score) of the graphs reflected the overall change of all metabolites in the metabolic pathway. Therein, a score of 1 indicated that the expression trend of all annotated differential metabolites in the pathway was up-regulated, and −1 indicated that the expression trend of all annotated differential metabolites in the pathway was down-regulated. The length of the line segment indicated the absolute value of the DA Score. The size of the dots indicated the number of annotated differential metabolites in the pathway, and a larger dot indicated a higher number of differential metabolites in the pathway [33]. The results showed that the fungi–bacteria mixed system compared to the bacterial system in membrane transport—glycerophospholipid metabolism, ABC transporter, nucleotide metabolism—pyrimidine metabolism, amino acid metabolism—glycine, serine, and threonine metabolism, the overall expression of metabolites tended to be up-regulated. Meanwhile, the overall expression of metabolites in amino acid metabolism—lysine biosynthesis, lysine degradation, metabolism of taurine and hypotaurine, metabolism of cofactors and vitamins—biotin metabolism tended to be up-regulated in the fungi–bacteria mixed system compared to the fungal system.

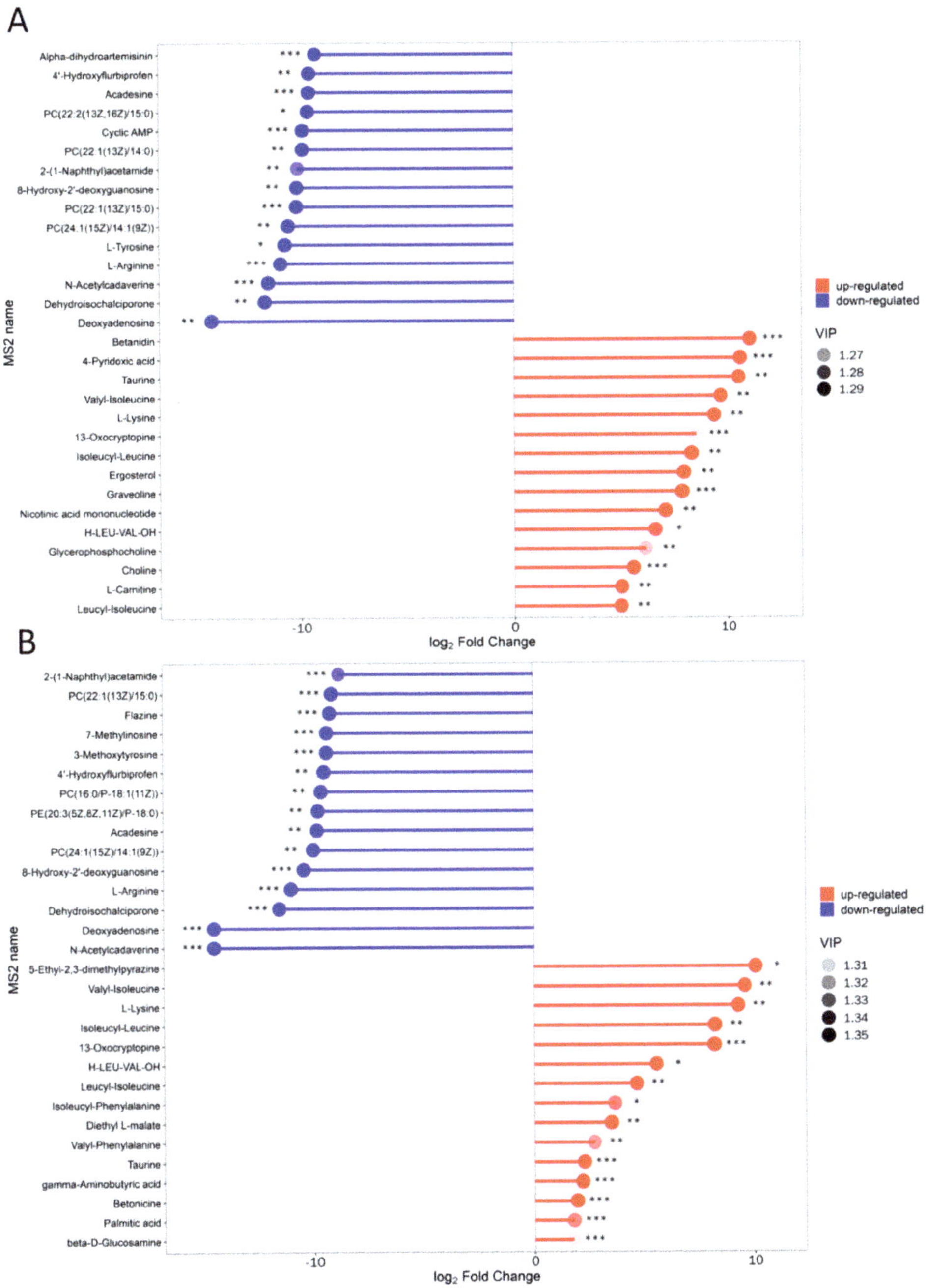

Figure 3. Matchstick analysis of different lignin degradation systems: (**A**) J-6 consortium vs. J-6 bacterial consortium; (**B**) J-6 consortium vs. J-6 fungal consortium. * stands for significance: * $0.01 < p < 0.05$, ** $0.001 < p < 0.01$, *** $p < 0.001$.

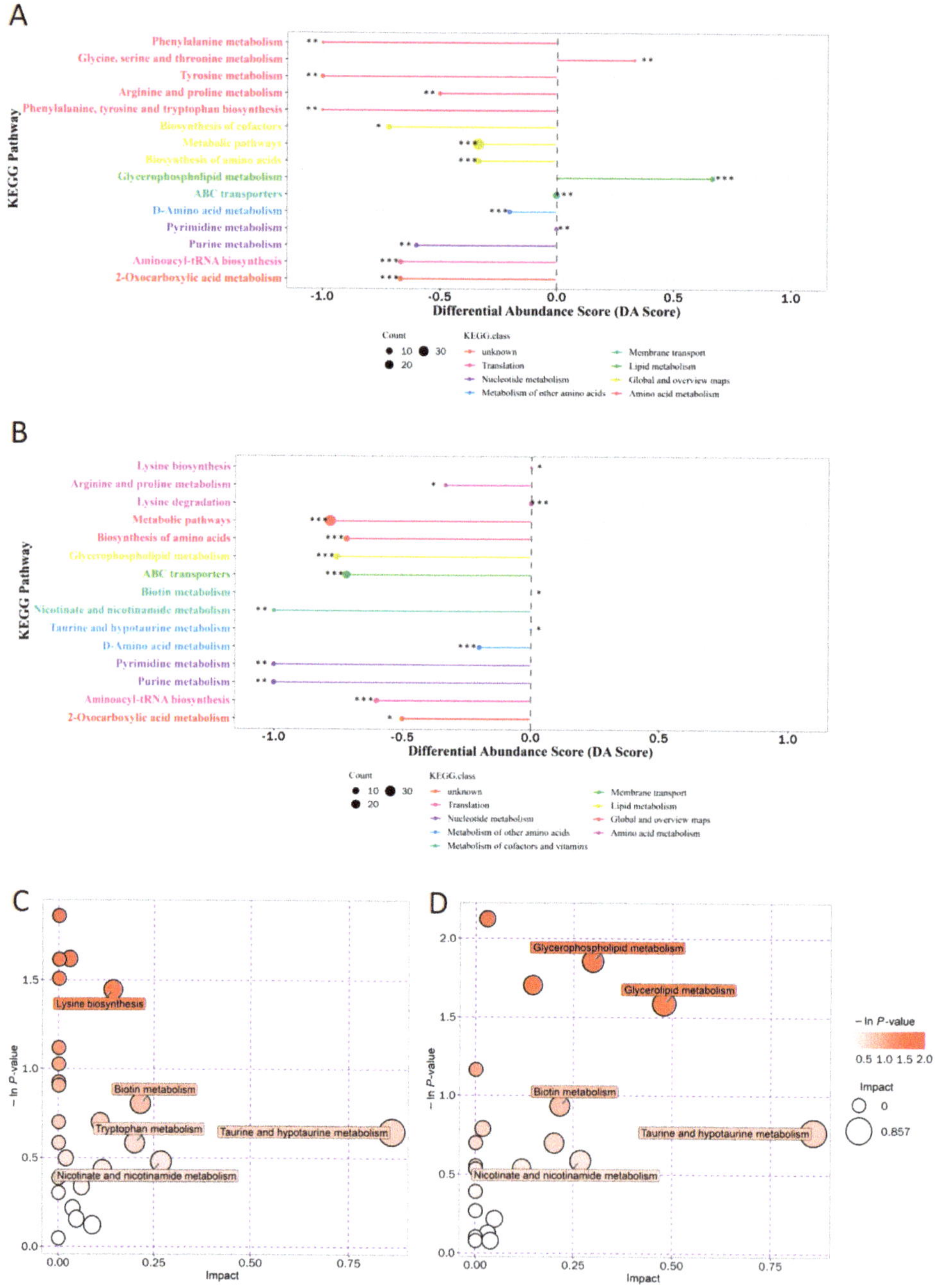

Figure 4. Functional annotation analysis of metabolomics: (**A**) differential abundance score for J-6 consortium vs. J-6 bacterial consortium; (**B**) differential abundance score for J-6 consortium vs. J-6 fungal consortium; (**C**) metabolic pathway analysis bubble plot for J-6 consortium vs. J-6 bacterial consortium; (**D**) metabolic pathway analysis bubble plot for J-6 consortium vs. J-6 fungal consortium. * stands for significance: * $0.01 < p < 0.05$, ** $0.001 < p < 0.01$, *** $p < 0.001$.

2.1.6. Metabolic Pathway Analysis of Differential Metabolites

The results of the metabolic pathway analysis were presented in bubble plots, represented in Figure 4C,D. Each bubble in the plot corresponded to a specific metabolic pathway. The horizontal position and size of the bubble indicated the impact factor of the pathway in the topology analysis, where larger bubbles meant higher impact factors. The vertical position and color of the bubble reflected the p-value of the enrichment analysis, where the color intensity represented the significance level, and darker-colored bubbles indicated smaller p-values, indicating a higher degree of enrichment significance [34]. Large dark bubbles (the pathways with higher enrichment) were selected as the focus of subsequent research. The comparison between the fungi–bacteria mixed system and the bacterial system showed differences in lysine biosynthesis, biotin metabolism, tryptophan metabolism, taurine, and hypotaurine metabolism, as well as nicotinate and nicotinamide metabolism. On the other hand, the comparison between the fungi–bacteria mixed system and the fungal system revealed distinctions in glycerophospholipid metabolism, glycerolipid metabolism, biotin metabolism, tryptophan metabolism, taurine, and hypotaurine metabolism, as well as nicotinate and nicotinamide metabolism.

2.1.7. Analysis of Interacting Substances

The fungal versus bacterial system and the bacterial versus fungal system were analyzed using the same methods as previously described. Based on the inference from the literature that utilizes similar methodologies to investigate the interaction of substances [16], it is likely that substances that are significantly up-regulated in both the fungi–bacteria mixed system versus the bacterial system and the fungal system versus the bacterial system are released by the fungi. Conversely, substances that are significantly up-regulated in both the fungi–bacteria mixed system versus the fungal system and the bacterial system versus the fungal system are likely released by the bacteria. And these substances may be the most important interacting substances between fungi and bacteria.

In order to screen for more accurate interacting substances, the differential metabolites obtained by the above metabolomics analysis were screened with FOLD_CHANGE (up-regulation value) greater than 2 (those lower than 2 were not screened), and at the same time satisfied the p-value value of less than 0.05. As shown in Table 2, the substances that were significantly up-regulated in the fungi–bacteria mixed system compared with the bacterial system and the fungal system compared with the bacterial system were betanidin, 4-pyridoxic acid, graveoline, ergosterol, nicotinic acid mononucleotide, glycerophosphocholine, and so on. The substances that were significantly up-regulated in the mixed system versus the fungal system and the bacterial system versus the fungal system were mainly L-phenylalanine and taurine.

In summary, through synthesizing the metabolic levels of each metabolite and their roles in the pathway, it could be hypothesized that the key compounds released by the fungi that might affect the bacterial system were ergosterol, betanidin, glycerophosphocholine, 4-pyridoxic acid, gravoline, and nicotinic acid mononucleotide. The key compounds released by the bacteria that might affect the fungal system were L-phenylalanine and taurine. Among these substances, betanidin and glycerophosphocholine may play a role in regulating osmosis in microbial cells. Betanidin serves as the synthetic precursor of betalain. Betalain can be involved in osmoregulatory processes that help sustain normal physiological functions. Additionally, it can act as a non-enzymatic antioxidant, aiding in the neutralization of excessive reactive oxygen species during unfavorable conditions to maintain standard cellular metabolic activities [35]. Glycerophosphocholine, recognized as a vital component of cell membrane structure, exhibits a dose-dependent elevation in cellular glycerophosphocholine levels in response to high osmolarity [36]. Glycerophosphocholine has been identified as the preferred osmoprotectant over other endogenous metabolites [37]. Nicotinic acid or its derivatives of nucleosides and mononucleotides are important intermediates in the synthesis of a variety of body-required bioproducts, thus, nicotinic acid mononucleotide may be associated with microbial energy-producing metabolism. Ergosterol and taurine may be associated with microbial stress adaptation and may enhance microbial resistance. Higher levels of ergosterol are linked to improved resistance against low temperature, freezing, low sugar, alcohol, and oxidative stress in the yeast. Meanwhile, the yeast with decreased ergosterol content exhibited reduced adaptation to high osmotic stress [38]. Chen et al. [39] found that taurine has an effect on regulating cellular metabolism and enhancement of stress resistance. Previously, several studies have demonstrated that phenylalanine affects metabolism through induction under stress conditions [40,41]. Therefore, L-phenylalanine may affect microbial metabolism through induction.

2.2. *Verification of Key Interacting Substances*

The effects of these key compounds on lignin degradation and microbial growth were experimentally verified. The compounds released by the fungi and added to the bacterial culture system were ergosterol, betanidin, nicotinic acid mononucleotide, 4-pyridoxalic acid, gravoline, and glycerophosphorylcholine. The compounds released by the bacteria and added to the fungal culture system were L-phenylalanine and taurine, respectively.

It could be seen from Figure 5 that in the experiment using alkali lignin as a degradation substrate within a certain concentration range, four interacting substances, i.e., taurine, L-phenylalanine, ergosterol, and betanidin, significantly promoted the degradation of lignin. The increase in fungi–bacteria interacting substances concentration correlated with an increase in the degradation efficiency of lignin. The lignin degradation efficiency reached the highest value when the concentration was 0.75–1 μg/L, and the degradation efficiency was 8.11% (taurine, 0.75 μg/L), 10.89% (L-phenylalanine, 1 μg/L), 29.94% (ergosterol, 1 μg/L), 25.12% (betanidin, 0.75 μg/L), respectively. Compared with the blank experimental group (bacterial consortium: 23.16%; fungal consortium: 6.24%), the degradation efficiency was improved. From Figure 6, it can be seen that when the lignin degradation efficiency was higher, the growth of the microorganisms was also better. In addition, the effects of these compounds and their analogs on lignin degradation systems have also been reported. For example, the research of Wang et al. [42] proved that ergosterol was a kind of ligninolytic-inducing compound in the incubation of *Phanerochaete sordida* YK-624 and they also identified and synthesized the ergosterol metabolites ergosta-4,7,22-triene-3,6-dione and ergosta-4,6,8(14),22-tetraen-3-one, which were compounds that also enhanced lignin-degrading activity. However, the other four interacting substances (4-pyridoxic acid, gravorin, choline glycerophosphate, and niacin mononucleotides) had no significant effect on the degradation of lignin. Compared with the experiments we did before, under the same conditions (optimal lignin degradation conditions), the J-6 microbial consortium can degrade lignin up to 54% efficiently [43]. In contrast, in this experiment, the J-6 bacterial consortium reached a degradation rate of 29.94% in the presence of ergosterol from fungi, and this degradation rate was the highest degradation rate among all groups in this experiment. It was hypothesized that fungi and bacteria in the microbial consortium may degrade lignin more efficiently by releasing interacting metabolites compared to the fungal consortium or bacterial consortium alone.

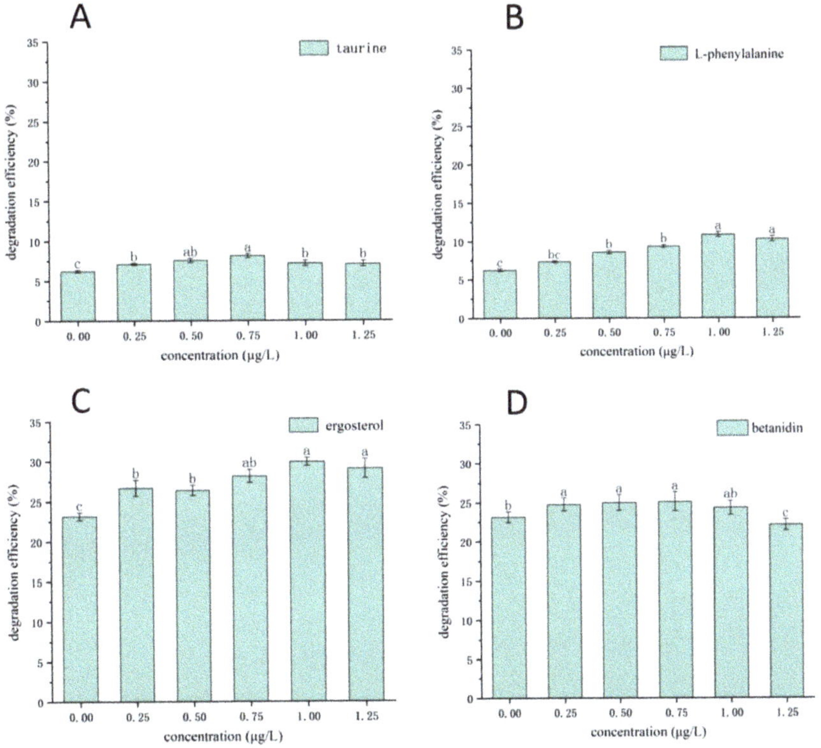

Figure 5. Verification of key interacting substances in the fungi–bacteria lignin degradation system: (**A**): lignin degradation efficiency of fungal culture systems adding taurine; (**B**) lignin degradation efficiency of fungal culture systems adding L-phenylalanine; (**C**) lignin degradation efficiency of bacterial culture systems adding ergosterol; (**D**) lignin degradation efficiency of bacterial culture systems adding betanidin). Different letters in the graph indicate significant differences ($p < 0.05$).

In the experiment with lignocellulose biomass as the substrate, the degradation efficiency of lignin did not change significantly with the increase in the concentration of fungi–bacteria interacting substances. It was hypothesized that the fungi or bacteria consortium alone might not be adapted to an environment with biomass as a substrate but can be adapted to an environment with alkaline lignin as a substrate. Therefore, when using biomass as a substrate, it was more necessary to use the fungi–bacteria mixed system for lignin degradation. This has also been confirmed in our previous research [11].

2.3. *Comparison of Lignin Degradation Products*

To further validate the roles of these key interacting substances in the degradation of lignin by microbial consortium, additional experiments were conducted. The lignin degradation products produced by a bacterial/fungal culture system incorporating ergosterol and L-phenylalanine, which significantly increased the efficiency of lignin degradation, were analyzed. Additionally, the lignin degradation products from a bacterial/fungal culture system where these two substances were not added were also analyzed. The results (Table 3) showed that different microbial consortia have different characteristics in the type of degradation products. For example, most of the degradation products of bacterial consortium are substances with high molecular weights, such as 13-oxocryptopine, and substances containing multiple benzene rings, such as 5,5′-dimethoxy-3,3′,7,7′-tetramethyl-

2,2′-binaphthalene-1,1′,4,4′-tetrone. However, with the addition of ergosterol, lignin degradation products of high molecular weight or containing multiple benzene rings are significantly reduced. In contrast, the degradation products of fungal consortium hardly contain compounds with multiple benzene rings but contain many metabolic intermediates such as L-alanyl-L-glutamine, L-citrulline, as well as many small molecules, such as 1,7-diphospho-1-epi-valienol and selenohomocystine. In addition, with the addition of L-phenylalanine, small molecule lignin degradation products increased, and there was not much change in metabolite types in general.

Due to the limited detection time of GC-MS analysis, not all lignin degradation products were detected, but the major products of lignin degradation were detected. For example, Zhang et al. [43] used GC-MS analysis to detect lignin degradation products such as phenol, benzaldehyde, 3-methyl, butylated hydroxytoluene, benzenebutyric acid,2,3-dimethoxy, etc., were also detected in this experiment.

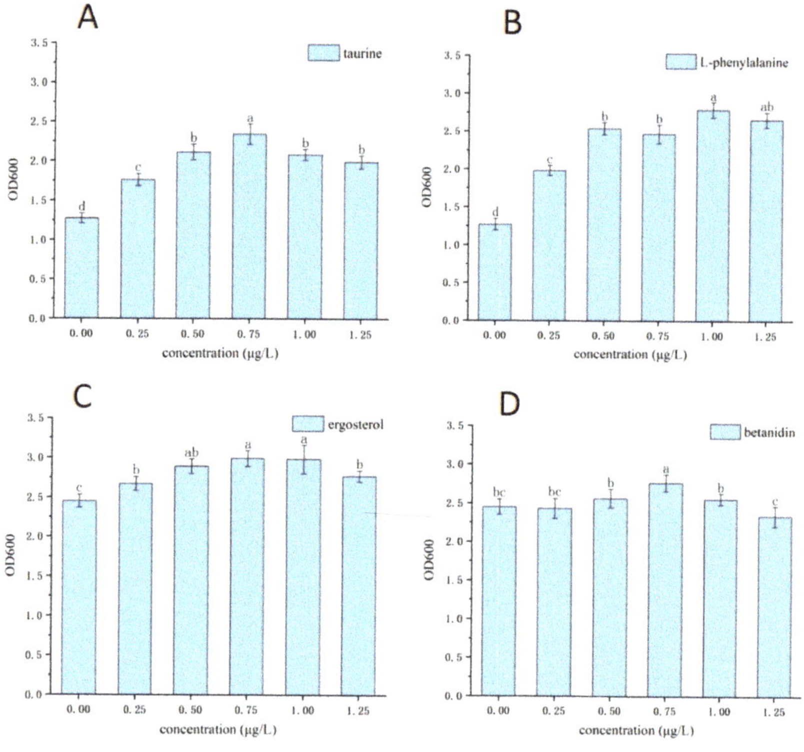

Figure 6. Verification of key interacting substances in the fungi–bacteria lignin degradation system: (**A**) biomass of fungal culture systems adding taurine; (**B**) biomass of fungal culture systems adding L-phenylalanine; (**C**) biomass of bacterial culture systems adding ergosterol; (**D**) biomass of bacterial culture systems adding betanidin). Different letters in the graph indicate significant differences ($p < 0.05$).

2.4. The Fungi–Bacteria Interaction Mechanism of Microbial Consortium

Through the metabolomics analysis of various data, eight key fungi–bacteria interacting substances, including taurine, L-phenylalanine, ergosterol, betanidin, 4-pyridoxic acid, gravoline, glycerophosphorylcholine, and nicotinic acid mononucleotide were screened. Through experimental verification, taurine, L-phenylalanine, ergosterol, and betanidin were found to most significantly enhance the efficiency of lignin degradation by a single consortium, especially the J-6 bacterial consortium, which achieved a 29.94% degradation rate with the addition of ergosterol, a 29.27% increase compared to the bacterial consortium without ergosterol. Moreover, as the efficiency of lignin degradation increased, so did the biomass. Subsequent analysis of degradation products from microbial consortia with and without the addition of these key interactive substances showed that with the addition of ergosterol, high molecular weight or polycyclic aromatic metabolites in the bacterial consortium significantly decreased. With the addition of L-phenylalanine, intermediate metabolites in the fungal consortium decreased while small molecule metabolites increased. This indicated that these two key interactive compounds played a significant role in achieving complete degradation of lignin.

Table 3. Identification of the products of the lignin degradation system.

No.	RT [1]	Compound	Bacterial Consortium		Bacterial Consortium + Ergosterol		Fungal Consortium		Fungal Consortium + L-Phenylalanine	
			1d	2d	1d	2d	1d	2d	1d	2d
1	10.45 min	Phenol		+		+		+		+
2	11.58 min	Ethyl 3-hydroxybutyrate			+		+	+	+	
3	13.27 min	Benzaldehyde, 3-methyl		+	+	+	+	+	+	+
4	18.32 min	5,5′-Dimethoxy-3,3′,7,7′-tetramethyl-2,2′-binaphthalene-1,1′,4,4′-tetrone	+	+						
5	19.81 min	Butylated Hydroxytoluene				+				+
6	22.79 min	Benzenebutyric acid, 2,3-dimethoxy-	+	+	+		+	+	+	
7	23.77 min	1,2-Benzenedicarboxylic acid, butyloctyl ester	+	+	+		+		+	
8	23.77 min	Phthalic acid, hept-3-yl isobutyl ester	+	+	+		+		+	
9	24.64 min	n-Hexadecanoic acid	+	+	+	+	+	+	+	+
10	26.51 min	Octadecanoic acid		+	+	+	+	+	+	+
11	32.69 min	1-Hydroxy-2-(2,3,4,6-tetra-O-acetyl-beta-D-glucopyranosyl)-9H-xanthene-3,6,7-triyl triacetate		+	+	+		+	+	+
12	34.27 min	L-Alanyl-L-Glutamine			+		+	+		+
13	35.75 min	1,7-diphospho-1-epi-valienol						+		+
14	35.82 min	Selenohomocystine				+		+	+	+
15	36.32 min	L-Citrulline					+	+	+	
16	36.33 min	3-Methoxytyrosine		+			+			+
17	36.55 min	4-(Trimethylammonio)butanoate				+		+	+	+
18	36.80 min	Chlomethoxyfen	+		+				+	
19	38.45 min	13-Oxocryptopine	+	+	+		+		+	

[1] RT: retention time.

Based on the above analysis, a hypothesis regarding the interaction mechanism of these key chemical currencies in the degradation of the lignin mixed system was formulated (Figure 7). Betanidin from fungi may play a role in regulating the osmotic pressure of bacterial cell membranes, while ergosterol could be involved in stress-adaptive regulation and enhancing stress resistance in bacteria. Additionally, L-phenylalanine from bacteria may be involved in inducing cellular metabolism in fungi, while taurine could play a role in stress-adaptive regulation and enhancing stress resistance in fungi.

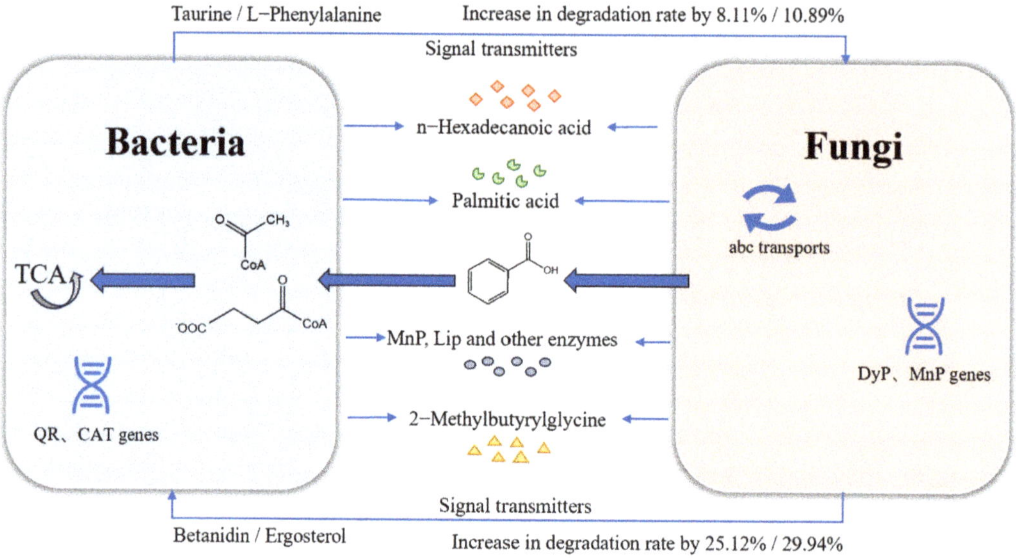

Figure 7. Degradation mechanism of lignin consortium system and key interacting substances interaction mechanism.

3. Materials and Methods

3.1. Materials

The microbial consortium J-6 was obtained from our previous research and stored at 4 °C in a refrigerator [43]. The fungi of J-6 were dominated by Saccharomycetales (98.92%) and the bacteria were mainly *Cupriavidus* sp. (29.84%), *Shinella* sp. (47.38%), and *Bosea* sp. (7.96%). This experiment was conducted using common lignocellulosic biomass in China, including bagasse (Guangxi, China), eucalyptus root (Guangxi, China), and corn straw (Heilongjiang, China). The lignin content of the biomass was from 16.37% to 26.31% [11]. In addition, the main experimental reagents, such as alkaline lignin, betanidin, ergosterol, taurine, L-phenylalanine, magnesium sulfate, and so on, were purchased from Shanghai Aladdin Biochemical Technology Co., Ltd. (Shanghai, China). The reagents were analytically pure.

3.2. Culture Medium

3.2.1. Basic Medium

The basic medium consisted of the following components: 2.5 g/L sodium chloride, 2.5 g/L $(NH_4)_2SO_4$, 1 g/L potassium dihydrogen phosphate, 0.2 g/L $MgSO_4 \cdot 7H_2O$, 0.1 g/L calcium chloride, 0.5 g/L glucose, 1 mL/L trace elements and 0.5 g/L alkaline lignin, which were acquired from our prior experiments [43]. Additionally, 1 mol/L NaOH or HCl solution was added to the mixture in order to regulate the pH level.

3.2.2. Degradation Medium

The composition of the degradation medium was as follows: biomass (1.6 g/L) or alkaline lignin (0.5 g/L), 0.2 g/L $MgSO_4$, 1.0 g/L KH_2PO_4, 2.5 g/L NaCl, 2.5 g/L $(NH_4)_2SO_4$, 0.1 g/L $CaCl_2$, 0.5 g/L glucose, and 1 mL/L trace elements. Trace elements solution formulation was as follows: 0.1 g/L $ZnSO_4 \cdot 7H_2O$, 0.16 g/L $CoCl_2 \cdot 6H_2O$, 0.15 g/L $CuSO_4 \cdot 5H_2O$, 0.1 g/L $MnSO_4 \cdot H_2O$, 0.02 g/L H_3BO_3, 0.8g/L $Na_2MoO_4 \cdot 8H_2O$, and 0.05 g/L $NiCl_2 \cdot 6H_2O$.

3.3. Isolation of Fungal and Bacterial Consortium

According to the available study [44], the isolation of fungal bacteria was done by the use of antifungal agents and antibiotics, and this isolation method was proven to be effective. This method was also applied in this study.

For bacterial isolation, the J-6 consortium suspension was diluted in 10 times the volume of sterile saline, and the diluted consortium suspension was shaken in a vortex apparatus for homogenization. The consortium suspension was moved to the basic medium with a 10% inoculum volume. At the same time, 0.05 g/L of nystatin and 0.2 g/L of cycloheximide were added to inhibit the growth of fungi. After that, it was put into a constant-temperature shaker to be cultured for 24 h at 37 °C and 200 rpm.

When the fungi were isolated, the suspension was inoculated into the culture medium according to the above method. Then, 0.1 g/L penicillin and 0.1 g/L streptomycin were added to inhibit the growth of bacteria before incubation, followed by being put into the constant-temperature shaker at 28 °C and 200 rpm for 3–7 d after the inoculation was completed.

The above isolated bacterial and fungal consortia were transformed and amplified three times, respectively. Then the colony composition was analyzed to ensure complete isolation by the sequencing method. After analyzing the community structure, it was determined that the fungi and bacteria had been completely isolated.

3.4. Metabolomics Analysis

3.4.1. Analytics Information and Raw Data Preprocessing

Due to the complexity of natural lignin, the use of stable lignin purification products is an effective and conventional method to study the microbial degradation behavior and mechanism of lignin, and there is great similarity between the two structures [45]. Therefore, in the process of metabolomics research in this study, the carbon source used was alkali lignin.

Three groups of control experiments were conducted separately, inoculating J-6 consortium (labeled 6), J-6 fungal consortium (labeled F), and J-6 bacterial consortium (labeled B) into the alkaline lignin degradation medium, and lignin degradation was conducted in a constant-temperature shaker at 30 °C and 200 rpm for 48 h to obtain the test samples. Then, 100 μ L of different samples were transferred to an EP tube (Eppendorf tube). After the addition of 400 μ L of extract solution (acetonitrile:methanol = 1:1), the samples were vortexed for 30 s, sonicated for 10 min in and ice-water bath, and incubated for 1 h at −40 °C. Then, the samples were centrifuged at 13,800× *g* for 15 min at 4 °C. The resulting supernatant was transferred to a fresh glass vial for LC-MS/MS analysis. Specifically, the ionization source of the Orbitrap platform was electrospray ionization. The positive and negative ion modes were combined to detect the metabolome.

This analysis consisted of four stages, which were basic data analysis, differential metabolite screening, advanced data analysis, and speculation of interacting substances. The original data was prepared and sorted through deviation filtering, missing value filtering, missing value filling, and data standardization.

3.4.2. Basic Data Analysis

Hierarchical clustering analysis (HCA) was performed on all samples and their metabolites, and the substances with high expression were screened. Metabolite classification statistics were carried out to classify the identified metabolites according to the chemical classification information. The metabolite classification and proportion pie chart were drawn, focusing on the types of metabolites that accounted for a relatively large proportion. Multivariate analysis (MVA) was conducted with principal component analysis (PCA), orthogonal partial least squares-discriminant analysis (OPLS-DA) and so on. Through PCA using SIMCA software (V16.0.2, Sartorius Stedim Data Analytics AB, Umea, Sweden), the overall distribution trend as well as the degree of difference of samples between groups, were obtained. To gather more accurate data regarding the differences between metabolite groups and the level of correlation of the experimental group, non-orthogonal and orthogonal variables were analyzed independently after the orthogonal variables in the metabolites that were not correlated with the categorical variables were filtered out using OPLS-DA in SIMCA software. Univariate analysis (UVA) was studied with a Student's *t*-test and analysis of variance (ANOVA) to obtain independent changes in metabolite levels.

3.4.3. Differential Metabolite Screening

Combining the results of univariate and multivariate statistical analyses, the differential metabolites were screened, and volcano plots were used to illustrate the findings of screening differential metabolites, which intuitively demonstrated the overall distribution of the metabolite differences between groups.

3.4.4. Advanced Data Analysis

Matchstick plot analysis of differential metabolites was carried out to compare each group, calculate the ratio corresponding to the quantitative value of differential metabolites, and screen out the different metabolites with greater variation through the matchstick diagram analysis. KEGG annotation of differential metabolites was to sort out various relevant information about pathways in the KEGG Pathway database and draw pathway maps. According to the enrichment results of differential metabolites in KEGG metabolic pathways, the general changes of all differential metabolites in a pathway were obtained. Besides, the pathways whose overall expression tended to be up-regulated were screened out. Metabolic pathway analysis of differential metabolites was performed to find out the target pathway where differential metabolites had a greater impact on the pathway through the comprehensive analysis of the pathway where differential metabolites reside (including enrichment analysis and topological analysis).

3.4.5. Speculation on Fungi–Bacteria Interacting Substances

The key fungi–bacteria interacting substances were analyzed by combining the metabolic levels of these metabolites and their roles and effects in the pathway.

3.5. *Complementation Experiments with Interacting Substances*

The isolated bacterial/fungal suspensions were inoculated into the medium containing the interaction compounds to be verified at a 10% inoculation rate, and the concentration gradients of the interaction compounds to be verified were set as 0 μg/L, 0.25 μg/L, 0.5 μg/L, 0.75 μg/L, 1 μg/L, and 1.25 μg/L, respectively. Three parallel experiments were set for each concentration gradient. After inoculation, the samples were incubated at 30 °C and 200 rpm in a constant-temperature shaker. Throughout the incubation period, samples were sampled periodically. The incubation time of the biomass degradation system was 14 d, and the incubation time of the alkaline lignin degradation system was 48 h (the incubation conditions and incubation time were determined according to the previous experiments).

3.6. *Analytical Methods*

LC-MS/MS analyses were performed using a UHPLC system (Vanquish, Thermo Fisher Scientific, Waltham, MA, USA) with a UPLC BEH Amide column (2.1 mm × 100 mm, 1.7 μm) coupled to Q Exactive HFX mass spectrometer (Orbitrap MS, Thermo Fisher Scientific, USA). The mobile phase consisted of 25 mmol/L ammonium acetate and 25 mmol/L ammonia hydroxide in water (pH = 9.75) (A) and acetonitrile (B). The auto-sampler temperature was 4 °C, and the injection volume was 2 μL.

According to the previously reported method, soluble lignin concentration was measured by absorbance at 280 nm [46]. The NREL (National Renewable Energy Laboratory) method of the U.S. Department of Energy was used to determine the lignin content in biomass [47]. The absorbance value of the microorganism suspension at 600 nm was measured using a visible spectrophotometer (TU-1901, Purkinje General Instrument, Beijing, China). The community structure analysis method was high-throughput sequencing [43].

The lignin degradation products were analyzed by gas chromatography–mass spectrometry (GC-MS) analysis, which was performed using previously reported methods [48].

3.7. Statistical Methods

The above experiments were replicated three times, and the average data were recorded. One-way ANOVA was employed to identify any significant bias in the results (Tukey test, $p < 0.05$).

4. Conclusions

Investigating the fungi–bacteria interaction mechanism during lignin degradation to improve the degradation efficiency of lignin is of great significance for global carbon cycling and value-added utilization of biological resources research. The results showed that the positive interaction between fungi and bacteria could improve the efficiency of lignin degradation. The metabolites released by fungi, especially betanidin and ergosterol, had an impact on bacterial metabolism, promoted the degradation of macromolecules, and significantly increased the lignin degradation efficiency. Metabolites released by bacteria, especially L-phenylalanine and taurine, played a key role in fungal metabolism, leading to more complete degradation. The interaction mechanism of chemical currencies exchange between fungi and bacteria during lignin degradation obtained in this study can provide theoretical guidance for microbial consortium degradation technology. In future research, it is necessary to utilize the results of this study to develop suitable biodegradation enhancement methods for different lignin degradation systems in order to optimize the lignin degradation process.

Supplementary Materials: The following supporting information can be downloaded at: https://www.mdpi.com/article/10.3390/molecules30030508/s1, Figure S1: TIC plot of positive ionization mode by UHPLC-QE-MS detection for (the representative sample).

Author Contributions: W.Z. conceptualized the study, participated in its design, and coordination. W.Z. and Y.W. drafted the manuscript. Y.W. and C.D. carried out most of the tests. Z.W. and Z.L. completed some supplementary experiments. All authors have read and agreed to the published version of the manuscript.

Funding: This research was supported by the Zhejiang Provincial Philosophy and Social Sciences Planning Project (project No. 24NDJC196YB), the Zhejiang Provincial Natural Science Foundation of China under Grant No. LQ20B060004, the National Natural Science Foundation of China (No. 21476190) and the Zhejiang Shuren University Basic Scientific Research Special Funds (2021XZ016).

Institutional Review Board Statement: Not applicable.

Informed Consent Statement: Not applicable.

Data Availability Statement: The data are available from the corresponding author on reasonable request.

Conflicts of Interest: The authors declare that they have no known competing financial interests or personal relationships that could have appeared to influence the work reported in this paper.

References

1. Shen, X.J.; Sun, R.C. Recent advances in lignocellulose prior-fractionation for biomaterials, biochemicals, and bioenergy. *Carbohydr. Polym.* **2021**, *261*, 117884. [CrossRef] [PubMed]
2. Kamimura, N.; Sakamoto, S.; Mitsuda, N.; Masai, E.; Kajita, S. Advances in microbial lignin degradation and its applications. *Curr. Opin. Biotechnol.* **2019**, *56*, 179–186. [CrossRef] [PubMed]
3. Sawatdeenarunat, C.; Surendra, K.C.; Takara, D.; Oechsner, H.; Khanal, S.K. Anaerobic digestion of lignocellulosic biomass: Challenges and opportunities. *Bioresour. Technol.* **2015**, *178*, 178–186. [CrossRef]
4. Schoenherr, S.; Ebrahimi, M.; Czermak, P. Lignin degradation processes and the purification of valuable products. *Lignin-Trends Appl.* **2018**, *5*, 9-8. [CrossRef]
5. Rahimi, A.; Ulbrich, A.; Coon, J.J.; Stahl, S.S. Formic-acid-induced depolymerization of oxidized lignin to aromatics. *Nature* **2014**, *515*, 249–252. [CrossRef]
6. Bouxin, F.P.; McVeigh, A.; Tran, F.; Westwood, N.J.; Jarvis, M.C.; Jackson, S.D. Catalytic depolymerisation of isolated lignins to fine chemicals using a Pt/alumina catalyst: Part 1—Impact of the lignin structure. *Green Chem.* **2015**, *17*, 1235–1242. [CrossRef]
7. Zhang, B.; Li, C.Z.; Dai, T.; Huber, G.W.; Wang, A.Q.; Zhang, T. Microwave-assisted fast conversion of lignin model compounds and organosolv lignin over methyltrioxorhenium in ionic liquids. *RSC Adv.* **2015**, *5*, 84967–84973. [CrossRef]
8. Johnston, S.R.; Lynne, B.; Weightman, A.J. Bacteria in decomposing wood and their interactions with wood-decay fungi. *FEMS Microbiol. Ecol.* **2016**, *92*, fiw179. [CrossRef]
9. Iimura, Y.; Abe, H.; Otsuka, Y.; Sato, Y.; Habe, H. Bacterial community coexisting with white-rot fungi in decayed wood in nature. *Curr. Microbiol.* **2021**, *78*, 3212–3217. [CrossRef]
10. Zhong, C.; Wang, C.M.; Wang, F.X.; Jia, H.H.; Wei, P.; Zhao, Y. Enhanced biogas production from wheat straw with the application of synergistic microbial consortium pretreatment. *RSC Adv.* **2016**, *6*, 60187–60195. [CrossRef]
11. Zhang, W.; Diao, C.Y.; Wang, L. Degradation of lignin in different lignocellulosic biomass by steam explosion combined with microbial consortium treatment. *Biotechnol. Biofuels Bioprod.* **2023**, *16*, 55. [CrossRef] [PubMed]
12. Tan, F.Y.; Cheng, J.; Zhang, Y.; Jiang, X.F.; Liu, Y.Q. Genomics analysis and degradation characteristics of lignin by Streptomyces thermocarboxydus strain DF3-3. *Biotechnol. Biofuels Bioprod.* **2022**, *15*, 78. [CrossRef] [PubMed]
13. Wilhelm, R.C.; Singh, R.; Eltis, L.D.; Mohn, W.W. Bacterial contributions to delignification and lignocellulose degradation in forest soils with metagenomic and quantitative stable isotope probing. *ISME J.* **2019**, *13*, 413–429. [CrossRef] [PubMed]
14. Lacerda, G.V., Jr.; Pastore, R.A.A.; Delforno, T.P.; Centurion, V.B.; Noronha, M.F.; Ventura, J.P.; Sartoratto, A.; Melo, I.S.; Oliveira, V.M. Taxonomic and functional dynamics of the soil microbiome from a tropical dry forest in kraft lignin-amended microcosms. *Appl. Soil Ecol.* **2023**, *183*, 104766. [CrossRef]
15. Amin, S.A.; Hmelo, L.R.; van Tol, H.M.; Durham, B.P.; Carlson, L.T.; Heal, K.R.; Morales, R.L.; Berthiaume, C.T.; Parker, M.S.; Djunaedi, B.; et al. Interaction and signalling between a cosmopolitan phytoplankton and associated bacteria. *Nature* **2015**, *522*, 98–101. [CrossRef]
16. Song, H.; Lavoie, M.; Fan, X.J.; Tan, H.N.; Liu, G.F.; Xu, P.F.; Fu, Z.W.; Paerl, H.W.; Qian, H.F. Allelopathic interactions of linoleic acid and nitric oxide increase the competitive ability of *Microcystis aeruginosa*. *ISME J.* **2017**, *11*, 1865–1876. [CrossRef]
17. Bruger, E.L.; Synder, D.J.; Cooper, V.S.; Waters, C.M. Quorum sensing provides a molecular mechanism for evolution to tune and maintain investment in cooperation. *ISME J.* **2021**, *15*, 1236–1247. [CrossRef]
18. Shibl, A.A.; Ochsenkühn, M.A.; Mohamed, A.R.; Isaac, A.; Coe, L.S.; Yun, Y.; Skrzypek, G.; Raina, J.-B.; Seymour, J.R.; Afzal, A.J.; et al. Molecular mechanisms of microbiome modulation by the eukaryotic secondary metabolite azelaic acid. *eLife* **2024**, *12*, RP88525. [CrossRef]
19. Li, Q.; Wu, X.Q.; Zhang, X.J. Research progress in microbial degradation of straw lignin. *Acta Microbiol. Sin.* **2023**, *63*, 4118–4132. [CrossRef]
20. Thornbury, M.; Sicheri, J.; Slaine, P.; Getz, L.J.; Finlayson-Trick, E.; Cook, J.; Guinard, C.; Boudreau, N.; Jakeman, D.; Rohde, J.; et al. Characterization of novel lignocellulose-degrading enzymes from the porcupine microbiome using synthetic metagenomics. *PLoS ONE* **2019**, *14*, e0209221. [CrossRef]
21. Sunil, S.A.; Anita, R.; Siu, K.S. Quantitative proteomic study of *Aspergillus fumigatus* secretome revealed deamidation of secretory enzymes. *J. Proteom.* **2015**, *119*, 154–168. [CrossRef]
22. Beale, D.J.; Pinu, F.R.; Kouremenos, K.A.; Poojary, M.M.; Narayana, V.K.; Boughton, B.A.; Kanojia, K.; Dayalan, S.; Jones, O.A.H.; Dias, D.A. Review of recent developments in GC–MS approaches to metabolomics-based research. *Metabolomics* **2018**, *14*, 152. [CrossRef] [PubMed]
23. Zhou, B.; Xiao, J.F.; Tulif, L.; Ressom, H.W. LC-MS-based metabolomics. *Mol. Biosyst.* **2012**, *8*, 470–481. [CrossRef] [PubMed]
24. Xiao, J.F.; Zhou, B.; Ressom, H.W. Metabolite identifi cation and quantitation in LC-MS/MS-based metabolomics. *TrAC-Trend Anal. Chem.* **2012**, *32*, 1–14. [CrossRef]
25. Saccenti, E.; Hoefsloot, H.C.J.; Smilde, A.K.; Westerhuis, J.A. MMWB Hendriks. Reflections on univariate and multivariate analysis of metabolomics data. *Metabolomics* **2014**, *10*, 361–374. [CrossRef]

26. Zhu, X.D.; Zhou, Z.J.; Guo, G.J.; Li, J.D.; Yan, H.; Li, F. Proteomics and metabolomics analysis of the lignin degradation mechanism of lignin-degrading fungus *Aspergillus fumigatus* G-13. *Anal. Methods* **2023**, *15*, 1062–1076. [CrossRef]
27. Kumar, M.; You, S.; Beiyuan, J.; Luo, G.; Gupta, J.; Kumar, S.; Singh, L.; Zhang, S.; Tsang, D.C.W. Lignin valorization by bacterial genus Pseudomonas: State-of-the-art review and prospects. *Bioresour. Technol.* **2021**, *320*, 124412. [CrossRef]
28. Li, F.; Zhao, Y.; Xue, L.; Ma, F.; Dai, S.Y.; Xie, S. Microbial lignin valorization through depolymerization to aromatics conversion. *Trends Biotechnol.* **2022**, *40*, 1469–1487. [CrossRef]
29. Luo, C.Y.; Yao, X.Z.; Chen, J.; Lyu, L.T. Analysis of Substance Differences before and after Fermentation of Black Tea Infusion by *Eurotium cristatum* Based on Non-Targeted Metabonomics. *Cienc. Tecnol. Alime* **2021**, *46*, 246–253. [CrossRef]
30. Guo, L.; Xiao, P.Y.; Li, L.S.; Chen, S.; Yuan, G. Mechanism of trehalose-enhanced metabolism of heterotrophic nitrification-aerobic denitrification community under high-salt stress. *Chin. J. Biotechnol.* **2022**, *38*, 4536–4552. [CrossRef]
31. Ye, X.L.; Yu, F.H.; Zhou, J.; Zhao, C.N.; Wu, J.; Ni, X. Analysis of the gut microbiota in children with gastroesophageal reflux disease using metagenomics and metabolomics. *Front. Cell. Infect. Microbiol.* **2023**, *13*, 1267192. [CrossRef] [PubMed]
32. Kanehisa, M.; Goto, S.; Furumichi, M.; Tanabe, M.; Hirakawa, M. KEGG for representation and analysis of molecular networks involving diseases and drugs. *Nucleic Acids Res.* **2010**, *38* (Suppl. S1), D355–D360. [CrossRef] [PubMed]
33. Song, H.; Kim, J.; Guk, J.H.; Kim, W.H.; Nam, H.; Suh, J.G.; Seong, J.K.; Cho, S. Metagenomic analysis of the gut microbiota of wild mice, a newly identified reservoir of Campylobacter. *Front. Cell. Infect. Microbiol.* **2021**, *10*, 596149. [CrossRef] [PubMed]
34. Yang, X.M.; Cui, F.X.; Zhan, X.K. Metabolomic analyses of potential regulatory substances of primordium development of *Lanmaoa asiatica*. *Res. Pap.* **2023**, *42*, 1185–1202. [CrossRef]
35. Polturak, G.; Aharoni, A. "La Vie en Rose": Biosynthesis, sources, and applications of betalain pigments. *Mol. Plant* **2018**, *11*, 7–22. [CrossRef]
36. Ridgway, N.D. The role of phosphatidylcholine and choline metabolites to cell proliferation and survival. *Crit. Rev. Biochem. Mol. Biol.* **2013**, *48*, 20–38. [CrossRef]
37. Chen, L.Y.; Li, J.; Guo, T.N.; Ghosh, S.; Koh, S.K.; Tian, D.C.; Zhang, L.; Jia, D.Y.; Beuerman, R.W.; Aebersold, R.; et al. Global metabonomic and proteomic analysis of human conjunctival epithelial cells (IOBA-NHC) in response to hyperosmotic stress. *J. Phys. Chem. Lett.* **2015**, *14*, 3982–3995. [CrossRef]
38. Montanes, F.M.; Pascual-Ahuir, A.; Proft, M. Repression of ergosterol biosynthesis is essential for stress resistance and is mediated by the Hog1 MAP kinase and the Mot3 and Rox1 transcription factors. *Mol. Microbiol.* **2011**, *79*, 1008–1023. [CrossRef]
39. Chen, B.; Abaydula, Y.; Li, D.L.; Tan, H.W.; Ma, X.L. Taurine ameliorates oxidative stress by regulating PI3K/Akt/GLUT4 pathway in HepG2 cells and diabetic rats. *J. Funct. Foods* **2021**, *85*, 104629, Erratum in *J. Funct. Foods* **2021**, *87*, 104861. [CrossRef]
40. Jan, R.; Khan, M.A.; Asaf, S.; Lee, I.J.; Bae, J.S.; Kim, K.M. Overexpression of OsCM alleviates BLB stress via phytohormonal accumulation and transcriptional modulation of defense-related genes in *Oryza sativa*. *Sci. Rep.* **2020**, *10*, 19520. [CrossRef]
41. Manela, N.; Oliva, M.; Ovadia, R.; Sikron-Persi, N.; Ayenew, B.; Fait, A.; Galili, G.; Perl, A.; Weiss, D.; Oren-Shamir, M. Phenylalanine and tyrosine levels are rate-limiting factors in production of health promoting metabolites in Vitis vinifera cv. Gamay Red cell suspension. *Front. Plant Sci.* **2015**, *6*, 538. [CrossRef] [PubMed]
42. Wang, J.Q.; Yin, R.; Hashizume, Y.; Todoroki, Y.; Mori, T.; Kawagishi, H.; Hirai, H. Ergosterol and Its Metabolites Induce Ligninolytic Activity in the Lignin-Degrading Fungus *Phanerochaete sordida* YK-624. *J. Fungi* **2023**, *9*, 951. [CrossRef] [PubMed]
43. Zhang, W.; Ren, X.; Lei, Q.; Wang, L. Screening and comparison of lignin degradation microbial consortia from wooden antiques. *Molecules* **2021**, *26*, 2862. [CrossRef] [PubMed]
44. Slosarcíková, P.; Novotny, C.; Malachová, K.; Válková, H.; Fojtík, J. Effect of yeasts on biodegradation potential of immobilized cultures of white rot fungi. *Sci. Total Environ.* **2017**, *589*, 146–152. [CrossRef]
45. Shi, Y.; Chai, L.Y.; Tang, C.J.; Yang, Z.H.; Zhang, H.; Chen, R.H.; Chen, Y.H.; Zheng, Y. Characterization and genomic analysis of kraft lignin biodegradation by the beta-proteobacterium *Cupriavidus basilensis* B-8. *Biotechnol. Biofuels* **2013**, *6*, 1. [CrossRef]
46. Chang, Y.-C.; Choi, D.; Takamizawa, K.; Kikuchi, S. Isolation of *Bacillus* sp. strains capable of decomposing alkali lignin and their application in combination with lactic acid bacteria for enhancing cellulase performance. *Bioresour. Technol.* **2014**, *152*, 429–436. [CrossRef]
47. Sluiter, A.; Hames, B.; Ruiz, R.; Scarlata, C.; Sluiter, J.; Templeton, D.; Crocker, D. *Determination of Structural Carbohydrates and Lignin in Biomass*; Technical Report NREL/TP-510-42618; Contract No. DE-AC36-08-GO28308; The National Renewable Energy Laboratory (NREL) of the US Department of Energy (USDOE): Golden, CO, USA, 2012. Available online: https://www.nrel.gov/docs/gen/fy13/42618.pdf (accessed on 14 January 2025).
48. Shi, Y.; Chai, L.; Tang, C.; Yang, Z.; Zheng, Y.; Chen, Y.; Jing, Q. Biochemical investigation of kraft lignin degradation by *Pandoraea* sp. B-6 isolated from bamboo slips. *Bioprocess Biosyst. Eng.* **2013**, *36*, 1957–1965. [CrossRef]

molecules

Article

Engineering PVA-CNF-MOF Composite Films for Active Packaging: Enhancing Mechanical Strength, Barrier Performance, and Stability for Fresh Produce Preservation

Sergio Carrasco [1], Juan Amaro-Gahete [2,3,*], Eduardo Espinosa [1], Almudena Benítez [4], Francisco J. Romero-Salguero [3] and Alejandro Rodríguez [1]

[1] BioPrEn Group (RNM940), Chemical Engineering Department, Instituto Químico para la Energía y el Medioambiente (IQUEMA), Faculty of Science, University of Córdoba, 14014 Córdoba, Spain; q62carcs@uco.es (S.C.); a02esvie@uco.es (E.E.); q42ropaa@uco.es (A.R.)

[2] Materiales Polifuncionales Basados en Carbono (UGR-Carbon), Department of Inorganic Chemistry, Unidad de Excelencia Química Aplicada a Biomedicina y Medioambiente, University of Granada (UEQ-UGR), 18071 Granada, Spain

[3] Deparment of Organic Chemistry, Instituto Químico para la Energía y el Medioambiente (IQUEMA), Faculty of Science, University of Córdoba, Campus de Rabanales, Ed. Marie Curie, 14071 Córdoba, Spain; qo2rosaf@uco.es

[4] Departament of Inorganic Chemistry and Chemical Engineering, Instituto Químico para la Energía y el Medioambiente (IQUEMA), Faculty of Science, University of Córdoba, 14071 Córdoba, Spain; q62betoa@uco.es

* Correspondence: j.amarogahete@ugr.es

Abstract: Food waste is a global challenge, with nearly 40% of food discarded annually, leading to economic losses, food insecurity, and environmental harm. Major factors driving spoilage include microbial contamination, enzymatic activity, oxidation, and excessive ethylene production. Active packaging offers a promising solution by extending shelf life through the selective absorption or release of specific substances. In this study, polyvinyl alcohol (PVA) films incorporating metal-organic frameworks (MOFs) were prepared via solvent casting to enhance their mechanical and barrier properties. Five MOFs (HKUST-1, MIL-88A, BASF-A520, UiO-66, and MOF-801) were embedded in the PVA matrix and analyzed for their physical, mechanical, and optical characteristics. The incorporation of TEMPO-oxidized cellulose nanofibers (CNF) improved MOF dispersion, significantly strengthening film performance. Among the formulations, PVA-CNF-MOF-801 exhibited the best performance, with a 130% increase in tensile strength, a 50% reduction in water vapor permeability, and a 168% improvement in UV protection compared with neat PVA films. Ethylene adsorption tests with climacteric fruits confirmed that CNF-containing films retained ethylene more effectively than those without CNFs, although the differences among the MOFs were minimal. These results highlight the potential of PVA-CNF-MOF composite films as sustainable active packaging materials, providing an effective strategy to reduce food waste and its environmental impact.

Keywords: metal-organic frameworks; cellulose nanofibers; PVA; PVA-CNF-MOF film; active packaging

1. Introduction

The global food supply chain faces considerable hurdles due to population growth, which is projected to reach 9.8 billion inhabitants by 2050 [1], compounded by intensifying environmental pressures driven by climate change, socio-economic disruptions in the post-COVID-19 era that have exposed critical vulnerabilities in global supply chains, and geopolitical instabilities such as the Russia-Ukraine conflict, which has markedly affected the availability and cost of key agricultural commodities [2–4]. Several studies suggest that the increase in population will require a greater food supply, which is expected to increase by 35% to 56% between 2010 and 2050 [5], leading to a rise in food waste [6]. According to the latest available data, nearly 40% of the food intended for human consumption (1.3 billion tons of edible food) ends up as waste [7]. This substantial waste, known as "food loss and waste" (FLW), represents a hazard to the economy, environment, and food security [8]. FLW, estimated to be worth 1 trillion US dollars [9], is very harmful to the environment due to decomposition in landfills, where part of it breaks down into greenhouse gases and methane, with a global warming potential 28–36 times higher than carbon dioxide [10]. In fact, FLW contributes to 8–10% of greenhouse gas emissions, with an estimated carbon footprint of 3.3 billion tons of CO_2 [11]. Consequently, the United Nations (UN) has implemented measures to reduce FLW as part of the Sustainable Development Goals (SDG 12) for 2030 [12]. Solutions to this problem include reducing the amount of waste generated, optimizing the food chain, increasing consumer awareness [13], and improving packaging systems that increase the shelf-life of food [6].

The longevity of food products can be influenced by numerous factors, including microbial contamination, enzymatic degradation, oxidation, or the presence of excessive ethylene [14]. To mitigate these challenges, a range of preservation methods have been devised to extend shelf life, such as coating preservation, vacuum preservation, and active packaging [15]. Among these options, active packaging is emerging as a newer approach that is gaining prominence over traditional packaging systems due to the ability of this type of packaging to control the amount of moisture, oxygen, or ethylene in contact with food. Additionally, active packaging can desorb substances with antimicrobial and antioxidant properties, thus enhancing food preservation. Its versatility and flexibility, driven by its ability to be customized for different applications, make it a favorable field for future research and innovation [15].

In this context, the utilization of metal-organic frameworks (MOFs) is gaining significance in active packaging materials. MOFs represent a class of hybrid porous materials composed of metal ions or clusters coordinated to organic ligands. These materials stand out for their high specific surface area and porosity, presenting a high variation in components in their structure, thus facilitating interaction with multiple materials [16]. One of the main problems faced by MOFs is that they are mainly presented in powder form, which hinders their processing and therefore their direct application or use on an industrial scale [17]. This can be solved by mixing MOFs with other types of materials, such as polymers, forming different types of composites, such as films, membranes, or foams, with the aim of increasing their application and improving the synergistic properties of both components [16,18].

MOFs have garnered significant attention due to their exceptional properties in thin film materials. These hybrid structures incorporate an organic ligand, which imparts functional versatility, and an inorganic substrate, which contributes to mechanical rigidity and stability. Although the integration of inorganic substrates ensures robust mechanical performance and long-term stability, it often involves high costs and intricate fabrication processes [16]. In contrast, polymer membranes, such as polyvinyl alcohol (PVA), offer a more economically viable and industrially feasible alternative. By combining MOFs with PVA, composite films are created that synergistically exploit the adaptable functionality and porosity of MOFs alongside the cost-effectiveness, flexibility, and processability of PVA. This approach not only reduces fabrication complexity but also broadens the scope for scalable and practical applications in various industries. [19]. However, challenges arise during the preparation process, as MOF particles may undergo coverage or aggregation, potentially diminishing their effectiveness. Notwithstanding these challenges, the versatility and potential applications of MOF/PVA composite films, ranging from catalysis to food packaging [20,21], position them as a promising area of research and development in materials science. Addressing issues related to particle coverage and aggregation could further enhance the performance and applicability of these composite materials across various industries.

In this aspect, the incorporation of TEMPO-oxidized cellulose nanofibers (CNFs), produced through a selective oxidation process using the TEMPO (2,2,6,6-tetramethylpiperidine-1-oxyl) catalyst, offers an intriguing solution as these nanofibers exhibit a high content of carboxyl groups and a large specific surface area, enabling improved interactions with the positive charge of metal ions. This strategy would improve the dispersion of MOF particles and ensure a better homogenization in the PVA composite film resulting in a more effective synthesis procedure. Also, CNF adhesion to PVA films has been shown to be efficient since it improves mechanical properties and thermal stability as well as reduces water vapor permeability and water uptake [22,23]. The integration of the mechanical properties of PVA-CNF films with the capacity of MOFs for the adsorption and desorption of bioactive compounds positions these composite materials as highly suitable for diverse applications, including active packaging [24]. Notably, certain MOFs have demonstrated exceptional suitability for ethylene retention [25], a key factor in the preservation of fruits and vegetables, thereby significantly extending their shelf life.

This study focuses on the production of MOF/PVA composite films through a simple solution casting process [26]. Five types of MOFs (MIL-88A, A520-BASF, HKUST-1, MOF-801, and UiO-66) were compared to evaluate their influence on the physicochemical properties of the resulting materials. MOF/PVA films were characterized to determine their physical properties (density, moisture content, water solubility, swelling behavior, and water vapor permeability), mechanical properties (tensile strength and Young's modulus), and optical properties (transparency and UV-light barrier). Their structural and morphological features were further analyzed using X-ray diffraction (XRD), Fourier transform infrared spectroscopy (FTIR), and scanning electron microscopy (SEM). To improve the interaction between the film components, TEMPO-oxidized cellulose nanofibers, obtained to our knowledge for the first time from cotton linters, were incorporated. In this way, MOF-PVA-CNF films were developed and characterized using a similar methodology. In addition to their role as physical barriers, these composite films were designed to provide active functionalities relevant to food preservation, including enhanced UV shielding, improved barrier properties against water vapor, and ethylene scavenging capacity, thereby contributing to the delayed ripening and reduced spoilage of fresh produce. Both MOF-PVA and MOF-PVA-CNF composite films were subsequently tested on climacteric fruits to assess their ethylene adsorption capacity, demonstrating their potential as sustainable active materials for food packaging.

2. Results and Discussion

2.1. MOF Characterization

Table 1 summarizes the structural and adsorption properties of the synthesized MOFs, highlighting their potential for ethylene capture based on previous studies [26]. Ethylene adsorption in these materials depends on their specific surface area, pore structure, and active sites. HKUST-1 exhibits the highest ethylene adsorption capacity per gram due to its large surface area. However, when adsorption is normalized per unit area, MIL-88A demonstrates the highest specific ethylene adsorption, surpassing even HKUST-1. This suggests that MIL-88A possesses a high density of active adsorption sites despite its lower total surface area. Ethylene adsorption in MIL-88A is primarily driven by metal-π interactions, in which open metal sites (Lewis acid sites) play a crucial role. A similar mechanism could be expected in other MOFs with accessible metal centers, such as HKUST-1 (Cu-based), BASF-A520 (Al-based), and UiO-66 and MOF-801 (Zr-based). However, their interaction efficiency may vary due to differences in Lewis acidity, pore structure, and metal-ligand coordination environments. While HKUST-1 and BASF-A520 contain transition metals that could contribute to strong ethylene interactions, UiO-66 and MOF-801 feature more stable Zr coordination, which might favor selective adsorption while maintaining good structural integrity.

Table 1. Summary of structural and adsorption properties of the MOFs synthesized for this study, highlighting their potential as candidates for ethylene capture based on prior studies [26]. BTC = 1,3,5-benzenetricarboxylic acid, Fum = Fumaric acid, BDC = 1,4-benzenedicarboxylic acid.

Material	Metallic Center	Organic Linker	S_{BET} (m^2/g)	Theoretical Pore Size (Å)	Ethylene Adsorption ($mmol \cdot g^{-1}$)
HKUST-1	Cu	BTC	2042	9.0	8.33
MIL-88A	Fe	Fum	359	6.0	1.63
BASF-A520	Al	Fum	911	5.8	3.90
UiO-66	Zr	BDC	1400	7.5, 12.0	2.61
MOF-801	Zr	Fum	856	4.8, 5.6, 7.4	2.31

From a food safety perspective, the biocompatibility and stability of MOFs must be considered. Zirconium-based MOFs (UiO-66 and MOF-801) are generally regarded as highly stable and suitable for applications involving food contact. Aluminum-based MOFs, such as BASF-A520, also exhibit good stability, which can be advantageous for long-term use. HKUST-1, being Cu-based, may require careful consideration regarding potential metal leaching, though its performance in adsorption remains high. Similarly, MIL-88A (Fe-based) presents an interesting balance between adsorption efficiency and biocompatibility, as iron is an essential element in biological systems. Additionally, the choice of organic linkers influences toxicity; thus, selecting chemically stable and non-reactive ligands ensures safe application in active food packaging. Fumaric acid-based linkers, present in MIL-88A, BASF-A520, and MOF-801, are generally considered less toxic, as fumarate is a naturally occurring metabolite involved in biological cycles such as the Krebs cycle in humans [27]. In contrast, aromatic carboxylate ligands, such as benzene-1,3,5-tricarboxylate (HKUST-1) and benzene-1,4-dicarboxylate (UiO-66), may require further evaluation due to potential concerns associated with aromatic compounds [28]. However, the overall toxicity MOFs is influenced not only by its linker but also by its metal center and structural stability, making case-by-case assessments essential for food-related applications.

When considering the integration of MOFs into a PVA matrix with CNF, several factors must be considered, including particle dispersion, interfacial compatibility, and potential hydrogen bonding or electrostatic interactions. MOFs with higher surface areas and more polar functional groups may exhibit better dispersion and interaction with PVA and CNF, leading to enhanced mechanical and barrier properties. In this regard, BASF-A520 [$Al(OH)(Fum) \cdot 3.5H_2O$], UiO-66 [$Zr_6O_4(OH)_4(BDC)_6$], and MOF-801 [$Zr_6O_4(OH)_4(Fum)_6$] could be strong candidates due to their structural stability, high surface area, and the presence of hydroxy functional groups coordinated to the metal cluster, which may favor interactions with PVA and CNF via hydrogen bonding. Conversely, MIL-88A [$Fe_3O(Fum)_3(H_2O)_2$], which contains fumaric acid linkers in its molecular structure, also exhibits high hydrophilicity, potentially improving dispersion within the polymeric matrix. However, the dynamic structural flexibility (breathing effect) of MIL-88A may influence its performance, depending on the processing conditions. HKUST-1 [$Cu_3(BTC)_2$] despite its high surface area, may require careful consideration due to its potential sensitivity to moisture and its aromatic linker, which could affect its compatibility with the hydrophilic nature of PVA and CNF. Overall, while all MOFs appear suitable for incorporation into the PVA-CNF system, Zr-based MOFs (UiO-66 and MOF-801) and fumaric acid-linked MOFs (MIL-88A and BASF-A520) may provide the best combination of stability, compatibility, and functional properties.

The N_2 adsorption-desorption isotherms presented in Figure 1a provided insights into the textural characteristics of the studied MOF samples (Table 2). The isotherms for all samples predominantly exhibited Type I isotherms, which are characteristic of microporous materials [29]. This is evidenced by the steep rise at low relative pressures (P/P_0), indicating significant micropore filling due to strong adsorbent-adsorbate interactions. However, the isotherms for HKUST-1 (a) and MIL-88A (b) also displayed characteristics of Type IV isotherms, typically associated with mesoporous materials [26,30]. These features were observed in the form of a gradual slope and the presence of hysteresis loops at higher relative pressures, suggesting a combined micro- and mesoporous structure. Specifically, the hysteresis observed in these two samples corresponds to Type IV hysteresis, indicative of capillary condensation occurring in mesopores. This suggested that HKUST-1 and MIL-88A exhibit a hybrid micro- and mesoporous structure. In contrast, the remaining samples lacked significant hysteresis, confirming their predominantly microporous nature [31–33]. The textural properties summarized in Table 2 support these observations. Among them, HKUST-1 and UiO-66 achieved the highest Langmuir surface areas (1526 and 1531 m^2/g, respectively) and micropore volumes (0.584 and 0.585 cm^3/g), strongly confirming their dominant microporous nature [34,35]. MIL-88A, although having the lowest Langmuir surface area (469 m^2/g) and total pore volume (0.198 cm^3/g), displayed an average pore width of 4.9 nm, indicative of some degree of mesoporosity within its structure [26]. In contrast, BASF-A520 presented the largest average pore width (10.6 nm) and a significant total pore volume (0.414 cm^3/g), consistent with part of its mesoporous framework [31]. Similarly, MOF-801 exhibited a combination of microporosity and mesoporosity, with a total pore volume of 0.333 cm^3/g and an average pore width of 4.9 nm. The particle size distribution determined by laser diffraction for each MOF is presented in Figure 1b, with the following average particle sizes (D [4.3]) determined for the samples: MIL-88A (4.0 μm), BASF-A520 (17.0 μm), HKUST-1 (31.5 μm), MOF-801 (1.0 μm), and UiO-66 (37.3 μm). These values may be overestimated due to particle aggregation and deviations from sphericity in ethanolic dispersion, as the laser diffraction method assumes an equivalent spherical diameter that may not fully represent the morphology of irregularly shaped or aggregated MOF particles. Such effects contribute to the broader particle size distributions observed in some of the samples.

Table 2. Porous textural properties of MOF samples.

MOF Sample	Langmuir Surface Area (m^2/g)	t-Plot Micropore Area (m^2/g)	t-Plot External Surface Area (m^2/g)	Total Pore Volume (cm^3/g)	t-Plot Micropore Volume (cm^3/g)	BJH Desorption Average Pore Width (nm)
HKUST-1	1526	1471.8	54.2	0.584	0.512	4.8
MIL-88A	469	429.9	39.4	0.198	0.146	4.9
BASF-A520	968	909.2	59.2	0.414	0.313	10.6
UiO-66	1531	1478.0	52.6	0.585	0.514	6.3
MOF-801	853	797.1	55.6	0.333	0.272	4.9

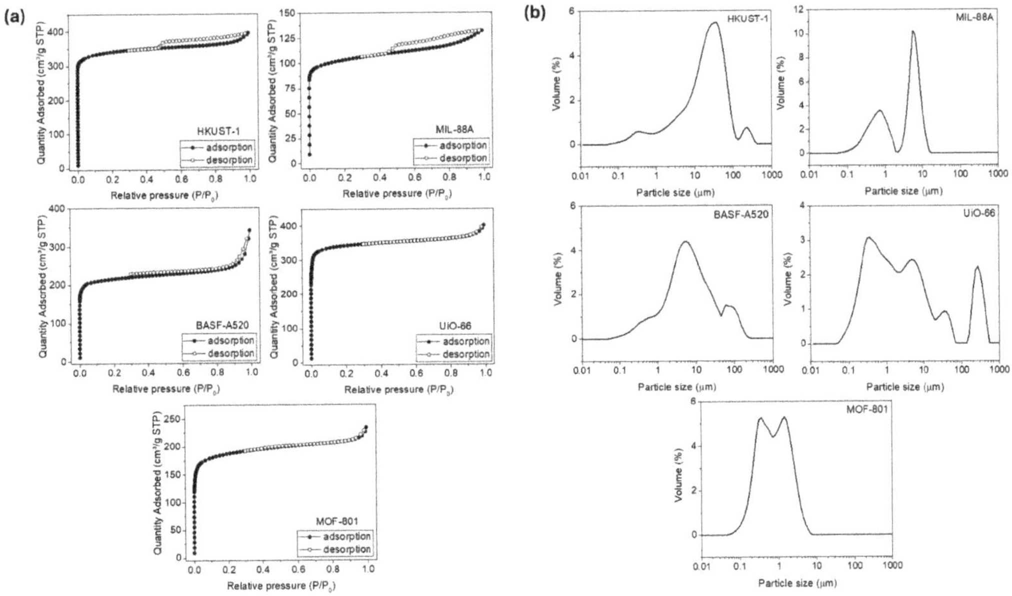

Figure 1. (**a**) N_2 adsorption-desorption isotherms at −196 °C and (**b**) particle size distribution of MOF samples: HKUST-1, MIL-88A, BASF-A520, UiO-66, and MOF-801.

Regarding the morphology observed by SEM, different shapes of MOFs particles were evidenced (Figure 2). HKUST-1 (Figure 2(a1,a2)) consisted of small particles approximately 2 μm in length and 1 μm in width, along with larger particles ranging between 20–30 μm in length and 15–20 μm in width, both displaying an octahedral shape [36]. The MIL-88A material (Figure 2(b1,b2)) exhibited a distinctive morphology characterized by particles with a prismatic central body and pyramidal tips at both ends [26]. The sample consisted of well-defined elongated crystals, forming large micrometric particles measuring approximately 5.0 μm in length and 2.0 μm in width. This unique structural arrangement highlights the highly ordered structure and anisotropic crystal growth. The BASF A520 sample (Figure 2(c1,c2)) showed a morphology consisting of quasi-spherical crystals approximately 100 nm in size [37]. These nanoscale crystals likely formed as a result of rapid crystallization and were observed to aggregate into larger, micron-sized particles. This hierarchical structure reflects the interplay between crystal growth kinetics and particle assembly during synthesis. SEM images of UiO-66 (Figure 2(d1,d2)) revealed a granular morphology, characterized by aggregates of small crystallites produced through the direct reaction between zirconium salt and terephthalic acid. The particles demonstrated a homogenous size distribution around 100 nm, in agreement with previous studies reported in the literature [38]. This homogeneity underscores the precision of the synthesis process and the reproducibility of the material's morphological features. Conversely, MOF-801 (Figure 2(e1,e2)) exhibited uniform octahedral shapes, approximately 400 nm in size.

The TGA profiles presented in Figure 3 provided valuable insights into the thermal stability of the MOFs studied. HKUST-1 displayed an initial weight loss below 200 °C, attributed to the desorption of water and residual solvents. The primary decomposition step occurs between 300 and 350 °C, corresponding to the thermal degradation of the copper-coordinated benzene tricarboxylate linkers. This behavior indicates moderate thermal stability, which is consistent with the previously reported thermal properties of copper-based MOFs [39]. MIL-88A showed an initial weight loss below 150 °C due to the release of adsorbed water and residual solvents. A significant degradation event occurred from 200 °C, associated with the breakdown of the organic linker. This thermal profile reflected the moderate thermal stability of iron-based MOFs with flexible frameworks, which is a characteristic feature of the MIL family [26]. BASF-A520, or aluminum fumarate, demonstrated a higher thermal stability compared to MIL-88A and HKUST-1. Minimal weight loss was observed below 350 °C, primarily due to solvent desorption. The major decomposition step took place between 400 and 500 °C, corresponding to the degradation of the fumarate linker. This behavior highlights the robustness of aluminum-based MOFs, which often exhibit significant thermal stability [40]. UiO-66 exhibited the highest thermal stability among the studied MOFs. Apart from minor solvent desorption below 400 °C, this material remained stable, with its primary decomposition event occurring between 450 and 550 °C. This corresponded to the breakdown of the terephthalate linker and is indicative of the exceptional thermal stability of zirconium-based MOFs, which benefit from their rigid and robust framework structure [41]. Lastly, MOF-801, another zirconium-based MOF, showed a thermal stability that was slightly lower than that of UiO-66. Negligible weight loss was observed below 300 °C, with the major degradation step occurring between 350 and 450 °C. This thermal behavior was attributed to the decomposition of the organic linker, further emphasizing the stability of zirconium-coordinated MOFs [42]. These TGA results reflect the influence of their metal centers and organic linkers on their thermal behavior, which underlines the importance of structural composition in determining the thermal properties of MOFs [43] for various applications related to active packaging.

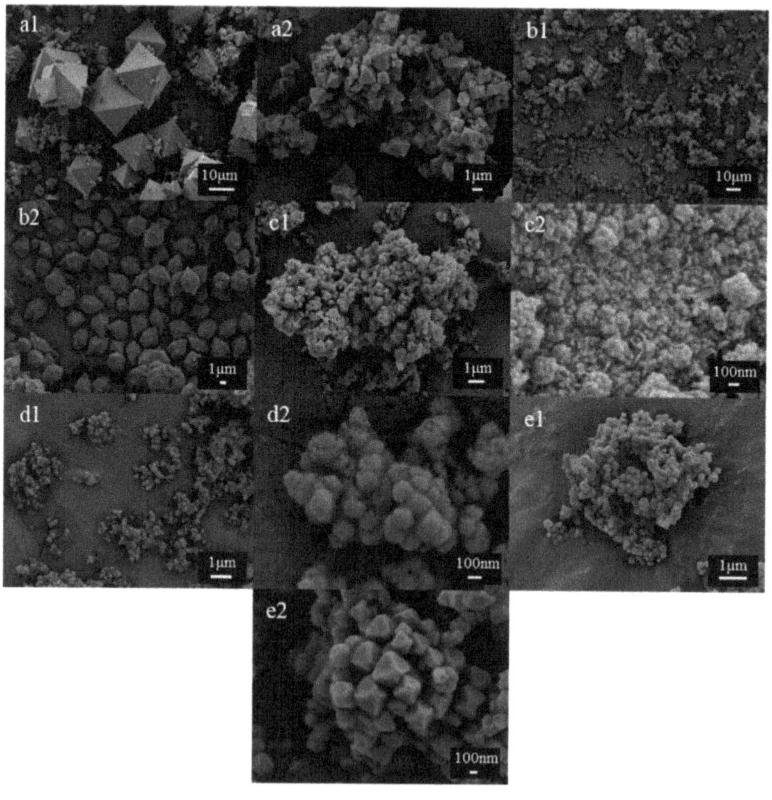

Figure 2. SEM images of MOFs obtained at different magnifications. These correspond to (**a1**,**a2**) HKUST-1, (**b1**,**b2**) MIL-88A, (**c1**,**c2**) BASF-A520, (**d1**,**d2**) UiO-66, and (**e1**,**e2**) MOF-801.

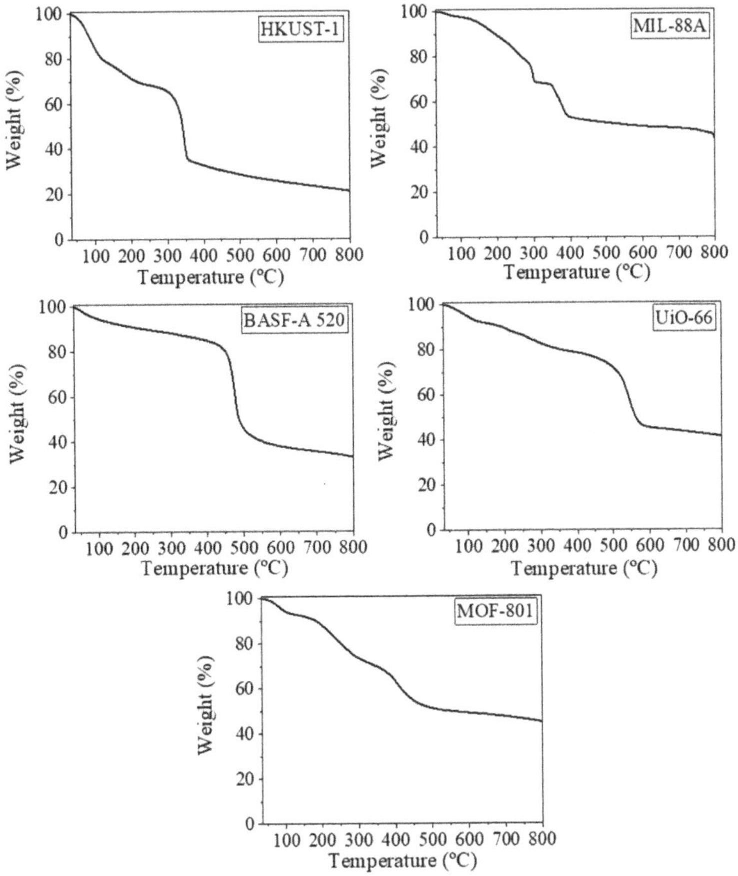

Figure 3. TGA curves of the MOF samples in air atmosphere.

2.2. Characterization of the PVA-MOF Films

The morphology of the cross-section and surface of the films was examined through SEM images, shown in Figure 4. The surface images of the films demonstrated the presence of MOFs within the PVA film matrix, indicating effective homogenization, as they were distributed across the entire surface. This behavior could be expected due to the strong interaction between the MOFs and the polyhydroxylated structure of PVA [16,44,45]. However, some aggregations of MOFs were observed on the film surface, which may negatively impact the film's properties. This phenomenon occurs because, at high concentrations, MOF particles tend to aggregate [46]. As observed in the characterization of the MOFs, these images highlighted the distinct morphology of each MOF used, both in terms of size and composition. It is also evident that there was heterogeneity in particle size among the different MOFs, as the particle size distribution results showed (Figure 1b). The cross-sectional images of the films displayed a homogeneous appearance, further indicating an effective MOF distribution throughout the matrix.

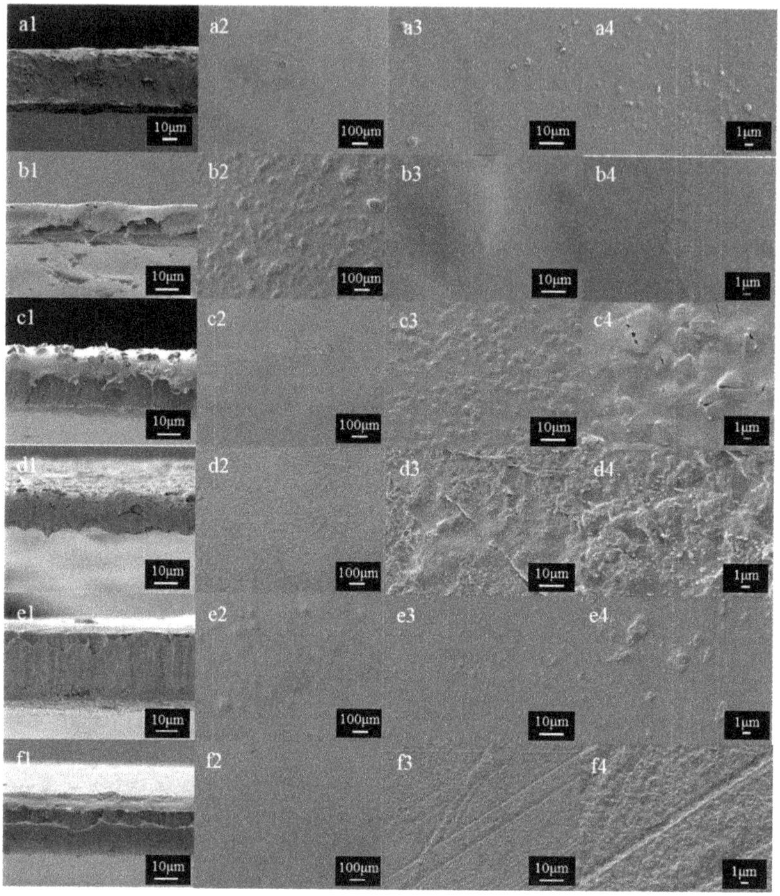

Figure 4. SEM images of the surface of different films obtained. These correspond to (**a1–a4**) PVA films, (**b1–b4**) PVA-HKUST-1, (**c1–c4**) PVA-MIL-88A, (**d1–d4**) PVA-BASF-A520, (**e1–e4**) PVA-UiO-66, and (**f1–f4**) PVA-MOF-801. (1) Corresponds to the cross section; (2) Corresponds to the 100× magnification; (3) Corresponds to the 1500× magnification; and (4) Corresponds to the 5000× magnification.

The results obtained from XRD and FTIR (Figures 5 and 6) confirmed the presence of the MOF in the matrix observed in Figure 4. The XRD patterns of the different composite films are shown in Figure 6. All the films showed a characteristic peak related to the semi-crystalline nature of PVA at 19.5°, attributed to the crystal plane (111) [47]. The diffraction peaks observed in the PVA-HKUST-1 film (Figure 5a) at 6.6°, 9.4°, 11.5°, and 13.3° were assigned to the (200), (220), (222) and (400) crystal planes from HKUST-1, respectively [48]. The PVA-MIL-88A film showed a diffraction peak at 10.1° which correspond to the (100) plane of the MOF (Figure 5b) [49]. The peak observed at 10.3° in PVA-BASF-A520 is related to the (100) plane of the MOF (Figure 5c) [50]. In Figure 5d, one of the most intense peaks from UiO-66 can be seen in the mixed film at 7.4° and 8.5°, which are related to the (111) and (200) crystal planes, respectively [51]. The XRD pattern of MOF-801 (Figure 5e) showed well defined peaks, which are also in the composite film. Some of them, at 10° and 21.7°, correspond to the (200) and (440) planes, respectively [52]. In the composite films, some characteristic peaks of the MOFs are clearly present in the diffractogram, although slightly attenuated, most likely due to the dominant amorphous/semi-crystalline nature of the PVA matrix and the lower relative MOF content.

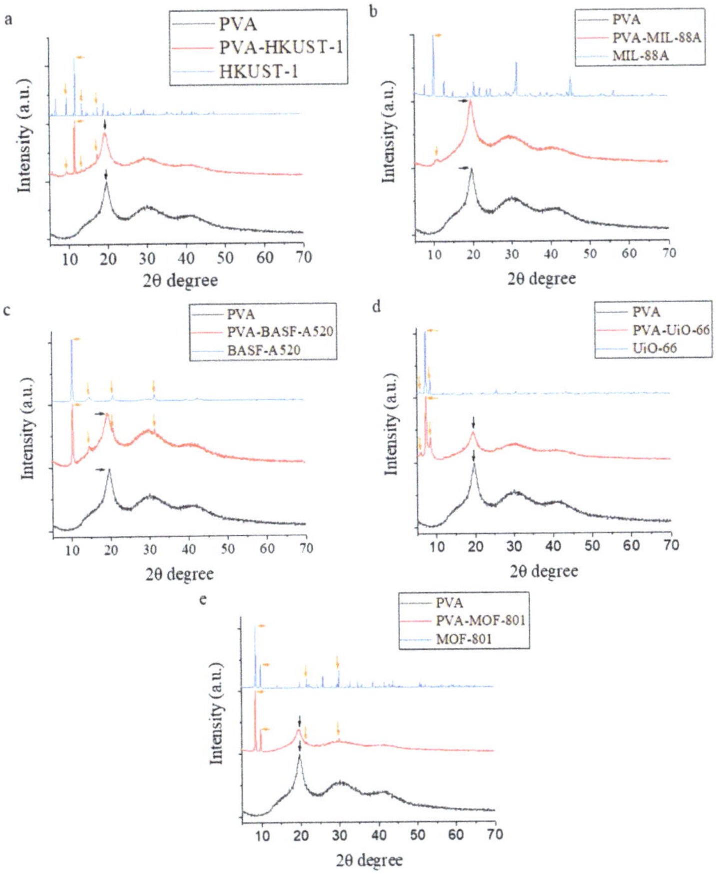

Figure 5. XRD pattern of the PVA-MOF composite films.

The chemical structure of the films, generated from different MOFs, was studied using FTIR analysis (Figure 6). The spectra obtained from the different films showed some characteristic peaks of the PVA, such as those observed at 3285 cm^{-1}, 2915 cm^{-1}, 1420 cm^{-1}, and 1080 cm^{-1}, attributed to the stretching of -OH groups, -CH bonds, C=O and C=C bonds, and C-O bonds, respectively [53]. Some of the most characteristics peaks of HKUST-1, such as those observed at 1650, 1450, and 1365 cm^{-1}, arising from asymmetric and symmetric stretching vibrations of carboxylate groups, are not observed in the PVA-HKUST-1 film [54]. The peaks observed at 1600 and 1380 cm^{-1} in the PVA-MIL-88A film correspond to symmetric and asymmetric vibration modes of the fumarate carboxyl group. Those observed at 990 and 530 cm^{-1} are attributed to C-H bond bending vibration and the stretching of the Fe-O bond, respectively [49]. In the PVA-BASF-A520 film, peaks at 690, 1400, and 1600 cm^{-1} are observed, corresponding to the Al-OH bond and symmetric and asymmetric vibrations of aluminum fumarate carboxyl groups, respectively. However, other characteristic peaks of BASF-A520 related to the O-H bond (980 cm^{-1}) and C-O bond (1160 cm^{-1}) cannot be observed in the film [55].The PVA-UiO-66 film presents characteristic peaks in the range of 1600–1400 cm^{-1}, associated with the stretching vibrations of C=C and C=O bonds [56]. MOF-801 presents a series of characteristic peaks, all observed analogously in the film. At 1575 cm^{-1}, a peak attributed to the asymmetric stretching of --C-O produced by the ligand carboxyl group can be observed. The symmetric stretching vibration of the C-H group is shown at 1400 cm^{-1}. Conversely, the peaks observed at 1215, 985, and 800 cm^{-1} may be ascribed to C-O stretching, CH_3 skeletal vibrations, and C=C-H out-of-plane bending, respectively. Additionally, the peaks at 655 and 475 cm^{-1} are indicative of vibrations associated with $Zr_6(OH)_4O_4$ and the asymmetric stretching of the Zr-(OC) group of MOF-801 [52]. Overall, no systematic shifts were detected in the main absorption bands of PVA upon MOF incorporation, which suggests that the polymer backbone is not chemically altered. However, the attenuation or absence of some MOF-specific peaks in the composite films indicates that certain vibrational modes may be overlapped or masked by the strong absorption of PVA. While clear evidence of chemical bonding cannot be drawn from the spectra, weak physical interactions such as hydrogen bonding between PVA hydroxyl groups and MOF carboxylate or hydroxyl moieties may contribute to the compatibility observed in the films.

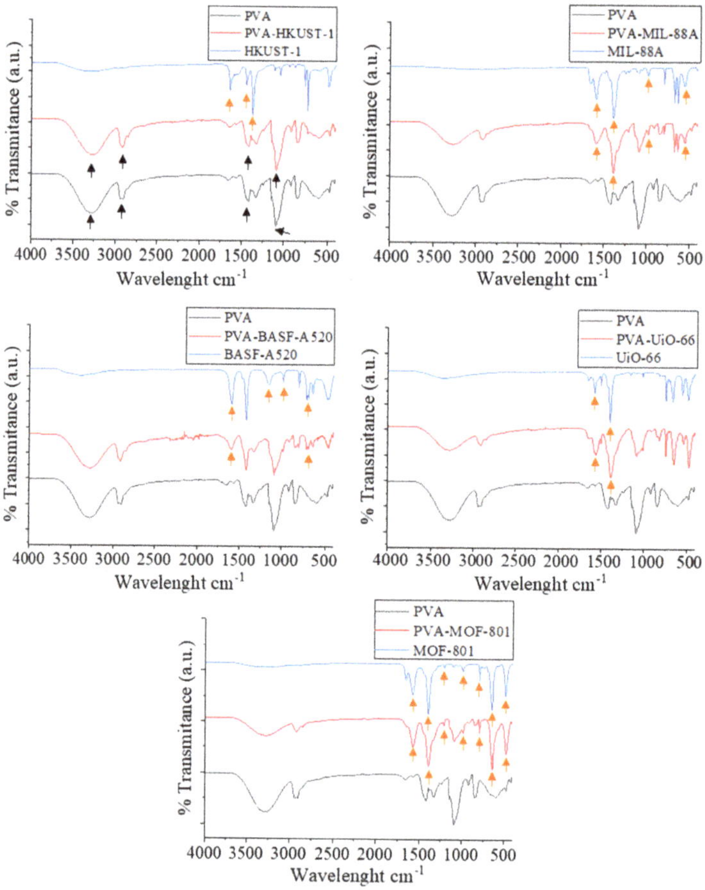

Figure 6. FTIR spectra of the different composite films.

The physical, optical, and mechanical properties of films are critical for their application as active food packaging materials. High transparency and efficient UV dispersion enhance consumer perception by improving food visibility while protecting food from oxidative damage [57]. Strong mechanical properties are essential to withstand stress during packaging and transportation, ensuring durability while maintaining the structural integrity of the package and protecting the food [58]. Furthermore, films with enhanced barrier properties, such as lower *WVP* and reduced moisture content, are particularly desirable, as they help preserve food quality and ensure the package's structural stability over time [59].

The results of the physical, optical, and mechanical characterization of the different PVA-MOF films are presented in Figures S1, S2 and S3, respectively, demonstrating that the addition of different MOF particles significantly influences the properties of the PVA film. Notably, distinct effects were observed among the various MOFs, attributed to their unique compositions, as well as differences in particle size, porosity, and specific surface area (Tables 1 and 2, and Figure 1). These variations in the overall properties arise from the incorporation of MOFs into the PVA matrix, which disrupts its uniform arrangement, preventing a denser and more consistent packing of the film [60]. The MOFs with smaller particle sizes tend to be distributed more homogeneously within the matrix, resulting in a lesser impact on the film. However, larger MOF particles promote the formation of stress zones and structural discontinuities that act as weak points within the material due to their less uniform distribution [61]. These larger particles create regions with reduced cohesion between the PVA and the added material, thereby decreasing matrix uniformity and facilitating the formation of microfractures and porosity channels. This behavior explains why the density of the PVA films, as well as the transparency of the films, decreases upon the incorporation of MOFs, with films containing smaller average particle sizes (such as MIL-88A and MOF-801) being less affected. Similarly, the films' moisture content was reduced due to the strong interactions between MOFs and the PVA polymer matrix [16], which limit water uptake by preventing hydrogen bonding with the PVA matrix. However, MOFs with higher porosity, such as HKUST-1, may allow for slightly more interactions with water molecules.

Another aspect to consider is the stability in aqueous media that MOFs present. This stability depends on several factors, including the nature of the metal center, the type of ligand, and the interaction between them. Metals with higher valence states, such as Zr^{4+}, tend to form stronger bonds with carboxylate groups due to their higher charge density, which allows for more stable coordination with the ligands. This results in a more robust structure that is less susceptible to hydrolysis. In contrast, carboxylate ligands, when coordinated with lower-charge metals like copper (Cu^{2+}), form weaker bonds, making these MOFs more prone to decomposition and dissolution in water [62]. These findings are consistent with the results obtained for water solubility, in which HKUST-1 (Cu^{2+}) exhibited significantly higher values compared to the others. On the other hand, the PVA-MIL-88A and PVA-A520-BASF films exhibited intermediate behavior as they do not interact as strongly as those with metals of higher valence states, like zirconium. Furthermore, particle size also affects solubility. Larger MOF particles, such as those in HKUST-1, facilitate water access to the PVA matrix by creating microfractures and porosity channels, thereby increasing overall solubility. The low water stability of HKUST-1, as well as the size of its particles and the associated issues, would explain why it exhibits higher *WVP* values compared to the other films.

The films containing MOFs exhibited a significant improvement in UV blocking properties compared to the PVA film. This behavior has been previously observed with various MOFs in prior studies and is attributed to proper particle dispersion within the matrix, good interaction with the matrix, their composition (metal center and ligand), and properties that enable UV light reflection (porous structure and surface area) [44,63,64].

Previous studies have reported improvements in the mechanical properties of PVA films with the addition of MOFs, likely due to strong interactions with the polyhydroxylated structure of the PVA matrix [16,44,45]. However, the results in Figure S3 deviate from these expectations, as most films showed no significant differences in mechanical properties, and in some cases, even a notable decrease is observed. This may be because prior studies achieved improvements using a lower MOF content (typically 0.5–2% by weight) [46]. As observed in the SEM images (Figure 4) and noted in previous sections, the MOF particle size and its tendency to aggregate play a critical role in influencing film properties. Smaller MOF particles were more uniformly dispersed in the PVA matrix [65], reducing the formation of microfractures and porosity channels. Conversely, larger MOF particles with higher aggregation tendencies create stress zones that weaken the structure. This is consistent with the mechanical characterization results, in which films with smaller-sized MOFs (MIL-88A and MOF-801) exhibit greater tensile strength and deformation resistance.

2.3. *Characterization of the CNFs*

Table 3 presents the results of the characterization of the cellulose nanofibers obtained from cotton linters, as well as data from other raw materials processed using similar methods. The results of the cotton linter nanofibers exhibit values that are very similar, and in some cases even superior, to those of the other nanofibers in terms of cationic demand, with the exception of those derived from orange peels. Specifically, the results closely resemble those of wheat straw nanofibers, which are commonly used as reinforcing agents in PVA films. This similarity is particularly evident in parameters such as cationic demand, specific surface area, and diameter. This suggests that cotton linter nanofibers could be suitable for use as reinforcing agents in PVA films. Therefore, incorporating these nanofibers could not only enhance the mechanical and barrier properties of PVA-CNF films but also facilitate a more uniform distribution of MOFs throughout the matrix. This improved distribution and enhanced performance make these films an even more promising material for active packaging applications. Additionally, the addition of CNF potentially could improve the overall distribution throughout the film by reducing MOF aggregation at high concentrations and thus leading to an overall improvement in properties [18].

Table 3. Results from the characterization of the TEMPO nanocellulose compared with other CNFs obtained from different sources.

Parameters	Yield (%)	Cationic Demand (μeq/g)	Carboxyl Content (μeq/g)	Specific Surface Area (m^2/g)	Diameter (nm)	Length (nm)
Cotton Linter (This work)	63.55 ± 7.89	1178.59 ± 59.02	534.02 ± 27.61	314 ± 32	8.01 ± 0.82	319.29 ± 30.66
Raspberry [57]	>95	1264 ± 32.6	240.2 ± 4.5	548	5	N.R.
Horticultural residues (bell pepper, tomato, eggplant) [66]	63.44 ± 4.52	1043.54 ± 18.2	148.12 ± 5.26	436	6	614
Vine shoots [66]	60.42 ± 5.56	1227.91 ± 18.8	168.93 ± 10.25	516	5	755
Orange peel wastes [23]	75.58 ± 2.21	2144.73 ± 29.64	559.73 ± 16.60	772	3	2383
Wheat straw [67]	98.71	1116.5 ± 43.10	367.0 ± 8.72	367.01	6.81	1395
Olive tree pruning [68]	26.44 ± 4.15	521.27 ± 9.33	311.95 ± 19.02	101.93	24	705

2.4. *Characterization of the PVA-CNF-MOF Films*

2.4.1. PVA-CNF Films

To assess the reinforcement capacity of our nanofibers, multiple PVA-CNF films were prepared at various nanofiber concentrations. The results are shown in Table S1.

The evaluation of mechanical properties reveals that both the tensile strength and Young's modulus increase as the CNF content rises, with a significant peak observed at the 5% incorporation level. This enhancement is largely attributed to the nanofibers' ability to create inter- and intramolecular hydrogen bonds, their more uniform dispersion within the matrix, and their high compatibility with the PVA. Nevertheless, when the CNF concentration surpasses 5%, the films experience a reduction in tensile strength and Young's modulus. This decline is likely caused by the tendency of nanofibers to aggregate at higher concentrations, leading to the formation of clusters within the matrix, which in turn weakens the material [23].

The *WVP* results indicate a significant improvement in the barrier properties of the films after the incorporation of nanofibers, with values decreasing from 0.1749 ± 0.0109 for pure PVA to 0.0622 ± 0.0062 ($g \cdot Pa^{-1} \cdot s^{-1} \cdot m^{-2} \cdot 10^{-7}$) for the 5% CNF composition. This effect is attributed to the so-called tortuous pathway, where the nanofibers act as fillers within the matrix, obstructing the optimal movement of water molecules, which become trapped. The films with the CNF show better results, as previously mentioned, due to a more homogeneous nanofiber distribution within the matrix, further enhancing the barrier properties. However, higher nanofiber concentrations lead to an expected increase in *WVP*, likely due to the formation of nanofiber aggregates that facilitate the passage of water molecules [23,69]. This behavior was expected to result in a lower moisture content, as well as a reduction in solubility and swelling. However, no significant differences were observed in moisture content, and a slight increase in swelling and moisture was noted in some film compositions. The transparency of the films is also related with the dispersion of the nanofibers in the PVA matrix [67]. As shown in the results, an increase in the nanofiber content leads to a decrease in transparency, indicating a homogeneous dispersion within the film. However, the 7% composition shows an increase in transparency, which may be due to the aggregation of nanofibers, facilitating the passage of light. The UV barrier results did not show any significant difference between the different films.

Based on the results obtained, it was determined that the composition with 5% cellulose nanofibers exhibited the greatest improvement in properties relevant to active packaging, such as *WVP* and overall mechanical properties. Therefore, this composition was selected for the fabrication of the PVA-CNF-MOF films.

2.4.2. Characterization of the PVA-CNF-MOF Films

Figure 7 shows SEM images of the cross-section and surface of the PVA-CNF-MOF films. These images are significantly different from the images obtained for the PVA-MOF films (Figure 4). A more homogeneous distribution and the absence of aggregations are evident on the film surfaces, even at higher magnifications. Similarly, the cross-section of the films presents a homogeneous appearance, further indicating effective MOF distribution throughout the matrix. This behavior is expected, as previously mentioned, since cellulose nanofibers enhance the dispersion of MOFs due to the strong interaction between the carboxylate groups of the nanofibers and the metal centers of the MOFs, serving as binding sites for the MOFs [18]. This is a crucial aspect, as it is expected to result in the improved physical and mechanical properties of the films.

The XRD and FTIR results are shown in Figures S4 and S5, respectively. In the FTIR spectrum, no characteristic peaks of cellulose were observed due to overlapping with the characteristic peaks of PVA, attributed to their similar structure and the low concentration of cellulose nanofibers present [70]. However, the presence of cellulose nanofibers can be identified in the XRD spectrum, which shows a diffraction peak at 22°, corresponding to the (200) crystallographic plane of cellulose [71]. In both spectra, the characteristic peaks of the respective MOFs, as well as those of PVA, are also observed, as previously described.

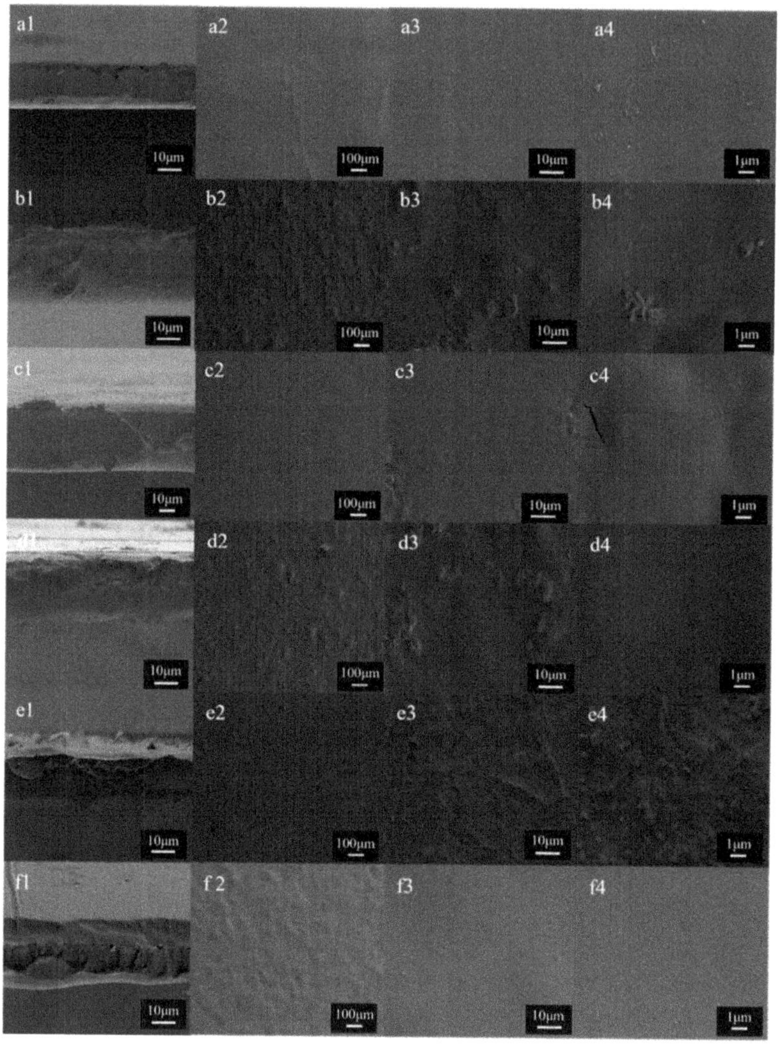

Figure 7. Surface images of various films obtained via SEM. These correspond to (**a1–a4**) PVA-CNF, (**b1–b4**) PVA-CNF-HKUST-1, (**c1–c4**) PVA-CNF-MIL-88A, (**d1–d4**) PVA-CNF-BASF-A520, (**e1–e4**) PVA-CNF-UiO-66, and (**f1–f4**) PVA-CNF-MOF-801 films. (1) Corresponds to the cross-section view; (2) Corresponds to the 100× magnification; (3) Corresponds to the 1500× magnification; and (4) Corresponds to the 10,000× magnification.

The expected improvement in the properties of the films with the inclusion of CNF is evident in the results obtained from the characterization of their physical, optical and mechanical properties (Figures 8–10). As previously mentioned, and as observed in the SEM images (Figure 7), the presence of cellulose nanofibers likely facilitates a more homogeneous dispersion of MOFs within the matrix. This prevents the formation of stress zones and structural discontinuities that could act as weak points in the material [61], while compacting the matrix, thereby reducing both porosity and accessibility. This phenomenon explains the slight increase in density and in mechanical properties (Figure 10), as well as the significant reduction in swelling values and *WVP* (Figure 8), which is also associated with the tortuous pathway mechanism previously described [23]. Additionally, films containing MOFs (HKUST-1, BASF-A520, and UiO-66) with larger particles showed a notable reduction in solubility values, as expected. Conversely, an increase in solubility is observed in films incorporating smaller MOFs. These unexpected behaviors could be attributed to the film formation process itself.

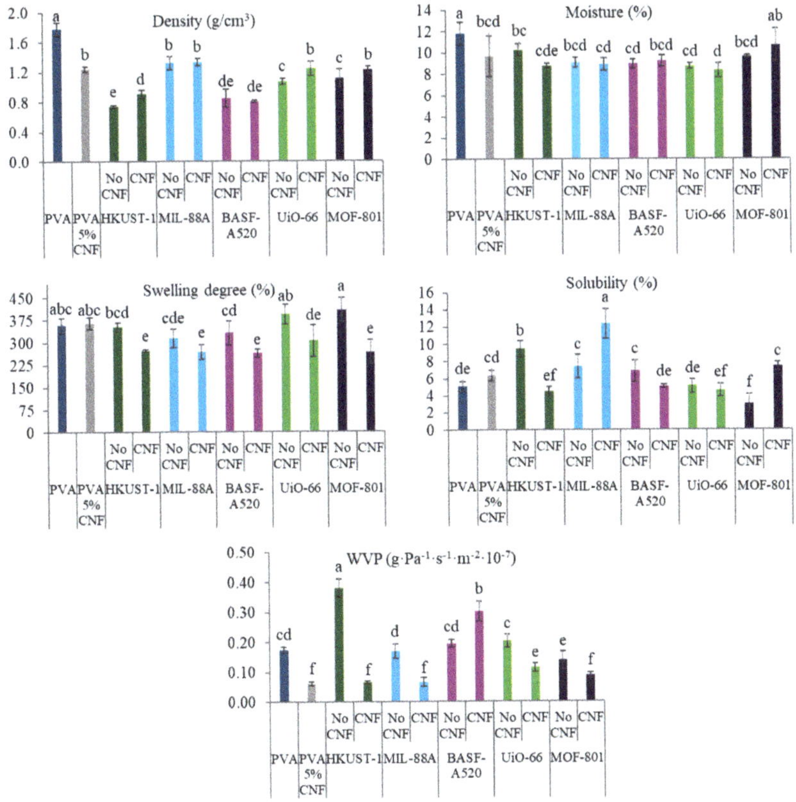

Figure 8. Physical properties of the PVA-MOF and PVA-CNF-MOF films. Different letters above bars indicate significant differences among formulations ($p \leq 0.05$).

The improved distribution provided by the CNF is also reflected in the optical properties (Figure 9). The enhanced dispersion of the MOFs results in a more uniform distribution on the film surface, which explains the observed variations compared to the films without CNF.

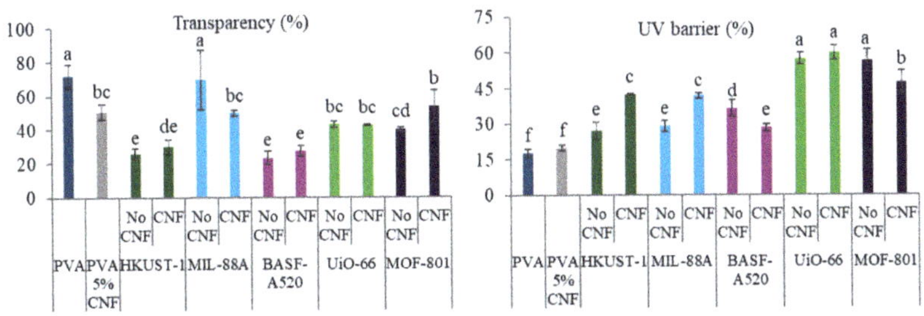

Figure 9. Optical properties of the PVA-MOF and PVA-CNF-MOF films. Different letters above bars indicate significant differences among formulations ($p \leq 0.05$).

The mechanical properties of the films, shown in Figure 10, exhibit a significant improvement with the incorporation of cellulose nanofibers, particularly highlighting the reinforcing effect in compositions containing HKUST-1 and MOF-801. The film with MOF-801 achieved exceptional values for its Young's modulus and tensile strength, reaching 5426 MPa and 69.12 MPa, respectively. These values represent an increase of approximately 290% in its Young's modulus and 130% in its tensile strength compared to the properties of pure PVA.

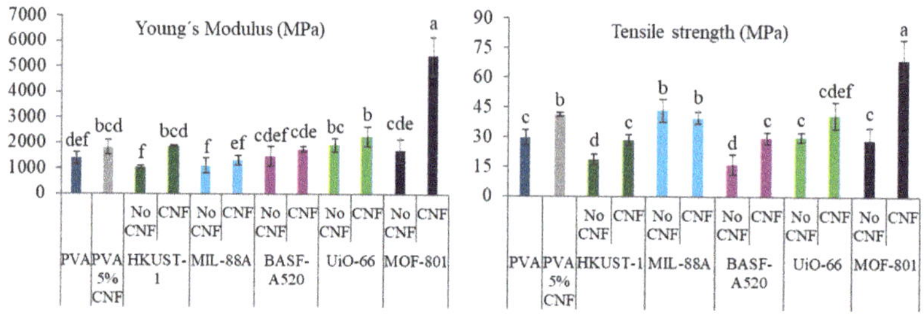

Figure 10. Mechanical properties of the PVA-MOF and PVA-CNF-MOF films. Different letters above bars indicate significant differences among formulations ($p \leq 0.05$).

2.5. Scavenging Activity

The results obtained from the proof-of-concept test to measure ethylene retention by the PVA-MOF and PVA-CNF-MOF films are shown in Table 4. This table presents the texture and browning rate results for the selected bananas. The texture results demonstrate the mechanical strength of the banana pulp after being stored in a controlled environment with the films. Some compositions showed significant improvements with the incorporation of CNF, such as in the case of HKUST-1, BASF-A520, and MOF-801, highlighting the importance of proper homogenization to improve the active surface of the films. Similarly, the browning rate also showed improvements in films with CNF compared to their counterparts without nanofibers. However, despite the observed improvements in films incorporating CNF and the different ethylene adsorption capacities of the MOFs (Table 1), only small differences were observed between them and the PVA film, in which the difference in the firmness (texture) of the bananas was notable.

Table 4. Results of browning rate and texture obtained from the banana characterization. Different letters under texture and browning rate parameters indicate significant differences among formulations ($p \leq 0.05$).

Parameters	Texture (g)		Browning Rate (%)	
Samples	PVA-MOF	PVA-CNF-MOF	PVA-MOF	PVA-CNF-MOF
0% MOF	77.80 ± 33.09 bcd	64.81 ± 22.29 de	40.82 ± 13.81 bc	44.37 ± 7.69 b
HKUST-1	53.19 ± 20.80 e	77.02 ± 35.47 bcd	37.01 ± 7.94 bcd	20.28 ± 0.00 d
MIL-88A	78.11 ± 21.56 bcd	71.97 ± 25.67 bcd	42.06 ± 7.78 bc	37.30 ± 1.96 bcd
BASF-A520	67.13 ± 19.75 cde	85.91 ± 16.19 ab	41.88 ± 18.70 bc	26.69 ± 4.66 bcd
UiO-66	98.11 ± 22.94 a	83.32 ± 21.56 abc	33.64 ± 7.04 bcd	28.33 ± 17.25 bcd
MOF-801	63.91 ± 16.41 de	87.15 ± 26.82 ab	64.29 ± 16.56 a	22.53 ± 29.50 cd

In future studies, using freshly harvested fruit under controlled storage conditions could help reduce variability and provide a clearer assessment of the effect of MOFs on the formulations. The variability observed in this study may be attributed to the natural heterogeneity of fresh produce, particularly when sourced from supermarkets. Differences in cultivation practices, prior storage conditions, environmental exposure, and handling can influence key parameters such as texture and browning rate, potentially masking the impact of MOFs on fruit preservation [72–75].

Nevertheless, PVA-CNF-MOF-801 films show great promise due to their excellent mechanical properties, high UV protection, and low *WVP*. The superior mechanical performance observed for PVA-CNF-MOF-801 films can be attributed to the intrinsic stability of the MOF-801 framework, its smaller particle size, and its favorable interaction with the polymeric and CNF matrix. These factors likely promoted better interfacial adhesion and stress transfer compared with other MOFs, thereby enhancing their tensile strength and Young's modulus. Additionally, PVA-CNF-MIL-88A films are also of interest, despite their mechanical properties being inferior to those of PVA-CNF-MOF-801 films. Both MOFs exhibit high biocompatibility, making them suitable for direct contact with food products.

3. Materials and Methods

3.1. Materials

Cellulose derived from cotton linter was supplied by Cotton South S.L (Granada, Spain). Sodium bromide (NaBr, 99% purity) was supplied by ACS reagent (Washington, DC, USA). Ethanol was supplied from Alcoholes del Sur S.A (Córdoba, Spain). Cupric nitrate hemipentahydrate ($Cu(NO_3)_2{\cdot}2.5H_2O$), benzene-1,3,5-tricarboxylic acid, iron(III) chloride hexahydrate ($FeCl_3{\cdot}6H_2O$), fumaric acid, aluminum sulfate hydrate ($Al_2(SO_4)_3{\cdot}18H_2O$), zirconium(IV) chloride ($ZrCl_4$), 1,4-benzenedicarboxylic acid, zirconium oxychloride octahydrate ($ZrOCl_2{\cdot}8H_2O$, purity $\geq$ 99.5%), formic acid, N,N-Dimethylformamide anhydrous (DMF, purity $\geq$ 99.8%), 2,2,6,6-tetramethylpiperidine-1-oxyl (TEMPO, free radical, 98% purity), Bis-(ethylenediamine)copper(II) hydroxide solution (CED), and poly(vinyl alcohol) (PVA) were supplied by Sigma Aldrich (St. Louis, MO, USA). Sodium hypochlorite solution (10% w/v, technical grade) and sodium hydroxide (pellets technical grade) were supplied from PanReac AppliChem ITW Reagents (Barcelona, Spain). Pes-Na (0.001 N) and Cationic Titrant (0.001 N) were provided by BTG Instruments AB (Espoo, Finland). Granulated calcium chloride anhydrous (extra pure) from Scharlau Chemie S.A (Barcelona, Spain) was employed.

All MOF materials were prepared following procedures previously reported in the literature: HKUST-1 [76], MIL-88A [26], BASF-A520 [31], UiO-66 [77], and MOF-801 [78].

3.2. Synthesis of the TEMPO-Oxidized Cellulose Nanofibers (CNFs)

Cellulose nanofibers were obtained by chemical pretreatment process involving TEMPO catalytic oxidation and subsequent high-pressure homogenization. This pretreatment involves the oxidation of cellulose through the addition of 5 mmol NaClO/g fiber in the presence of NaBr (0.1 g/g fiber) and TEMPO (0.016 g/g fiber), followed by the addition of 0.5 M NaOH until the reaction reaches stable pH values of 10.2. Once the reaction time is completed (2 h), 100 mL of ethanol was added, the pulp was filtered and a 1.5% suspension in distilled water was prepared and stirred for 24 h. After this time, high-pressure homogenization was carried out using an optimized pass sequence [67] in the GEA Niro Panda Plus 2000 high-pressure homogenizer (GEA Spain Alcobendas, Alcobendas, Madrid, Spain). The nanofibers generated were stored under refrigeration at 4 °C.

3.3. Synthesis of the PVA-CNF-MOF Films

Films with a grammage of 35 g/m^2 were fabricated using a solvent casting technique. For this purpose, an aqueous solution of PVA was prepared at a concentration of 30 g/L. The mixture was then stirred for 3 h at 85 °C.

To create MOF/PVA films, different types of MOFs were added up to 10% with respect to the dry weight of PVA. The MOF/PVA mixtures were then diluted to a total volume of

100 mL and mechanically agitated for 3 h to ensure proper homogenization. After the time elapsed, the solutions were poured into Petri dishes with an area of 144 cm^2 and dried in a climate-controlled environment at 25 °C with 50% relative humidity.

The reinforcing effect of CNF on PVA films was investigated by adding different proportions of nanofibers (1, 3, 5, and 7% based on the dry weight of PVA). For this purpose, the determined amount of CNF was diluted in 50 mL of distilled water and dispersed by applying a homogenizer IKA T18 digital Ultra Turrax (IKA Works Spain, Limited Company, Barcelona, Spain). Once uniformly dispersed, they were mixed with the predetermined amount of PVA, further diluted to 100 mL, shaken and then poured into Petri dishes.

MOF-PVA-CNF films with the optimal nanofiber content were prepared following a similar procedure as described above. In this case, CNFs were pre-mixed with the various MOFs at different concentrations (1, 3, 5, and 10% *w*/*w* based on PVA) and homogenized prior to blending with PVA. All film preparations were conducted in triplicate to evaluate their reproducibility.

A schematic illustration summarizing the workflow for producing PVA-MOF and PVA-CNF-MOF films is provided in Figure 11.

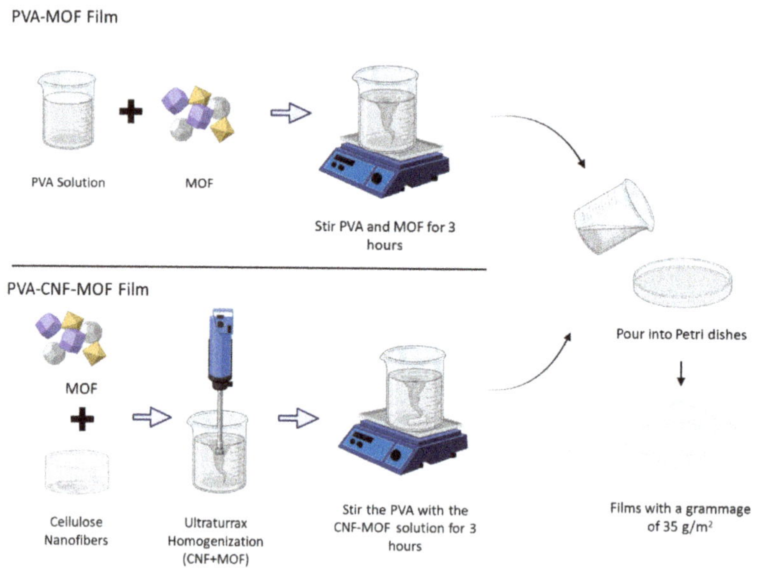

Figure 11. Step-by-step schematic of the synthesis procedures for PVA-MOF and PVA-CNF-MOF films.

3.4. Structural, Textural, Chemical and Morphological Characterization Techniques

X-ray diffraction (XRD) patterns were obtained using a Bruker D8 Discover diffractometer (Bruker, Berlin, Germany) equipped with a LynxEye linear position-sensitive detector and a Cu-Kα radiation source (λ = 1.5418 Å), operating in reflection geometry mode. A Ni filter was used to suppress the Kβ radiation, and both $K\alpha_1$ and $K\alpha_2$ components were included in the analysis. The scans were performed over a 2θ range of 5–70°, with a step size of 0.02°, at a scan rate of 0.04°/min, using fixed divergence and receiving slits. The total acquisition time per sample was approximately 55 min. Samples were mounted on standard Bruker polymer-based transparent sample holders, which are optimized for thin film analysis and provide minimal background signal. The textural properties of the MOFs were analyzed through N_2 adsorption-desorption isotherms in a Micromeritics ASAP 2020 instrument (Micromeritics, Norcross, GA, USA) at −196 °C. Before the measurements, a degasification step was carried out at 120 °C overnight under vacuum conditions (10^{-6} mbar) to remove residual adsorbed gases and impurities. Thermogravimetric analysis (TGA) measurements were performed using a Mettler Toledo-TGA/DSC (Mettler Toledo, Greifensee, Switzerland) under air atmosphere by heating the samples from 25 to 800 °C at 10 °C min^{-1}. Particle size analysis was performed using a Mastersizer S laser diffraction system (Malvern Instruments, Malvern, UK). The sample was dispersed in ethanol and subjected to ultrasonic treatment for 10 min to prevent particle aggregation. Volume Moment Mean (D [4.3], De Brouckere Mean Diameter) was obtained for each MOF material. The incorporation of MOFs and the CNF in the films was evaluated by Fourier transform infrared spectroscopy (FTIR) using a Perkin-Elmer Spectrum Two FTIR-ATR spectrometer (PerkinElmer U.S. LLC, Shelton, CT, USA). The spectrum was analyzed with a resolution of 4 cm^{-1} in a range of 400–4000 cm^{-1}. The different morphologies of the MOFs, PVA-MOF films, and PVA-CNF-MOF films were examined by Scanning Electron Microscopy (SEM) technique on a JEOL JSM 7800 F Scanning Electron Microscope (JEOL Ltd., Akishima, Japan). For this purpose, the samples were coated with gold to improve the conductivity. The accelerating voltage applied to the samples was 5 KV, with a working distance of 10 mm.

3.5. Characterization of the CNFs

The CNFs were characterized in terms of their nanofibrillation yield, cationic demand, carboxylation rate, and degree of polymerization. The nanofibrillation yield was determined by preparing a 0.1% suspension of the nanofibers, followed by centrifugation for 12 min at 10,000 rpm. For the cationic demand determination, 0.1% suspension was prepared with 0.2 dry grams. Once homogenized, a 15 mL aliquot was mixed with 15 mL of a cationic polymer (cationic titrant) and centrifuged at 4000 rpm for 90 min. Subsequently, a Mütek PCD 05 particle charge detector (Malvern Panalytical Ltd., Malvern, UK) was used to determine the cationic demand of the supernatant aliquots.

The carboxyl content was calculated by conductimetry of a 0.3 g dry sample of nanofiber, diluted to 55 g with distilled water and adding 5 mL of 0.1 M NaCl to increase the conductivity, followed by titration with a standardized 0.01 M NaOH solution.

The results obtained for the carboxyl rate and cation demand were used for the calculation of the specific surface area (σ) and the diameter of the nanofibers (d), making use of results and equations reported in the literature (Equations (1) and (2)) [79]:

$$\sigma_{nanofibers}\left(\mathrm{m}^2/\mathrm{g}\right) = (CD - CC)\cdot\sigma_{cationic\ titrant} \tag{1}$$

where CD is the cationic demand and CC is the carboxyl content.

$$d(\mathrm{nm}) = \frac{4}{\sigma_{nanofibers}\cdot 1600\cdot 10^3} \tag{2}$$

The measurement of the degree of polymerization (DP) was determined by calculating the intrinsic viscosity according to [80]. For this purpose, a given amount of nanofiber was dissolved in 50 mL of 0.5 M CED solution and measured using a Manual SCAN/ISO 5351 Viscometers Type T (PSL Rheotek, Granger, IN, USA). The results of the degree of polymerization were used to calculate fiber length using the following equation (Equation (3)) [81]:

$$Length\ (\mathrm{nm}) = 4.286\cdot DP - 757 \tag{3}$$

3.6. Mechanical Properties of the PVA-CNF-MOF Films

Young's modulus and tensile strength of the films were determined using an LF Plus Lloyd Instrument testing machine with a 1 kN cell load. The analyses were performed on film samples measuring 10 cm in length and 1.5 cm in width accordance with ASTM D882.

For each sample the test was repeated 10 times, and the results are expressed as the mean value accompanied by its corresponding standard deviation.

3.7. Physical Properties of the PVA-CNF-MOF Films

The films obtained were measured in terms of weight, area and thickness to calculate the density of the films. The moisture content of the films was calculated by determining weight differences of the films before and after putting them in an oven at 105 °C for 24 h. These same fragments are used to determine the swelling and water solubility of the films. To determine swelling, the dried fragments were placed in 50 mL of distilled water for 24 h. After this time, they were extracted, the excess water was removed by depositing them on a filter paper, and the weight was recorded. These fragments were returned to the oven for 24 h to calculate the water solubility. All these measurements were carried out in triplicate, and their result is shown as the mean with the standard deviation.

Water vapor permeability (*WVP*) of the films was determined according to ASTM E96/E96M-10 standard test method. For this purpose, TQC SHEEN Permeability Cup aluminum capsules with a surface area of 10 cm^2 were used. The capsules were loaded with $CaCl_2$ to serve as the desiccant material. These capsules, containing the film and salt, were then introduced into a climatic chamber at 25 °C and 50% RH, where the weight was recorded at different hours. The *WVP* result was calculated using the following equations (Equations (4) and (5)):

$$WVP = \frac{WVTR}{\Delta p} = \frac{WVTR}{S(R_1 - R_2)} \tag{4}$$

In this equation, Δp represents the vapor pressure difference, S denotes the saturation vapor pressure at the test temperature, and R_1 and R_2 indicate the relative humidities at the source and vapor sink, respectively, presented as fraction.

$$WVTR = \frac{G}{tA} \tag{5}$$

where G is the weight variation (g), t is the time elapsed during the measurement (h), and A is the test area (ft^2).

3.8. Optical Properties of the PVA-CNF-MOF Films

The ultraviolet barrier and transparency of the films were determined by measuring their transmittance at 280 and 660 nm, respectively, using a Jenway 7315 Advance UV/Visible Spectrophotometer (Fisher Scientific International, Inc., Hampton, VA, USA). The transmittance (T) values were then processed using the following equations (Equations (6) and (7)):

$$Transparency\ (\%) = \frac{log\%T_{660}}{thickness\ (mm)} \tag{6}$$

$$UV - light\ barrier\ (\%) = 100 - \left(\frac{\%T_{280}}{\%T_{660}} \cdot 100\right) \tag{7}$$

3.9. Ethylene Scavenging Application

The ethylene retention capacity of the films with 10% MOF were tested through an experiment using bananas. Bananas from the same bunch were selected and placed in zip-lock polyethylene bags, with films inside. The bags containing the bananas were stored in a climatic chamber at 25 °C and 50% relative humidity for eight days. Throughout the experiment, the bananas were photographed daily to monitor the progression of browning, which was quantified using the following equation (Equation (8)):

$$R_b = S_b / S_t \quad (8)$$

where S_b represents the brown area, and S_t is the total area of the banana peel [82].

The firmness of the bananas was measured on the ninth day using a TA. XT Plus texture analyzer equipped with a P6 cylindrical probe (6 mm diameter). Samples with a thickness of 1.5 cm were taken from different parts of the banana (front, rear, and middle sections), and measurements were performed in triplicate. The test was conducted with a testing speed of 2.00 mm/s, a post-test speed of 2.00 mm/s, and a plotting parameter of 5.00 mm/s over a distance of 5 mm [83].

3.10. Statistical Analysis

The experimental results are expressed as the mean $\pm$ standard deviation. Analysis of variance (ANOVA) followed by the Duncan post hoc test were performed. Samples were considered to be statistically different at $p \leq 0.05$.

4. Conclusions

This study evaluated the potential of PVA-MOF and PVA-MOF-CNF films for active packaging applications. The incorporation of TEMPO-oxidized cellulose nanofibers (CNFs) significantly enhanced the homogeneity of the MOF distribution, reducing aggregation and improving the interaction between components. This improvement led to a more structurally and functionally cohesive matrix, resulting in a superior overall performance compared to their counterparts without nanofibers. Notably, the films exhibited substantial improvements in mechanical properties, including significant increases in Young's modulus and tensile strength. Additionally, the barrier properties were enhanced, with a marked reduction in *WVP* and improved UV shielding, which was attributed to the MOFs. These enhancements are critical for ensuring the performance of the films in protecting and extending the shelf life of perishable foods, underscoring their suitability for active packaging applications.

While the PVA-CNF-MOF films showed improved ethylene retention compared to PVA-MOF films, the differences in ethylene retention among the PVA-CNF-MOF variants were statistically very small. This limitation highlights the influence of external factors on product variability, emphasizing the need for subsequent research to refine experimental conditions and isolate the effects of the materials themselves. To address this, future studies should consider using freshly harvested fruit stored under controlled conditions to minimize variability and provide a clearer assessment of MOF performance.

Despite this limitation, PVA-CNF-MOF-801 demonstrated the most promising results among the compositions tested in the ethylene scavenging analysis. This film exhibited high biocompatibility, making it suitable for direct contact with food products, in this case, bananas. This characteristic, combined with their enhanced mechanical and barrier properties, reinforces the potential of PVA-CNF-MOF films as viable materials for food packaging applications. In the future, their use could be expanded to other ethylene-producing commodities, such as tomatoes, avocados, or apples, thereby contributing to sustainable strategies for extending the shelf-life of fresh produce.

Supplementary Materials: The following supporting information can be downloaded at: https://www.mdpi.com/article/10.3390/molecules30193971/s1, Figure S1: Physical properties (density, moisture, swelling degree, solubility and *WVP*) of the composite films. Different letters above bars indicate significant differences among formulations ($p \leq 0.05$); Figure S2: Transparency and UV barrier of the different composite films. Different letters above bars indicate significant differences among formulations ($p \leq 0.05$); Figure S3: (a) Young's Modulus, (b) tensile strength and (c) stress-strain curves of PVA films with the different MOFs. Different letters above bars indicate significant differences among formulations ($p \leq 0.05$); Figure S4: XRD pattern of the PVA-CNF-MOF films; Figure S5: FTIR of the PVA-CNF-MOF composite films; Table S1: Results of the characterization of the PVA-CNF composite films in terms of mechanical, physical and optical properties.

Author Contributions: Conceptualization, J.A.-G., E.E., F.J.R.-S. and A.R.; methodology, S.C., J.A.-G., A.B., E.E., F.J.R.-S. and A.R.; formal analysis, J.A.-G., E.E., A.B., F.J.R.-S. and A.R.; investigation, S.C., J.A.-G., E.E., A.B., F.J.R.-S. and A.R.; resources, J.A.-G., E.E., A.B., F.J.R.-S. and A.R.; data curation, S.C., J.A.-G., E.E. and A.B.; writing—original draft preparation, S.C., J.A.-G., E.E. and A.B.; writing—review and editing, S.C., J.A.-G., E.E., A.B., F.J.R.-S. and A.R.; visualization, J.A.-G., E.E., A.B., F.J.R.-S. and A.R.; supervision, J.A.-G., E.E., A.B., F.J.R.-S. and A.R.; validation, J.A.-G., E.E., A.B., F.J.R.-S. and A.R.; project administration, J.A.-G., E.E., A.B., F.J.R.-S. and A.R.; funding acquisition, J.A.-G., E.E., A.B., F.J.R.-S. and A.R. All authors have read and agreed to the published version of the manuscript.

Funding: This work was supported by the Ministry of Science, Innovation and Universities of the Spanish Government for the projects PID2022/142657OB-I00/MCIN/AEI/10.13039/501100011033/FEDER, EU; PID2023-152932OB-I00/MICIU/AEI/10.13039/501100011033, ERDF/EU; and the BIOMEDPACK project PRIMA-S2-2023/PCI2024-153453/MICIU/AEI/10.13039/501100011033/EU. Additional financial support was provided by the Andalusian Regional Government through Project ProyExcel_00492 and by FEDER funds linked to the research groups FQM-346, FQM-175, and RNM-940. The authors also acknowledge funding from Cotton South, S.L., under the framework of the project *Producción de celulosa de alto grado de polimerización. Obtención de NFC* (Reference: 12021221). J.A.-G. acknowledges the support of a "Juan de la Cierva" fellowship (Reference: JDC2022-048903-I), funded by MCIN/AEI/10.13039/501100011033 and the European Union "NextGenerationEU"/PRTR. Likewise, A.B. acknowledges funding from the "Juan de la Cierva-Incorporación" fellowship (Reference: IJC2020-045041-I) under the same funding program.

Institutional Review Board Statement: Not applicable.

Informed Consent Statement: Not applicable.

Data Availability Statement: The data underlying this study are fully presented within the manuscript. Further information or specific datasets used in the analysis are available from the corresponding author upon request.

Acknowledgments: The authors gratefully acknowledge the valuable support provided by the "Instituto Químico para la Energía y el Medioambiente (IQUEMA)" and the "Servicios Centrales de Apoyo a la Investigación (SCAI)" at the University of Córdoba, Córdoba, Spain.

Conflicts of Interest: The authors declare no conflicts of interest.

References

1. United Nations. World Population Projected to Reach 9.8 Billion in 2050, and 11.2 Billion in 2100. Available online: https://www.un.org/en/desa/world-population-projected-reach-98-billion-2050-and-112-billion-2100 (accessed on 11 March 2024).
2. Lin, F.; Li, X.; Jia, N.; Feng, F.; Huang, H.; Huang, J.; Fan, S.; Ciais, P.; Song, X.-P. The Impact of Russia-Ukraine Conflict on Global Food Security. *Glob. Food Secur.* **2023**, *36*, 100661. [CrossRef]
3. Heydari, M. Cultivating Sustainable Global Food Supply Chains: A Multifaceted Approach to Mitigating Food Loss and Waste for Climate Resilience. *J. Clean. Prod.* **2024**, *442*, 141037. [CrossRef]
4. Akram, M.W.; Akram, N.; Shahzad, F.; Rehman, K.U.; Andleeb, S. Blockchain Technology in a Crisis: Advantages, Challenges, and Lessons Learned for Enhancing Food Supply Chains during the COVID-19 Pandemic. *J. Clean. Prod.* **2024**, *434*, 140034. [CrossRef]
5. van Dijk, M.; Morley, T.; Rau, M.L.; Saghai, Y. A Meta-Analysis of Projected Global Food Demand and Population at Risk of Hunger for the Period 2010–2050. *Nat. Food* **2021**, *2*, 494–501. [CrossRef] [PubMed]
6. Verghese, K.; Lewis, H.; Lockrey, S.; Williams, H. Packaging's Role in Minimizing Food Loss and Waste Across the Supply Chain: Packaging's Role in Minimizing Food Waste Across the Supply Chain. *Packag. Technol. Sci.* **2015**, *28*, 603–620. [CrossRef]

7. Champions 12.3. SDG Target 12.3 on Food Loss and Waste: 2023 Progress Report. Available online: https://champions123.org/publication/sdg-target-123-2023-progress-report (accessed on 11 March 2024).
8. Abiad, M.G.; Meho, L.I. Food Loss and Food Waste Research in the Arab World: A Systematic Review. *Food Sec.* **2018**, *10*, 311–322. [CrossRef]
9. *Food and Agriculture Organization (FAO) Food Wastage Footprint: Full Cost-Accounting: Final Report*; FAO: Rome, Italy, 2014; ISBN 978-92-5-108512-7.
10. United States Environmental Protection Agency (US EPA). Understanding Global Warming Potentials. Available online: https://www.epa.gov/ghgemissions/understanding-global-warming-potentials (accessed on 19 March 2024).
11. United Nations Environment Programme (UNEP). Promoting Sustainable Lifestyles. Available online: http://www.unep.org/regions/north-america/regional-initiatives/promoting-sustainable-lifestyles (accessed on 19 March 2024).
12. United Nations. Goal 12 | Department of Economic and Social Affairs. Available online: https://sdgs.un.org/goals/goal12 (accessed on 19 March 2024).
13. Stöckli, S.; Niklaus, E.; Dorn, M. Call for Testing Interventions to Prevent Consumer Food Waste. *Resour. Conserv. Recycl.* **2018**, *136*, 445–462. [CrossRef]
14. Sultana, A.; Kathuria, A.; Gaikwad, K.K. Metal–Organic Frameworks for Active Food Packaging. A Review. *Environ. Chem. Lett.* **2022**, *20*, 1479–1495. [CrossRef]
15. Du, H.; Sun, X.; Chong, X.; Yang, M.; Zhu, Z.; Wen, Y. A Review on Smart Active Packaging Systems for Food Preservation: Applications and Future Trends. *Trends Food Sci. Technol.* **2023**, *141*, 104200. [CrossRef]
16. Liu, B.; Zhang, S.; Li, M.; Wang, Y.; Mei, D. Metal–Organic Framework/Polyvinyl Alcohol Composite Films for Multiple Applications Prepared by Different Methods. *Membranes* **2023**, *13*, 755. [CrossRef]
17. Liu, X.-M.; Xie, L.-H.; Wu, Y. Recent Advances in the Shaping of Metal–Organic Frameworks. *Inorg. Chem. Front.* **2020**, *7*, 2840–2866. [CrossRef]
18. Abdelhamid, H.N.; Mathew, A.P. Cellulose–Metal Organic Frameworks (CelloMOFs) Hybrid Materials and Their Multifaceted Applications: A Review. *Coord. Chem. Rev.* **2022**, *451*, 214263. [CrossRef]
19. Jia, S.; Ji, D.; Wang, L.; Qin, X.; Ramakrishna, S. Metal–Organic Framework Membranes: Advances, Fabrication, and Applications. *Small Struct.* **2022**, *3*, 2100222. [CrossRef]
20. Ceia, T.F.; Silva, A.G.; Ribeiro, C.S.; Pinto, J.V.; Casimiro, M.H.; Ramos, A.M.; Vital, J. PVA Composite Catalytic Membranes for Hyacinth Flavour Synthesis in a Pervaporation Membrane Reactor. *Catal. Today* **2014**, *236*, 98–107. [CrossRef]
21. Xie, M.; Wang, J.; Zhao, H. A PVA Film for Detecting Lipid Oxidation Intended for Food Application. *Sens. Actuators B Chem.* **2018**, *273*, 260–263. [CrossRef]
22. Chou, C.-T.; Shi, S.-C.; Chen, C.-K. Sandwich-Structured, Hydrophobic, Nanocellulose-Reinforced Polyvinyl Alcohol as an Alternative Straw Material. *Polymers* **2021**, *13*, 4447. [CrossRef]
23. Espinosa, E.; Rincón, E.; Morcillo-Martín, R.; Rabasco-Vílchez, L.; Rodríguez, A. Orange Peel Waste Biorefinery in Multi-Component Cascade Approach: Polyphenolic Compounds and Nanocellulose for Food Packaging. *Ind. Crops Prod.* **2022**, *187*, 115413. [CrossRef]
24. Sharanyakanth, P.S.; Radhakrishnan, M. Synthesis of Metal-Organic Frameworks (MOFs) and Its Application in Food Packaging: A Critical Review. *Trends Food Sci. Technol.* **2020**, *104*, 102–116. [CrossRef]
25. Zhang, B.; Luo, Y.; Kanyuck, K.; Bauchan, G.; Mowery, J.; Zavalij, P. Development of Metal–Organic Framework for Gaseous Plant Hormone Encapsulation To Manage Ripening of Climacteric Produce. *J. Agric. Food Chem.* **2016**, *64*, 5164–5170. [CrossRef]
26. Amaro-Gahete, J.; Klee, R.; Esquivel, D.; Ruiz, J.R.; Jiménez-Sanchidrián, C.; Romero-Salguero, F.J. Fast Ultrasound-Assisted Synthesis of Highly Crystalline MIL-88A Particles and Their Application as Ethylene Adsorbents. *Ultrason. Sonochem.* **2019**, *50*, 59–66. [CrossRef]
27. Wiśniewska, P.; Haponiuk, J.; Saeb, M.R.; Rabiee, N.; Bencherif, S.A. Mitigating Metal-Organic Framework (MOF) Toxicity for Biomedical Applications. *Chem. Eng. J.* **2023**, *471*, 144400. [CrossRef]
28. Kumar, P.; Anand, B.; Tsang, Y.F.; Kim, K.-H.; Khullar, S.; Wang, B. Regeneration, Degradation, and Toxicity Effect of MOFs: Opportunities and Challenges. *Environ. Res.* **2019**, *176*, 108488. [CrossRef]
29. Anovitz, L.M.; Cole, D.R. Characterization and Analysis of Porosity and Pore Structures. *Rev. Mineral. Geochem.* **2015**, *80*, 61–164. [CrossRef]
30. Yañez-Aulestia, A.; Trejos, V.M.; Esparza-Schulz, J.M.; Ibarra, I.A.; Sánchez-González, E. Chemically Modified HKUST-1(Cu) for Gas Adsorption and Separation: Mixed-Metal and Hierarchical Porosity. *ACS Appl. Mater. Interfaces* **2024**, *16*, 65581–65591. [CrossRef] [PubMed]
31. Rojas-Luna, R.; Amaro-Gahete, J.; Gil-Gavilán, D.G.; Castillo-Rodríguez, M.; Jiménez-Sanchidrián, C.; Ruiz, J.R.; Esquivel, D.; Romero-Salguero, F.J. Visible-Light-Harvesting Basolite-A520 Metal Organic Framework for Photocatalytic Hydrogen Evolution. *Microporous Mesoporous Mater.* **2023**, *355*, 112565. [CrossRef]

32. Azarifar, D.; Ghorbani-Vaghei, R.; Daliran, S.; Oveisi, A.R. A Multifunctional Zirconium-Based Metal–Organic Framework for the One-Pot Tandem Photooxidative Passerini Three-Component Reaction of Alcohols. *ChemCatChem* **2017**, *9*, 1992–2000. [CrossRef]
33. Li, C.-N.; Wang, S.-M.; Tao, Z.-P.; Liu, L.; Xu, W.-G.; Gu, X.-J.; Han, Z.-B. Green Synthesis of MOF-801(Zr/Ce/Hf) for CO_2/N_2 and CO_2/CH_4 Separation. *Inorg. Chem.* **2023**, *62*, 7853–7860. [CrossRef]
34. Lorzing, G.R.; Balto, K.P.; Antonio, A.M.; Trump, B.A.; Brown, C.M.; Bloch, E.D. Elucidating the Structure of the Metal-Organic Framework Ru-HKUST-1. *Chem. Mater.* **2020**, *32*, 7710–7715. [CrossRef]
35. Evans, T.G.; Salinger, J.L.; Bingel, L.W.; Walton, K.S. Determining Surface Areas and Pore Volumes of Metal-Organic Frameworks. *J. Vis. Exp.* **2024**, *205*, e65716. [CrossRef]
36. Qiao, C.; Jia, W.; Zhong, Q.; Liu, B.; Zhang, Y.; Meng, C.; Tian, F. MOF-Derived Cu-Nanoparticle Embedded in Porous Carbon for the Efficient Hydrogenation of Nitroaromatic Compounds. *Catal. Lett.* **2020**, *150*, 3394–3401. [CrossRef]
37. Rivera-Torrente, M.; Kroon, D.; Coulet, M.-V.; Marquez, C.; Nikolopoulos, N.; Hardian, R.; Bourrelly, S.; De Vos, D.; Whiting, G.T.; Weckhuysen, B.M. Understanding the Effects of Binders in Gas Sorption and Acidity of Aluminium Fumarate Extrudates. *Chem.-Eur. J.* **2022**, *28*, e202103420. [CrossRef]
38. Tang, X.; Luo, Y.; Zhang, Z.; Ding, W.; Liu, D.; Wang, J.; Guo, L.; Wen, M. Effects of Functional Groups of $–NH_2$ and $–NO_2$ on Water Adsorption Ability of Zr-Based MOFs (UiO-66). *Chem. Phys.* **2021**, *543*, 111093. [CrossRef]
39. Al-Janabi, N.; Hill, P.; Torrente-Murciano, L.; Garforth, A.; Gorgojo, P.; Siperstein, F.; Fan, X. Mapping the Cu-BTC Metal–Organic Framework (HKUST-1) Stability Envelope in the Presence of Water Vapour for CO_2 Adsorption from Flue Gases. *Chem. Eng. J.* **2015**, *281*, 669–677. [CrossRef]
40. Karmakar, S.; Dechnik, J.; Janiak, C.; De, S. Aluminium Fumarate Metal-Organic Framework: A Super Adsorbent for Fluoride from Water. *J. Hazard. Mater.* **2016**, *303*, 10–20. [CrossRef] [PubMed]
41. Athar, M.; Rzepka, P.; Thoeny, D.; Ranocchiari, M.; Bokhoven, J.A. van Thermal Degradation of Defective High-Surface-Area UiO-66 in Different Gaseous Environments. *RSC Adv.* **2021**, *11*, 38849–38855. [CrossRef] [PubMed]
42. Aghajani Hashjin, M.; Zarshad, S.; Motejadded Emrooz, H.B.; Sadeghzadeh, S. Enhanced Atmospheric Water Harvesting Efficiency through Green-Synthesized MOF-801: A Comparative Study with Solvothermal Synthesis. *Sci. Rep.* **2023**, *13*, 16983. [CrossRef] [PubMed]
43. Mouchaham, G.; Wang, S.; Serre, C. The Stability of Metal–Organic Frameworks. In *Metal-Organic Frameworks*; John Wiley & Sons, Ltd.: Hoboken, NJ, USA, 2018; pp. 1–28, ISBN 978-3-527-80909-7.
44. Dai, Y.; Tang, Q.; Zhang, Z.; Yu, C.; Li, H.; Xu, L.; Zhang, S.; Zou, Z. Enhanced Mechanical, Thermal, and UV-Shielding Properties of Poly(Vinyl Alcohol)/Metal–Organic Framework Nanocomposites. *RSC Adv.* **2018**, *8*, 38681–38688. [CrossRef]
45. Lu, C.; Xiao, H.; Chen, X. MOFs/PVA Hybrid Membranes with Enhanced Mechanical and Ion-Conductive Properties. *e-Polymers* **2021**, *21*, 160–165. [CrossRef]
46. Cirri, A.; Silakov, A.; Jensen, L.; Lear, B.J. Probing Ligand-Induced Modulation of Metallic States in Small Gold Nanoparticles Using Conduction Electron Spin Resonance. *Phys. Chem. Chem. Phys.* **2016**, *18*, 25443–25451. [CrossRef]
47. Aziz, S.B.; Dannoun, E.M.A.; Tahir, D.A.; Hussen, S.A.; Abdulwahid, R.T.; Nofal, M.M.; Abdullah, R.M.; Hussein, A.M.; Brevik, I. Synthesis of PVA/CeO2 Based Nanocomposites with Tuned Refractive Index and Reduced Absorption Edge: Structural and Optical Studies. *Materials* **2021**, *14*, 1570. [CrossRef]
48. Zhou, L.; Niu, Z.; Jin, X.; Tang, L.; Zhu, L. Effect of Lithium Doping on the Structures and CO_2 Adsorption Properties of Metal-Organic Frameworks HKUST-1. *ChemistrySelect* **2018**, *3*, 12865–12870. [CrossRef]
49. Ren, G.; Zhao, K.; Zhao, L. A Fenton-like Method Using ZnO Doped MIL-88A for Degradation of Methylene Blue Dyes. *RSC Adv.* **2020**, *10*, 39973–39980. [CrossRef] [PubMed]
50. Zayan, S.; Elshazly, A.; Elkady, M. In Situ Polymerization of Polypyrrole @ Aluminum Fumarate Metal–Organic Framework Hybrid Nanocomposites for the Application of Wastewater Treatment. *Polymers* **2020**, *12*, 1764. [CrossRef] [PubMed]
51. Moghaddam, Z.S.; Kaykhaii, M.; Khajeh, M.; Oveisi, A.R. Synthesis of UiO-66-OH Zirconium Metal-Organic Framework and Its Application for Selective Extraction and Trace Determination of Thorium in Water Samples by Spectrophotometry. *Spectrochim. Acta A Mol. Biomol. Spectrosc.* **2018**, *194*, 76–82. [CrossRef] [PubMed]
52. Shaik, M.R.; Adil, S.F.; Alothman, Z.A.; Alduhaish, O.M. Fumarate Based Metal–Organic Framework: An Effective Catalyst for the Transesterification of Used Vegetable Oil. *Crystals* **2022**, *12*, 151. [CrossRef]
53. Singh, S.; Gaikwad, K.K.; Lee, Y.S. Antimicrobial and Antioxidant Properties of Polyvinyl Alcohol Bio Composite Films Containing Seaweed Extracted Cellulose Nano-Crystal and Basil Leaves Extract. *Int. J. Biol. Macromol.* **2018**, *107*, 1879–1887. [CrossRef]
54. Lin, R.; Ge, L.; Diao, H.; Rudolph, V.; Zhu, Z. Ionic Liquids as the MOFs/Polymer Interfacial Binder for Efficient Membrane Separation. *ACS Appl. Mater. Interfaces* **2016**, *8*, 32041–32049. [CrossRef]
55. Fouladi, M.; Kavousi Heidari, M.; Tavakoli, O. Performance Comparison of Thin-Film Nanocomposite Polyamide Nanofiltration Membranes for Heavy Metal/Salt Wastewater Treatment. *J. Nanopart. Res.* **2023**, *25*, 77. [CrossRef]
56. Aghajanzadeh, M.; Zamani, M.; Molavi, H.; Khieri Manjili, H.; Danafar, H.; Shojaei, A. Preparation of Metal–Organic Frameworks UiO-66 for Adsorptive Removal of Methotrexate from Aqueous Solution. *J. Inorg. Organomet. Polym.* **2018**, *28*, 177–186. [CrossRef]

57. Morcillo-Martín, R.; Rabasco-Vílchez, L.; Espinosa, E.; Pérez-Rodríguez, F.; Rodríguez, A. Raspberry (*Rubus idaeus* L.) Waste-Derived Nanocellulose for Circular Application in Edible Films and Coatings. *LWT* **2023**, *188*, 115438. [CrossRef]
58. López, A.R.C.; Castillo, H.S.V.; Trochez, J.J.P.; Buitrago, C.E.M.; López, R.A.G. Mechanical properties of films obtained from thermoplastic starch pellets stored under different conditions. *Acta Agronómica* **2023**, *72*, 225–233. [CrossRef]
59. Chen, C.W.; Xie, J.; Yang, F.X.; Zhang, H.L.; Xu, Z.W.; Liu, J.L.; Chen, Y.J. Development of Moisture-Absorbing and Antioxidant Active Packaging Film Based on Poly(Vinyl Alcohol) Incorporated with Green Tea Extract and Its Effect on the Quality of Dried Eel. *J. Food Process. Preserv.* **2018**, *42*, e13374. [CrossRef]
60. Zhang, M.; Wang, G.; Zhang, X.; Zheng, Y.; Lee, S.; Wang, D.; Yang, Y. Polyvinyl Alcohol/Chitosan and Polyvinyl Alcohol/Ag@MOF Bilayer Hydrogel for Tissue Engineering Applications. *Polymers* **2021**, *13*, 3151. [CrossRef] [PubMed]
61. Sabetghadam, A.; Seoane, B.; Keskin, D.; Duim, N.; Rodenas, T.; Shahid, S.; Sorribas, S.; Guillouzer, C.L.; Clet, G.; Tellez, C.; et al. Metal Organic Framework Crystals in Mixed-Matrix Membranes: Impact of the Filler Morphology on the Gas Separation Performance. *Adv. Funct. Mater.* **2016**, *26*, 3154–3163. [CrossRef]
62. An, Y.; Lv, X.; Jiang, W.; Wang, L.; Shi, Y.; Hang, X.; Pang, H. The Stability of MOFs in Aqueous Solutions—Research Progress and Prospects. *Green Chem. Eng.* **2024**, *5*, 187–204. [CrossRef]
63. Wang, X.; Zhou, X.; Yang, K.; Li, Q.; Wan, R.; Hu, G.; Ye, J.; Zhang, Y.; He, J.; Gu, H.; et al. Peroxidase- and UV-Triggered Oxidase Mimetic Activities of the UiO-66-NH_2/Chitosan Composite Membrane for Antibacterial Properties. *Biomater. Sci.* **2021**, *9*, 2647–2657. [CrossRef]
64. Sun, L.; An, X.; Qian, X. Nano-MIL-88A(Fe) Enabled Clear Cellulose Films with Excellent UV-Shielding Performance and Robust Environment Resistance. *Nanomaterials* **2022**, *12*, 1891. [CrossRef]
65. Sánchez-Laínez, J.; Zornoza, B.; Mayoral, Á.; Berenguer-Murcia, Á.; Cazorla-Amorós, D.; Téllez, C.; Coronas, J. Beyond the H_2/CO_2 Upper Bound: One-Step Crystallization and Separation of Nano-Sized ZIF-11 by Centrifugation and Its Application in Mixed Matrix Membranes. *J. Mater. Chem. A* **2015**, *3*, 6549–6556. [CrossRef]
66. De Haro-Niza, J.; Rincón, E.; Gonzalez, Z.; Espinosa, E.; Rodríguez, A. Nanocellulose from Spanish Harvesting Residues to Improve the Sustainability and Functionality of Linerboard Recycling Processes. *Nanomaterials* **2022**, *12*, 4447. [CrossRef]
67. Espinosa, E.; Bascón-Villegas, I.; Rosal, A.; Pérez-Rodríguez, F.; Chinga-Carrasco, G.; Rodríguez, A. PVA/(Ligno)Nanocellulose Biocomposite Films. Effect of Residual Lignin Content on Structural, Mechanical, Barrier and Antioxidant Properties. *Int. J. Biol. Macromol.* **2019**, *141*, 197–206. [CrossRef]
68. Sánchez-Gutiérrez, M.; Espinosa, E.; Bascón-Villegas, I.; Pérez-Rodríguez, F.; Carrasco, E.; Rodríguez, A. Production of Cellulose Nanofibers from Olive Tree Harvest—A Residue with Wide Applications. *Agronomy* **2020**, *10*, 696. [CrossRef]
69. Heidarian, P.; Behzad, T.; Sadeghi, M. Investigation of Cross-Linked PVA/Starch Biocomposites Reinforced by Cellulose Nanofibrils Isolated from Aspen Wood Sawdust. *Cellulose* **2017**, *24*, 3323–3339. [CrossRef]
70. Ghorbel, N.; Kallel, A.; Boufi, S. Molecular Dynamics of Poly(Vinyl Alcohol)/Cellulose Nanofibrils Nanocomposites Highlighted by Dielectric Relaxation Spectroscopy. *Compos. Part A Appl. Sci. Manuf.* **2019**, *124*, 105465. [CrossRef]
71. Huang, S.; Wang, X.; Zhang, Y.; Meng, Y.; Hua, F.; Xia, X. Cellulose Nanofibers/Polyvinyl Alcohol Blends as an Efficient Coating to Improve the Hydrophobic and Oleophobic Properties of Paper. *Sci. Rep.* **2022**, *12*, 16148. [CrossRef] [PubMed]
72. Martínez, Z.N.; Menacho, P.Z.; Pachón-Ariza, F.A. Food Loss in a Hungry World, a Problem? *Agron. Colomb.* **2014**, *32*, 283–293. [CrossRef]
73. Papargyropoulou, E.; Lozano, R.; Steinberger, J.K.; Wright, N.; Ujang, Z.b. The Food Waste Hierarchy as a Framework for the Management of Food Surplus and Food Waste. *J. Clean. Prod.* **2014**, *76*, 106–115. [CrossRef]
74. Hodges, R.J.; Buzby, J.C.; Bennett, B. Postharvest Losses and Waste in Developed and Less Developed Countries: Opportunities to Improve Resource Use. *J. Agric. Sci.* **2011**, *149*, 37–45. [CrossRef]
75. Ishangulyyev, R.; Kim, S.; Lee, S.H. Understanding Food Loss and Waste—Why Are We Losing and Wasting Food? *Foods* **2019**, *8*, 297. [CrossRef]
76. Lin, K.-S.; Adhikari, A.K.; Ku, C.-N.; Chiang, C.-L.; Kuo, H. Synthesis and Characterization of Porous HKUST-1 Metal Organic Frameworks for Hydrogen Storage. *Int. J. Hydrogen Energy* **2012**, *37*, 13865–13871. [CrossRef]
77. Trepte, K.; Schaber, J.; Schwalbe, S.; Drache, F.; Senkovska, I.; Kaskel, S.; Kortus, J.; Brunner, E.; Seifert, G. The Origin of the Measured Chemical Shift of 129Xe in UiO-66 and UiO-67 Revealed by DFT Investigations. *Phys. Chem. Chem. Phys.* **2017**, *19*, 10020–10027. [CrossRef]
78. Tan, T.L.; Krusnamurthy, P.A.; Nakajima, H.; Rashid, S.A. Adsorptive, Kinetics and Regeneration Studies of Fluoride Removal from Water Using Zirconium-Based Metal Organic Frameworks. *RSC Adv.* **2020**, *10*, 18740–18752. [CrossRef]
79. Espinosa Víctor, E.; Tarrés, Q.; Delgado-Aguilar, M.; González, I.; Mutjé, P.; Rodríguez, A. Suitability of Wheat Straw Semichemical Pulp for the Fabrication of Lignocellulosic Nanofibres and Their Application to Papermaking Slurries. *Cellulose* **2016**, *23*, 837–852. [CrossRef]

80. *UNE-ISO 5351:2019*; Pastas. Determinación de La Viscosidad Intrínseca en Disolución de Cobre Etilen-Diamina (CED). UNE: Madrid, Spain, 2019. Available online: https://www.une.org/encuentra-tu-norma/busca-tu-norma/norma?c=N0061871 (accessed on 13 March 2024).
81. Shinoda, R.; Saito, T.; Okita, Y.; Isogai, A. Relationship between Length and Degree of Polymerization of TEMPO-Oxidized Cellulose Nanofibrils. *Biomacromolecules* **2012**, *13*, 842–849. [CrossRef]
82. Xiao, F.; Xiao, Y.; Ji, W.; Li, L.; Zhang, Y.; Chen, M.; Wang, H. Photocatalytic Chitosan-Based Bactericidal Films Incorporated with WO3/AgBr/Ag and Activated Carbon for Ethylene Removal and Application to Banana Preservation. *Carbohydr. Polym.* **2024**, *328*, 121681. [CrossRef]
83. Wang, H.; Zhou, X.; Hu, W.; Hu, X.; Murtaza, A.; Wang, L.; Xu, X.; Pan, S. Mechanism of the Abnormal Softening of Banana Pulp Induced by High Temperature and High Humidity During Postharvest Storage. *Food Bioprocess Technol.* **2024**, *17*, 3577–3587. [CrossRef]

MDPI AG
Grosspeteranlage 5
4052 Basel
Switzerland
Tel.: +41 61 683 77 34

Molecules Editorial Office
E-mail: molecules@mdpi.com
www.mdpi.com/journal/molecules

www.ingramcontent.com/pod-product-compliance
Lightning Source LLC
Chambersburg PA
CBHW061457210926
51850CB00008B/1742